D1483491

ANTI-EQUILIBRIUM

ON ECONOMIC SYSTEMS THEORY AND THE
TASKS OF RESEARCH

ANTI-EQUILIBRIUM

On economic systems theory and the tasks of research

by

JÁNOS KORNAI

Professor of Economics
Institute of Economics
Hungarian Academy of Sciences

NORTH-HOLLAND PUBLISHING COMPANY – AMSTERDAM • OXFORD
AMERICAN ELSEVIER PUBLISHING COMPANY, INC. – NEW YORK

330. 182
K84a
Cop. 2

© North-Holland Publishing Company – 1971

All rights reserved. No part of this publication may by reproduced, stored in a retrieval system, or transmitted, in any form or by any means, electronic, mechanical, photocopying, recording or otherwise, without the prior permission of the copyright owner.

Library of Congress Catalog Card Number: 75-134644
North-Holland ISBN: 0 7204 3055 0 (hardbound)
            0 7204 3805 5 (paperback)
American Elsevier ISBN: 0 444 10122 5

Published by:

North-Holland Publishing Company – Amsterdam
North-Holland Publishing Company, Ltd. – Oxford

Sole distributors for the U.S.A. and Canada:

American Elsevier Publishing Company, Inc.
52 Vanderbilt Avenue
New York, N.Y. 10017

First edition 1971
Second printing 1975

Printed in The Netherlands

CONTENTS

PART I: STARTING POINTS

UNIVERSITY LIBRARIES
CARNEGIE-MELLON UNIVERSITY
PITTSBURGH, PENNSYLVANIA 15213

CONTENTS

PART II: CONCEPTS AND QUESTIONS

LIST OF TABLES

LIST OF FIGURES

FOREWORD

This book is a "semi-finished product"—a combination of a theoretical treatise and a plan of work.[1] Although, I believe the concepts here outlined to be more or less fully developed and of current interest, I realize that a more complete articulation will require a considerable amount of research.

In the concluding remarks, after the reader has become acquainted with my ideas, I will give in greater detail my reasons for publishing a semi-finished product. In this foreword let me give a single motive only.

In 1968, the methods of economic administration and management in Hungary underwent a thorough reform. The systems of economic planning, enterprise control, material incentives, prices and incomes were essentially transformed. A number of economic processes which were formerly the subject of strict central control were, to a large extent, decentralized, and the role of profit as one of the material incentives was considerably increased.

The preparatory work for the reforms was carried out by several working panels, both theoretical economists and practical experts. It was a task of unparalleled interest and novelty to survey the economic system as a whole, and its major component factors and to predict the expected overall effect of the proposed changes.

Those participating in this work could rely on their daily experiences and their practical knowledge of the operation of economic systems in Hungary and abroad, but could hardly build upon any economic theory in the strict sense of the word. No scientifically established theorems were available to guide them in their work. It would certainly have caused great amusement if, at some committee meeting, a mathematical economist had claimed on scientific grounds that an atomized market with perfect competition was needed to ensure that prices would become the sole regulator of the economy and that an optimum equilibrium would be achieved.

---

[1] The research work on which this book is based started in 1965. Since that time, I have written three essays which may be considered preparatory drafts of the present work, namely "A gazdaság működésének szimulációs modelljei" (Simulation models of the functioning of the economy) [129], mimeographed, 1966; "Anti-Equilibrium" 1st version [130], in manuscript, 1967; and "Anti-Equilibrum" 2nd version [131], mimeographed, 1968.

Theory has proven unsuited to practical application. It was my exasperation with its inadequate and unworkable character that inspired this book.

This exasperation finds expression also in the sharp tone of the work. In some places, criticism turns into outright attack. Some comments may be considered unfair. After perusing the first draft of the text, I worried for a long time about whether or not it should be softened, and whether some opinions should be presented in a more diplomatic manner. But in the end I decided not to change the style. Some readers will, no doubt, dislike the tone. However, a sharp tone may prove useful, as certain maladies can be better remedied by shock treatment than by sedatives.

*   *   *

I take this opportunity to express my thanks to all those who have assisted me in my work.

My first acknowledgement is to *Tamás Lipták*. In the course of our long-standing relationship, he aided my research work with innumerable suggestions, encouragement and advice. He also assisted me in preparing the final text to this book.

I have received much assistance from the Institute of Economics of the Hungarian Academy of Sciences, under whose sponsorship my research work was carried out. Drafts of this book were discussed at several Institute meetings. I owe special thanks to *István Friss*, director of the Institute, as well as to *Tamás Bauer, András Bródy, Péter Erdős, Róbert Hoch, György Kondor, Tamás Nagy, Béla Martos, Judit Rimler* and many other members of the Institute's scientific staff for their valuable comments on the manuscript. My colleagues, *Andrea Deák* and *Attila Soós* have aided in the compilation of several Tables in this book, and in the preparation of the manuscript for publication. I am indebted to them for their assistance.

I have received helpful suggestions from *Pál Benedek* (Lórand Eötvös University of Sciences), *Balint Dömölki* ("Infelor" Enterprise for System Technique), *Tamás Frey* (Automation Research Institute, Hungarian Academy of Sciences), *József Tankó* (Compution Technical Centre, Hungarian Academy of Sciences), and *Márton Tardos* (Market Research Institute).

In 1968, I spent four months in the U.S.A., at the invitation of the Institute of Mathematical Studies in the Social Sciences, Stanford University, California. I wish to express my gratitude to Stanford University and particularly to *K. J. Arrow*, head of the Institute at that time, who was kind enough to give me many opportunities to clarify my ideas in the course of dicussions.

In 1970, I spent 6 months at the Cowles Foundation for Research in Economics at Yale University, as a guest under a visitors' program financed by the Ford Foundation. In addition to the opportunity to discuss final revisions of this book with members of the Cowles Foundation staff and other members of the faculty in the Economics Department, the Cowles Foundation provided for the editorial assistance, and typing of the manuscript. I am extremely grateful to the Cowles Foundation, to Yale University and to the Ford Foundation for giving me this excellent opportunity to further the improvement of my book.

I have received valuable comments on the 1968 or on the 1970 version of the book from *W. Fellner*, *T. C. Koopmans*, and *J. M. Montias* (Yale University), *J. Marschak* (University of California, Los Angeles), *R. Radner* and *T. Marschak* (University of California, Berkeley), *T. Vietorisz* (New School, New York) and *C. C. von Weizsäcker* (University of Heidelberg, Federal Republic of Germany and Massachusetts Institute of Technology, U.S.A.). Their suggestions were very helpful, but of course they bear no responsibility for the errors remaining in the book.

To the reader acquainted with modern economic theory it will be clear that the above list contains, among others, the names of economists who have played an outstanding role in the evolution of the theories criticized in my work. Their dispassionate objective treatment of my work commands particular respect. Their encouragement was one of the factors that prompted me to publish this work.

Last but not least, I wish to thank the translators, *P. Morvay* and *G. Hajdu*, and the editors, *R. W. Nelson* and *J. Yellen* (Yale University) for their devoted work. Their very conscientious efforts were helpful in bringing the book to the English-speaking reader. I am grateful also to *G. Ames*, *M. Crocker*, *L. De-Andrus* and *C. Villano*, (Cowles Foundation) and *A. Bakos*, *I. Lukács* and *A. Zerkovitz* (Budapest) for their contributions in secretarial work and typing.

Budapest—New Haven, 1970, June

JÁNOS KORNAI

ON THE STRUCTURE OF THE BOOK

S.1. Principles of Notation

The notation system of the book has been established on the basis of the following principles:

1. *Sets* will be denoted by Latin capitals in script-type character. When dealing with a set of abstract elements, the elements themselves will be denoted by the lower case of the letter used for the set. (E.g.: $\mathcal{G}$ is the set of products; g_1, g_2 ... are the individual products.)

2. As regards letter type, no distinction will be made between *matrices*, *vectors* and *real numbers*. From the explanations accompanying the formulae it will always be clear whether we are dealing with a matrix, a vector or a real number.

Matrices will always be denoted by capital letters, vectors by lower-case letters. The vectors and elements of a matrix are denoted by the same letter and provided with subscripts. E.g. the vectors of matrix X will be denoted x_i, its elements x_{ij}.

No distinction in denotation will be made between row and column vectors. From the relationship described it will always be clear whether we are dealing with a row or a column vector.

3. Notation has generally been chosen to correspond to the *initial letter* of the concept represented. (E.g. E stands for economy, P for preference, U for utility, etc.) But this principle could not always be consistently implemented.

4. Although a single concept will be denoted by the same symbol throughout the book, certain symbols must, unavoidably, be used for more than one concept.

S.2. Principle Notations

The list below is not complete, including only the principal symbols.

a = alternative
a^* = decision
$\mathcal{A}$ = set of possible alternatives
$\mathcal{B}$ = set of implementable alternatives
$\tilde{\mathcal{B}}$ = set of explored alternatives
C = abbreviation referring to the control sphere

c	= control unit
$\mathcal{C}$	= set of control units
$\mathcal{D}$	= set of acceptable alternatives (acceptable compromises)
E	= economic system
$\mathcal{F}$	= set of eligible alternatives
g	= product
$\mathcal{G}$	= set of products
K	= number of indicator types
p	= price
p	= decision problem
P	= preference ordering
$\mathcal{P}$	= set of decision problems
q	= quality parameter
Q	= abbreviation referring to quality
R	= abbreviation referring to the real sphere
r	= real unit
$\mathcal{R}$	= set of real units
s	= elementary selling intention
S	= complex selling intention
s	= information type
$\mathcal{S}$	= set of information types
t	= time as a variable
T	= time period
u	= information flow
U	= utility function
v	= memory content
V	= abbreviation referring to volume
w	= intensity
x	= product flow
y	= product stock
z	= extent of promoting process
α	= aspiration level
γ	= degree of consistency
δ	= ambition of decision
ε	= tension of aspiration
ζ	= tension of decision
$\varkappa$	= correction of aspiration
μ	= fulfilment ratio
ξ	= decision distribution function
φ	= control response function
Φ	= system of control response functions

2*

ψ = real response function
Ψ = system of real response functions
ω = result
Ω = relative strength of market forces

S.3. Definitions

The concepts introduced in the book, or used in a special sense are double spaced and defined when they first appear. The definitions have a double serial number, the first indicating the chapter and the second the order within the chapter. (E.g. 5.1, 6.1.)

We distingusih between two categories of defined concepts. First category: the *main concepts*. These are used repeatedly, they constitute the basic vocabulary of the "language" of the conceptual framework introduced in the book. The other category: *auxiliary concepts*. These are introduced to help define the main concepts.

In order to facilitate orientation for the reader, the serial number of the definitions of main concepts is always marked with an asterisk.

Sometimes a concept is introduced and explained, at the outset, only briefly or not at all. The exact definition is given later when other necessary concepts are available. In these cases reference is made to the location and serial number of the definition.

S.4. Statements

The book contains certain assertions which are set in italics and denoted by double serial numbers.

Some of these assertions, based on well known empirical facts, can be considered proved. Others qualify rather as hypotheses; their verification, refutation or correction is the task of further research. In numbering them, however, no distinction has been made between the proved statements and the hypotheses.

In each case the extent to which the assertion has been proven is indicated.

PART I

STARTING POINTS

1. INTRODUCTION: SCOPE OF THE SUBJECT

The subject of this book is far-reaching. Because it touches on a number of fundamental problems of economic theory and extends into other disciplines as well, it is necessary to delineate its scope fairly precisely.

1.1. The Systems-Theoretical Approach

Physical reality can be investigated by the physicist and chemist, by the researcher in thermodynamics, optics or mechanics, by the macrophysicist and microphysicist, each from his own point of view. The living organism, too, is viewed differently by the geneticist, the physiologist, the bio-chemist and the specialist in cellular research.

Economic reality likewise lends itself to analysis from various viewpoints. The approach of this book I propose to call *economic systems theory*, as suggested in the book's sub-title.

The economy is treated as a system composed of various elements: firms, households, government offices, social institutions, etc. Among the elements of the system there exist a variety of connections and interrelations. The behaviour of these elements is characterized by definite regularities. In the system's functioning over time, a prominent part is played by those processes, which *control* other processes, as, for example, the process, which controls production, trade and consumption.

The concepts ("system", "connection", "behavioural regularity", and "control process") will be defined in detail in Chapter 4.

At this point I wish only briefly to mention the systems-theoretical approach in order to make clear what will remain *outside* the scope of the work.

I will not engage in macroeconomic analysis in the usual sense of the word. Description of the national economy as a whole might be based on aggregates such as gross national income, total investment, and total consumption, abstracting from the fact that the economy is composed of smaller units or parts. Although analysis of this sort has proven both fruitful and important, this type of macroeconomic investigation will not be pursued in my book.

When describing the physical, technical interrelations of the economy, the connections between inputs and outputs, many economists disregard the specific problems involved in controlling these processes. This is the case, for example, with Leontief models and other production function theories. I will not, however, adopt this often justifiable and useful approach because my concern centers *on the control and on the flow of information.*

1.2. A Critique of Mathematical Economics

My primary concern is the critical evaluation of certain theories of mathematical economics; the works of the non-mathematical or "literary" economists will be dealt with only marginally. This is because I am particularly attracted by the problems of mathematical economics, and not because I consider this approach to be more deserving of criticism than the "literary" one.

I consider myself a mathematical economist; thus my critical remarks come not from "outside" but from "inside" the circle. These remarks, therefore, may be regarded in many cases as selfcriticism as well as criticism. It is my conviction that the further progress of economic theory will depend, if not exclusively at least significantly, on the advances made in the field of mathematical economics. It is in this area that I hope my work can make a contribution.

Further development of mathematical economics requires, at this point, a careful revision of the discipline's theoretical methods and its basic assumptions and axioms, as well as a consideration of the relevance of its theorems to economic reality. My book is intended to lend impetus to such a revision. This explains the seemingly paradoxical fact that, while engaged in mathematical economics, I largely avoid mathematical formalism. I wish to place emphasis on the economic interpretation of the mathematical formulae. It is here that I believe the short-comings to be particularly serious. Most frequently, authors confine themselves to the presentation of mathematical deductions leaving it to the reader to "decipher" the mathematics and attach an interpretation to the formalized assumptions and conclusions. Many readers, however, are willing to take it for granted, on the author's word of honor, that the mathematical assertions are economically meaningful. Among other things, I would like to dispute this naive belief in my book.

My book does not contain rigorous mathematical reasoning, finished models that lead from formalized assumptions to mathematically proven theorems. (This also indicates the semi-finished product character of the book, mentioned in the Foreword.) If, in some places I use mathematical symbols, I have several purposes in so doing. For one thing, it may facilitate a more compact, more exact formulation of definitions and, most important, presentation of interrelations among various concepts. Also, by using symbolic description, the tasks which have yet to be performed in the fields of measurement, observation and quantification may become clearer. Finally, formalization of some concepts and interrelations may elicit ideas and serve as a starting point for further research and construction of mathematical models.

1.3. Criticism and Positive Theory

Up to the present, economics has produced only a single finished theory describing the operation of the economy from a systems-theoretical point of view. This conceptual framework, typically expounded in formal mathematical models, is called *general equilibrium theory* and derives from the teachings of Walras.[1]

Since this is the *only* major formalized approach to economic systems theory, I propose to focus my attention on this theory. My book is basically a critical essay as indicated by its title: "Anti-Equilibrium", i.e. a criticism of equilibrium theory.

Although my polemics are directed primarily at general equilibrium theory per se, the book also deals with numerous other works and theories which employ the concepts, assumptions and schemas of thought characteristic of general equilibrium theory in its "pure" form.

The weaknesses of my position in this polemic are clear. In contrast to the traditional and classical thinking on this subject, some of my own remarks are neither sufficiently developed nor proven; they are, rather, intuitive. Frequently, I can only indicate the empirical observations needed to make certain statements convincing. It is a weakness of my argument that I cannot offer a complete and sophisticated *positive* theory to replace the criticized one. Only in some instances can I even present the outlines of a new theory, and perhaps indicate the direction to be followed in arriving at a new theory.

Few completely new results but many doubts in connection with earlier results characterize the present state of my work. I am raising questions rather than providing answers. Admittedly, it may be questioned whether I have the moral right to criticize sharply when I am not in a position to present a new theory superior to the old one. The only justification for presenting my doubts about, and arguments against, the general equilibrium theory is the fact that even overwhelmingly negative criticism may prove useful. Many examples could be cited from the history of science to demonstrate that important new ideas might never have emerged, had not profound doubts arisen about accepted ones.

[1] To avoid any misunderstanding, it must be noted that in speaking of equilibrium theory I always refer to the Walras school and not to any of the theories dealing with problems of budgetary equilibrium or balance-of-payments equilibrium.

1.4. The Present-Day Economy

My subject should also be delineated in one further respect; I propose to deal exclusively with contemporary economic systems, considering, however, a wide variety of types.

The present era is characterized by the co-existence of socialist and capitalist economic systems. Careful investigation of these systems leads one to the conclusion that, even within the group of socialist countries, one finds there existing systems which differ considerably from one another; the differences between the Soviet and Chinese, the Cuban and Hungarian, the Yugoslav and Polish economies are significant. Similarly, marked differences exist between the Dutch and Swiss, and the U.S. and the Japanese economies.

It is the fashion nowadays to speak about the "convergence" of different economic systems, of their gradual approach to each other. In capitalist countries the role of centralization, government intervention and planning appears to be increasing, while in socialist countries, decentralized decision-making and market allocation are growing in importance. There are economists and politicians who may be over-emphasizing these tendencies. Some, with the best of intentions, wish to lend support to the idea of peaceful coexistence with this argument; others, less well-intentioned, wish to play down differences which actually exist among the different systems. At the same time, there are economists and politicians in both socialist and capitalist parts of the world who refuse even to take notice of these tendencies. They recognize only the essential differences between these two modes of economic organization and deny the existence of certain features common to every economic system, and particularly to those of the present era.

In this work, I do not wish to subscribe to any of these biased views. It is my belief that there are certain phenomena to be found in every economic system which are *entirely general* in character. Some are closely connected to the era in which they occur; they characterize the economy of the 19th century, or of the first or second half of the 20th century. Some are related to the level of economic and technological development of a country and depend on whether the country in question is wealthy, average, or backward and poor. Finally, and most importantly, there are some phenomena distinctly connected with the political institutions, property relations and forms of political power of a given society.

An objective researcher must note and point out both the similarities and the dissimilarities of the different systems.

2. THEORY, FORMAL MODEL AND REALITY

Let me begin my criticism of equilibrium theory with a discussion of some epistemological and methodological issues in the philosophy of science concerning the relationship between model and reality.[1]

I shall touch only on those problems which I regard as most important for our particular point of view, for economic systems theory.

Therefore the professional philosopher will find no new epistemological ideas in this chapter. My remarks reflect, rather, the methodological worries of the mathematical economist who, only as a result of his own failures, perceives those truths which may have been evident to others.

2.1. What Do We Call a "Theory"?

One of the venerable classics of modern equilibrium theory is Debreu's *Theory of Value*. The subject of the book is "the explanation of the price of commodities."[2] Accordingly, it aims at the *explanation* of a critically important aspect of economic *reality*.

The treatment of the subject is axiomatic in character. Starting from basic concepts and axioms, the author gives an exact definition of every additional concept and deduces his theorems logically and rigourously.

The work does not consider the relationship between its basic assumptions and axioms on the one hand and reality on the other; the question of whether the former reflect the latter exactly, approximately, or not at all is ignored. Nor is the question of how the deduced theorems relate to reality ever discussed.

One might say that it is exactly this abstraction from concrete economic reality that lends, in the author's view, a *theoretical* character to his intellectual system. "Allegiance to rigour dictates the axiomatic form of the analysis

---

[1] In the preparation of Chapter 2, I have made use of *Kade*'s [109] work. Furthermore, I have taken into account the methodological debates now in progress concerning mathematical economics and modern economic theory in general. An outstanding document in this debate is *Koopmans*' essay, *The Construction of Economic Knowledge* [124]. See also *Friedman* [63], *Machlup* [155], *Nagel* [185], *Papandreou* [200], and *Samuelson* [215].

[2] See *Debreu* [50], p. VII.

where the theory, in the strict sense, is logically entirely disconnected from its interpretation."[3] In the introduction to the work Debreu indicates that whenever his analysis is completely divorced from the economic interpretation, the reader's attention will be drawn to this fact by the use of expressions such as "in theoretical language" or "for theoretical purposes", etc.

The whole spirit of Debreu's work, as suggested by the above quote, implies the following definition of the term "theory":

A theory is the set of theorems logically deducible from precisely formulated assumptions which are mutually consistent.

This conception of scientific "theories" has been widely accepted by mathematical economists. Accordingly, *every* theorem, or set of theorems, would deserve the rank of a "theory", provided that it is deducible from suitably formulated assumptions, regardless of the real content of either the assumptions or the conclusions.

In my opinion, the failure to clarify what can be called a "theory" in logic, in mathematics and in the real sciences, has influenced the evolution of general equilibrium analysis.

In the logical-mathematical sciences, "truth" is a logical criterion. A conclusion is "true" if it follows from the premises by means of deductions which are, after all, tautological. (It is beside the point that this may include rather intricate reasoning; in the last analysis everything can be traced back to tautological deductions.) Accordingly, in logical-mathematical sciences, "truth" means merely logical implication and nothing else.

Naturally, empiricism and observation have played an important part in the development of the logical-mathematical sciences. Experiences gained in architecture and warfare probably contributed to the discovery of the Pythagorean theorem. However, to prove this proposition it is not necessary to observe a hundred or a thousand right-angled triangles; its truth can be demonstrated by means of a logical argument.

In the real (natural and social) sciences, on the other hand, the only criterion of "truth" is experience, the comparison of assertions with reality.

I have emphasized the role of empiricism as a vehicle of progress in connection with the logical-mathematical sciences; conversely, logic and mathematics have promoted progress in the real sciences. The development of the latter is inconceivable without the use of mathematics, both for the statistical evaluation of observations and for the deduction and verification of the propositions themselves. Mathematics is an indispensable form of expression and tool of research in the real sciences.

It is one thing, however, to suggest that logic and mathematics serve a crucial

---

[3] *Debreu* [50], p. VIII.

role in the real sciences and quite another to assert that the same criterion of truth is relevant to both types of theory. In the real sciences, the criterion is *not* whether the proposition is logically true and tautologically deducible from earlier assumptions. The criterion of "truth" is, whether or not the proposition corresponds to reality.

In drawing a sharp distinction between the two criteria of "truth", I do not wish to become entangled in the debates concerning the philosophical foundations of mathematics and logic. I do not even wish to raise the question of whether or not, "in the final analysis", mathematics and logic also reflect actual reality. I believe that they do. Mathematical-logical sciences could not build their theories on axiomatical foundations which are in contradiction to reality.

I am ready to leave the answering of "in-the-final-analysis" type questions to the professional philosopher. What I am interested in is not the philosophical foundation of mathematics but rather the comparison between two types of scientist or, if you wish, *two types of scientific conscience*. The mathematician may sleep soundly if he believes that there are no inconsistencies among his axioms and that the deductions from the axioms to the theorems are correct; the mathematical-logical verification of his theorems will be complete. Those engaged in the real sciences cannot, however, rest content with that alone. Their consciences can only be clear if their propositions correspond to reality.

"Theory" thus requires a dual definition.

Definition 2.1. In the l o g i c a l - m a t h e m a t i c a l s c i e n c e s a theory is a theorem or body of theorems logically deducible from a set of mutually consistent axioms.

Definition 2.2*. In the r e a l s c i e n c e s a theory is a systematic description of the essential interrelations between the variables of reality. That is, only those theorems and propositions (deduced from assumptions not in conflict with reality) which describe the real world more or less accurately may be considered acceptable.

The above definition of theory in the real sciences has been generally accepted by those engaged in the sciences. Let me quote only a single authority, *Albert Einstein:*

"Physics constitutes a logical system of thought which is in a state of evolution, whose basis cannot be distilled, as it were, from experience by an inductive method, but can only be arrived at by free invention. The justification (truth content) of the system rests in the verification of the derived propositions by sense experiences.[4]

"The skeptic will say": It may well be true that this system of equations is

---

[4]"Physik und Realität", see *Einstein* [56], p. 322.

reasonable from a logical standpoint. But this does not prove that it corresponds to nature". You are right, dear skeptic. Experience alone can decide on truth".[5]

Definition 2.2 above does not mean that the rank of theory should be reserved exclusively for completely accurate and perfectly verified propositions.

It follows from our definition that a theory may be inaccurate and only approximate in character until a more accurate one can be established.

A theory may be temporarily unverified, and therefore hypothetical. But only hypotheses which have not been shown to contradict reality can be employed. (This prohibition must, of course, be interpreted cautiously. The requirement is obviously not that the theory must correspond to "sound reason" and "everyday experience" if a more profound and novel observation will lead to a proof of something differing from everyday experience.)

Theory may include temporarily unverifiable propositions, which can be neither confirmed nor refuted on the basis of our present knowledge, with the expectation that eventually they will prove capable of verification. In the first formulation therefore, one may rest content with a description of the phenomena which later must be observed if the theory is to become more accurate and suited to concretization.

There are many possible ways of verifying a scientific proposition. The most convincing one involves reliable observation of a great number of economic facts and testing of hypotheses with mathematical-statistical tools. This, however, is not always *neccessary*. It may be unnecessary to engage in mathematical-statistical analysis when the truth of a proposition is fully supported by *well-known empirical facts*. (In the course of further treatment of this subject, some of my propositions will be supported repeatedly by reference to generally known empirical facts.)

The researcher will often have to rely on indirect methods of verification such as carrying out experiments to simulate some definite situation, or interviewing men involved in practical affairs to determine how they would behave in certain hypothetical situations.

The difficulties of verification also compel us not to apply unrealistically high standards. The establishment of theories, which includes cognition often constitutes a lengthy process, which follows a roundabout path with many detours. Most living theories—even in the most highly developed natural sciences—are composed of a mixture of perfectly verified propositions and hypothetical, inaccurate approximations and conjectures.

Given these definitions, let us again consider the case of economic theory. Economics is not a logical-mathematical discipline but rather a real science, the fundamental task of which is to explain reality. *No system of ideas can be*

[5] "Über die allgemeine Gravitationslehre", see *Einstein* [56], p. 355.

called a "theory" explaining economic reality unless it satisfies Definition 2.2,
i.e. *the requirements of real-science theory.*

It is here that one of the main problems of modern mathematical equilibrium theory (and of some other branches of mathematical economics, too) lies. On the one hand, equilibrium theory claims to offer an *explanation* of reality. On the other hand, *it does not insist on verification.* It is content if its intellectual systems qualify as theories according to Definition 2.1; but a real science is obliged to meet the requirements of Definition 2.2.

Let there be no mistake; it is not the axiomatic treatment of the subject that I am criticizing here. Several branches of physics have reached the stage where an axiomatic exposition of their laws is possible; this is a sign of the discipline's maturity. But physics can, nonetheless. exist only as a real science, as should be abundantly clear from the above quote of Einstein.

The trouble with much of mathematical economics, and, particularly, equilibrium theory, is that it does not follow the developed and formalized natural sciences in requiring verification, but treats its discipline, instead, as a special branch of mathematics or logic.

2.2. Theory versus Intellectual Experiment

A distinction must be made between theory on the one hand and intellectual experiment on the other.

The conclusions arrived at by logical reasoning from arbitrary assumptions are called an *intellectual experiment.*

In intellectual experiments, the point of departure involves the posing of a question of the form "What would happen if. . . ?" The assumptions themselves may be realistic, but they might just as well be deliberately unrealistic.

We often have a number of well verified assumptions but feel uncertain about some particular one. We then work out alternative formulations of this assumption, and follow each to its logical conclusion. As a result, we derive a number of alternative propositions. Comparing these with each other and with reality, we may trace the argument back to the uncertain assumption.

Our starting assumptions may be based exclusively on unobservable phenomena. Thus, we may employ unverified but not impossible or absurd assumptions, and compare the conclusions drawn from them with reality.

We may start from a definitely absurd assumption, in order to demonstrate the absurdity of a conclusion and to exclude, thereby, certain lines of argument from further consideration.

The above examples show that in the course of creative work, an element in the process of cognition, an intellectual experiment based partly or entirely on unrealistic assumptions may play a useful role. However, acknowledging the usefulness of such an intellectual experiment is not the same thing as using it as a *substitute* for a genuine real-science theory.

Some intellectual experiments lead to theories in the real sciences, others do not. We may start from arbitrary assumptions; it is thus permissible to start exclusively from unrealistic assumptions in an intellectual experiment. Intellectual experiment is therefore a broad collective term, including not only those formalized systems of thought which may be called real scientific theories. The latter constitute a special case of the former.

2.3. Economic Real-Science Theory versus Decision Theory

A traditional mistake in economic thought, the effect of which has made itself felt from the days of the early English classicists to the present, involves confusing economic theory with decision theory.

Economic theory is a real science, the subject of which is the explanation of economic reality.

Decision theory, on the other hand constitutes, according to the definitions given in Section 2.1, a logical mathematical science.

The subject matter of decision theory is the following: Assume that we have a well defined situation with a number of alternative possibilities for action. We possess reliable, or unreliable, information concerning the consequences of alternative actions. Some precisely described criterion of choice is also given. Decision theory attempts to determine the most suitable action given the criterion of choice.

In regard to the *solution* of a decision theory problem, the sole criterion of "truth" is whether it is *logically* true. It should be verifiable (or refutable) by means of logical-mathematical methods that in the *given* situation, with the *given* possibilities for action with *given* consequences, the recommended action was really the most suitable according to the *given* criterion of choice. In the last sentence, the word "given" has been *underlined* four times to indicate that decision theory is not concerned with whether the situation described really reflects the actual situation, whether the possibilities for action taken into account are really possible, whether all alternatives have been taken into consideration, whether the consequences assumed in the model would in fact occur, or whether the decision criterion really expresses the interests of the decision-maker. It is concerned exclusively with the determination of the rational decision once all this is given. This is exclusively a logical-mathematical problem. It is not necessary to observe empirically how many of a hundred clever men would actually choose the solution termed rational by theory in the situation described in the decision model. The rationality of the solution is verified not empirically but in a purely logical way.

This does not imply that decision theory has nothing to do with experience. It is closely linked with the latter in two respects.

First, it comes into contact with reality in the course of its *application*. When one's intention is to employ some proposition of decision theory, it is necessary to know if the particular decision model gives a correct description of the problem under consideration. If this is not the case, then a logically correct answer obtained with the aid of the decision theory model may be misleading in practice. This applies, however, not only to decision theory, as a special logical-mathematical science, but to all branches of logical-mathematical science. Exactly the same is true with the differential and integral calculus. Its theorems have been proven logically and not empirically. They will not be disproven because some engineer applies theorems valid for continuous functions without success to a problem characterized by a non-continuous relationship.

Moreover, while the results of decision theory find application in practice, conversely, *practice may pose questions* to decision theory, inspiring it to generate new models and to work out new propositions.

Even though we recognize the close connection between decision theory and practice, the statement remains valid that decision theory constitutes a theory according to Definition 2.1 and not according to Definition 2.2, i.e. it is a logical-mathematical science and not a real science.

However, for at least a century, most economists have confused decision theory with economic theory as a result of the following psychological supposition: "Let us assume that man behaves rationally." In this case, every model which gives *recommendations* for strictly rational behaviour is properly regarded as constituting a *description of reality*.

This psychological assumption has been used in several forms. Historically, it dates back to the introduction of the "*homo oeconomicus*" concept in economic thinking, to the ideal type of the man whose every step is dictated by economic rationality. The philosophical background of this school of thought was provided by the English utilitarians (Bentham and others). Ever since, it has formed the nucleus of every economic model which intends to describe economic *reality* under the assumption that the economic units "optimize", i.e. minimize or maximize some function. Not only consumers and families, but also firms and even governments and states, are supposed to do this.

Some apply the assumption of rationality with more restraint by saying: "Let us assume *in the first approximation* that those concerned (the consumer, the firm, etc.) are behaving rationally." The use of the term "first approximation" is usually meant to indicate the fact that the author himself has no doubt about the inaccuracy and inadequateness of his assumption. However, matters are typically left at that, and no "second approximation" is attempted.

Actually, people will behave differently, some in a consistently clever way, some less intelligently, and others generally stupidly; for the majority, strict rationality is mixed with irrationality. Examining the various types of human behaviour as *mass phenomena*, certain *stochastic regularities* can be observed; it is the distribution of reactions to definite impulses that shows certain stable properties. However, all this cannot be characterized with the simple statement that man strives at "optimization".

The regularities of man's behaviour and actual economic performance can be observed and described by the real science of economics. However, observation and description cannot be replaced by the statement that man actually behaves as he *should*, that is, that he always heeds the strictly rational advice of optimization models.

From this it does not follow that decision theory cannot be applied to economic problems. It can be used for the purpose of *advising*, for working out recommendations concerning rational future action, and for planning and programming (provided, of course, that the right decision model, a model not far removed from the actual decision and planning problem is used).[6] The generalization of those experiences gained in the course of economic policy making may form part of an economic theory (extending the boundaries of the discipline beyond that of the real sciences in the strict sense of the word).

The possibilities for application of decision theory must be acknowledged; the importance of mathematical decision theory from the point of view of economic thinking cannot be overestimated. *The "transplantation" of the models of decision theory cannot, however, serve as a substitute for a scientific economic theory describing reality.* The widespread practice of "transplantations" of this type diverts attention from the essential problem of providing a correct description and explanation of economic events.

Decision theory includes a large number of models and theoretical propositions. One of these will be treated separately in some detail in two later chapters, namely, the theory of preference orderings and utility functions.

---

[6] The reliability of this statement may be enhanced by the fact that the author has by now been engaged for twelve years now in the utilization of mathematical models of decision theory for economic advice and planning. This proves, if nothing else, at least that I am *convinced* of the practical usefulness of decision theory for economic application. (See [128], [132], [133], [133a].) Acknowledgement of this usefulness does not, however, imply confusion about the scope of decision theory and scientific economics.

2.4. Economic Theory versus Econometrics

Econometrics constitutes an *auxiliary* science related to economic real-science theory, but it is not the economic real-science theory per se.

Econometrics[7] is the mathematical representation of interdependencies of economic variables. The parameters of the mathematical representation are estimated on the basis of statistical observation, by means of mathematical, statistical methods.

From many points of view, econometrics satisfies the requirements of Definition 2.2; it describes the real movements of the economy, verifies its assumptions on the basis of statistical observations, compares its findings with actual reality, and so forth.

Why is it that econometrics can be regarded only as an auxiliary science, and not as the economic real-science theory per se?

Let us survey, for example, the econometric models describing the movement of a national economy as a whole.[8] There are simultaneous equation systems composed of economic relationships. Both essential and less essential relationships figure here *coordinately;* both are members with "equal rights" of a common equation system which attempts to describe the operation of a particular economy, claiming approximate validity for a specific country in a specific period. However, every real scientific theory possesses an intricate and complete theoretical structure.

A *theoretical structure* is a hierarchically constructed intellectual system, a "multi-level" structure of laws. There are laws of a higher order, of more general validity, which are valid in a wider sphere, over a more extensive region or for a longer period of time, and there are others of a lower order that are more specialized in character, describe a narrower range of phenomena and are less generally applicable over space and in time.

In the case of a hierarchically constructed theory, the general laws may not be directly verifiable and it is only the validity of the special laws that can be confirmed by observations. The general law is confirmed if it offers a common explanation of several special laws and the latter are consistently deducible from the former.

---

[7] I use the expression "econometrics" in the narrower sense generally adopted in the West, i.e. I consider here only numerical models which make use of parameters estimated with mathematical-statistical methods and explain the functioning of the economy. Therefore, e.g. the mathematical economic models described in a pure symbolic form, used only for deduction of theorems but not specified numerically are not considered as part of econometrics.

[8] See e.g. the description of the Klein–Goldberger model: *Klein* [120], or publications on the Hungarian statistical macro-model, e.g. *Halabuk*'s paper [79].

3*

The most highly developed natural sciences—especially some branches of theoretical physics—have already achieved a theoretical structure involving a hierarchical organization of propositions which are either basic equations and laws or special laws that derive from the basic ones.

This hierarchical character, this type of multi-level structure, is absent from econometrics. All of its achievements, therefore, may be considered important as *preparations* for the formation of a scientific theory of economic systems, but not as a fully developed theory.

2.5. The Significance of the Formal Model

From what has been said above, the role assigned to formal models in economic theory should become clear.

(1) Formal models help to build up a theoretical structure. With the aid of such models, it is possible to determine whether contradictions exist among the various assumptions and propositions, whether the assumptions are necessary and sufficient and whether the hierarchy can be arranged in terms of more general and more specific laws.

(2) With the aid of the formal model, theorems may be derived through deductive reasoning which could not have been discovered in an inductive fashion on the basis of existing knowledge. Deduction cannot, however, replace the empirical verification which must be performed sooner or later.

(3) The generation and description of the formal model can indicate what observations are necessary and suggest which economic variables and parameters are actually relevant to the relationship under examination.

This and nothing more is the role of the formal model. But even this is of enormous importance, a fact that should satisfy every mathematical economist.

A formalized model *may* constitute a method of description in a *real-science theory*. But this is not necessarily the case. It may also be a tool in an intellectual experiment starting exclusively or partially from unrealistic assumptions, and, as such, a useful tool of cognition. However, the generation of a formal model will not, in itself, create a theory. The formal model may constitute a tool of theoretical research in the real sciences (thus, also in economic science), but the *generation of a model does not in itself guarantee that a real-science theory has in fact been achieved.*

2.6. The Indicators of the Maturity of a Real Science

The maturity of a real science may be characterized by means of various indicators. Three of these will be discussed:

(a) What stage has been reached in the verification of the theory, in its comparison with reality? How reliable is the verification?

(b) What stage has been reached in the formalization of the theory? How precise is the description of the theory?

(c) Has the theoretical structure of the theory and the hierarchy of its propositions been established?

I propose to call a real-science theory mature if it satisfies, to a large extent, requirements (a), (b), and (c).

By the above standards, I would consider highly mature, for example, numerous branches of theoretical physics.

How mature, then, is the theory of economic systems?

Unfortunately, *no mature economic systems theory exists.*

Modern mathematical equilibrium theory only satisfies criteria (b) and (c) because it constitutes an exact, formalized and hierarchically organized theoretical structure. It does not satisfy at all, however, the principal requirement of a real-science theory (Definition 2.2); its propositions have not been verified. Moreover, in several cases, both its assumptions and propositions are obviously in conflict with reality. *Equilibrium theory is merely an intellectual experiment.*

Later on, I will endeavour to substantiate this statement with arguments.

Instead of real maturity, we see that general equilibrium theory possesses only pseudo-maturity. For the description of the economic system, mathematical economics has succeeded in constructing a formalized theoretical structure, *thus giving an impression of maturity*, but one of the main criteria of maturity, namely, verification, has hardly been satisfied. In comparison to the vast amount of work devoted to the construction of the abstract theory, the amount of effort which has been applied, up to now, in checking the assumptions and statements seems inconsequential.

3. THE BASIC CONCEPTS OF GENERAL EQUILIBRIUM THEORY

3.1. Character of the Survey

Let me begin the survey of general equilibrium theory (GE theory) with a summary of its *basic assumptions*. Then, I shall review the *main concepts* and finally turn to the *principal questions* which the theory's propositions are intended to answer.

In the course of the survey I will rely, as far as possible, on GE theory's own conceptual framework; the concepts I intend to propose will be defined in a later part of the book.

The choice of a particular author on whose model my investigations should be based is difficult. GE theory has not yet attained a final synthetized form. It is a theory of long standing which presently finds expression in numerous models differing from each other to a greater or lesser extent. Important new publications, which claim to contribute to the development of the theory, appear almost monthly. Thus, it would be arbitrary to confine the analysis and criticism to the work of a specific author. However, I cannot provide a complete synthesis. This should, in my opinion, be undertaken by the adherents of the theory and not by its critics. Therefore, I will proceed in the following way.

With some arbitrariness, I have worked out a "prototype" of GE theory. In other words, I have drawn up a "meta-model" of the family of GE models, one that includes the characteristic features of most members of the family.

Adherents of the GE theory, will be able to deny that their model or theory corresponds *exactly* with that described in Sections 3.2–3.4 of this book. However, they must admit that it shares many characteristics in common with my meta-model.

My survey is based primarily on the most prominent representatives of the GE school, the *Arrow-Debreu* models of the 1950's.[1] It does not, however, extend to the modifications worked out in the 1960's. The more recent achievements of GE theory will be treated in the portion of the book dealing with history of the theory.

I shall not rely on the subtle formulations of the leading representatives of GE theory. On the contrary, my survey is closer in character to simple text-

---

[1] See e.g. *Debreu* [50], 1959.

book expositions than to the more sophisticated formulations of scholars employing refined tools. However, as I intend to criticize a *school* and not merely particular works, I do not regard this approach as inappropriate.[2]

3.2. Basic Assumptions

In this section, I will survey the basic assumptions (axioms, postulates) of GE theory. This can be done in several ways. It would be possible to describe the system of assumptions in terms of the smallest complete set of axioms. However, it will serve the purpose of my book better to treat every assumption with distinct economic content separately.

The basic assumptions will be expressed only in verbal form in order to bring out their economic content more fully.

The list of basic assumptions will not be complete; only the most characteristic ones will be discussed. I shall omit those which are without any special economic significance.

GE theory does not use all of its basic assumptions in the proof of each of its theorems. In the present survey I do not indicate which assumptions are required for proof of each theorem.

Some of the basic assumptions are not always *explicitly* made by the authors, but are recognizable as *implicit* assumptions if one examines their work.

Twelve basic assumptions will be outlined here. For future reference, each assumption will be given a short name. The assumption is set in italics in the text below, and commented upon wherever necessary.

1.A. *(Static character) The subject of investigation is the behaviour of the econopoint in time under investigation.*

1.B. *(Stationary character) The subject of investigation is the behaviour of the economic system over time, assuming, however, a number of elements in the model to be constant.*

The model can be interpreted in two ways. In the case of interpretation 1.A the model is strictly static; it offers only a snapshot of the economy, leaving both past and future outside the scope of investigation. In the case of interpretation 1.B the model is dynamic, with the qualification that the framework of the activities, their "setting" as it were, remains unchanged and that, as a result, the various processes within the economy are stationary ones. In the course of discussion of the basic assumptions it will become clear which of the components of GE theory are assumed to be constant over time.

---

[2] The basic assumptions and main theorems of equilibrium theory are summarized in a formal manner by *Debreu* [50], *Koopmans* [124], pp. 265–285, and others.

A more popular summary of equilibrium theory is given by *Baumol* in [27], Chapter 13.

In the following list of assumptions, interpretation 1.B is considered as an alternative to 1.A. Accordingly, some restrictions which are listed as basic assumptions, apply only if the model represents a system operating over time.

2. *(Constancy of the set of organizations) The economic system consists of a definite number of organizations; their number and set remain unchanged over time.*

The economics system is composed of interrelated elements; its elements are organizations. The complete list of organizations is called the set of organizations.

The second part of basic assumption 2 is only relevant in the case of the 1.B stationary interpretation of the model. It means that GE theory regards the set of organizations as unchanged over time.

3. *(Producers and consumers) The economic system consists exclusively of two types of organizations: producers and consumers.*

Three components of basic assumption 3 should be emphasized:

There is no other organization that plays a part in the economy.

The organization behaves in a unified manner; GE theory is not concerned with investigating conflicts within an organization.

No subordination or superordination exists among the organizations; all organizations have equal status.

4. *(Constancy of the set of products) The economic system produces a finite number of products; their number and set is unchanged over time.*

Let us imagine that each product is given a separate serial number and that the products are listed according to their serial numbers. Products with identical technical and functional properties which appear, however, at *different points* in space or at *different times* within a finite period of time would be given separate serial numbers. In specifying the quantity of each individual product, we accordingly specify its location in space and time as well.

The complete list of products will be termed the set of products.

The second part of assumption 4 is relevant only in the case of the 1.B, stationary interpretation of the model. GE theory regards the set of products as unchanged over time.

5. *(Simultaneous operation)*

5.A. *The activities of the producer can be described by a vector where the positive components give the inputs and the negative components the outputs. On the output-side: producing intention (production plan, production program), actual production, selling intention (supply) and actual sale are identical. On the input-side: factor use intention (the plan of factor use), actual use, buying intention (demand) and actual purchase are identical.*

5.B. *The consumer's activities can be described by a vector of identical dimensions where the positive components are the consumptions. The consumer's desired*

*consumption equals his planned purchases and actual c̶ ̲sumption equals de-
mand.*

5.C. *There is no time lag between the production, sale, and consumption of a
product.*

5.D. *At a definite point in time, production and consumption are accompanied
by a given price system; there is no time lag between the processes of production
and consumption on the one hand and their effects on the price system on the
other.*

According to basic assumption 5, the economy described by GE theory
operates without either material or monetary inventories and reserves.

Basic assumption 5 is particularly relevant in the case of the 1.B, stationary
interpretation.

. 6. *(Convexity of the production set) The set of feasible productions is convex.*[3]

I do not intend to indicate here the sub-assumptions which are necessary
and sufficient for the assertion of basic assumption 6.[4] We will content ourselves
with a looser treatment, listing only those sub-assumptions usually considered
necessary for "convexity" by the majority of GE authors.

6.A. *There are no indivisible products; the volume of production can be
described by means of continuous variables.*

6.B. *There are no indivisible resources; the volume and capacity of all resources
can be described by means of continuous variables.*

6.C. *The relationships between inputs and outputs can be described by means
of continuous and differentiable functions.*

6.D. *There are no increasing returns to scale. This means that if all inputs of
some product are increased in a given proportion, the volume of output cannot
increase in greater proportion.*

6.E. *The marginal rate of substitution between the factors of production is non-
increasing. This means that if the input of one factor is increased and that of
another is decreased, a constant or increasing quantity of the former factor will be
needed to replace successive equal units of the latter.*

We should emphasize that in the last decade there have been a number of
attempts to weaken the convexity (resp. concavity) assumptions 6, 7 and 8.
A few such models are known which prove — sometimes in a changed form — the
individual theorems of the GE theory, in the case of weaker assumptions than
those under 6.A–6.E (E.g. demanding only quasi-convexity, or quasi-concavity,

---

[3] It is not necessary to emphasize here the other properties of the production and
consumption sets (such as the closed character etc.) because from the economic point
of view only convexity is an essential restriction.

[4] A complete survey may be found in an article by *Martos* [166a], as well as in the
studies of *Arrow—Enthoven* [13] and *Arrow—Hurwicz—Uzawa* [16].

etc.)[5] These important contributions to theory will be discussed later in the book.

7. *(Profit maximization) The producer maximizes the difference between total revenue and total expenditure, i.e. his profit.*

According to basic assumption 7 the producer has a preference ordering on the set of production alternatives.[6] When choosing between two possible production alternatives with given prices, he can state unequivocally whether he prefers one to the other or is indifferent between them. The sole criterion of choice is his profit.

In view of the fact that returns are non-increasing (see assumption 6.D) and also because prices are given, the profit function is concave. In other words, a given increase in production results in a proportional or less than proportional increase in profit.

8. *(Maximization of consumer utility) The set of feasible consumptions is convex. The consumer has a preference ordering over this set and maximizes his utility function. The consumer utility function is concave.*

From the economic point of view, basic assumption 8 implies the following:

Since there are no indivisible products (as stated in 6.A) quantities consumed can be described by means of continuous variables.

The various products may be combined arbitrarily in the consumption bundle.

When choosing between two possible consumption alternatives, the consumer can state unequivocally whether he prefers one to the other or is indifferent between them. This can be represented by means of a utility function.

The concavity of the function implies that if all items in the consumption bundle are increased in a given proportion, utility cannot increase in greater proportion.

9. *(The constancy of production and consumption sets and of preference ordering) Neither the set of feasible productions described in assumption 6 nor the set of feasible consumptions described in assumption 8 change over time. Nor do the preference orderings of economic organizations change over time.*

Assumption 9 is relevant only in connection with the stationary interpretation, 1.B; it means that GE theory is not concerned with the effects which technical progress and the changes in resources over time exert on production.

---

[5] In this book we shall always say "convexity assumption" when meaning the convexity assumptions 6, 7 and 8. This however is to be understood always with the restrictions mentioned above (for example, in some relations it is only quasi-convexity).

[6] Strictly speaking, the same ought to be said here as in basic assumption 8; the decision-maker has a complete *pre*ordering of preferences. Instead, we content ourselves with the term "ordering." The exact definitions will be given in Chapter 10 which deals with preference ordering.

Similarly, the theory disregards the changes in demand over time, induced by technical, cultural and social factors.

10. *(Exclusivity of price information flows) Prices constitute the sole flow of information between the organizations of the economic system. At any given point in time, each product has a single uniform price.*

No direct or indirect flow of information of non-price character exists among the organizations of the economy.

11. *(Anonymity of market relations) In the economy, total production of a product is confronted with total consumption; producers are indifferent among consumers in the selling of output and vice versa.*

In the economic system, neither the seller nor the buyer draws any distinctions among his trading partners; no individual relations exist between producers and consumers. This can be interpreted as meaning that anonymous producers turn over their products to a black box, the market, which transmits them to the anonymous consumers.

12. *(Lack of uncertainty) Each organization of the economy knows its own set of possibilities and its preference ordering over it.*

The GE theory assumes that there is no uncertainty whatever in the economic system. Every organization knows exactly its own possibilities as well as all other relevant factors of the environment. In particular, all market clearing prices are known with certainty. On the basis of this knowledge, each organization makes its choices in accordance with its own preference ordering.

In the case of interpretation 1.B this means that there are no uncertainties in the plans and programs concerning the future.

In the last few years there have been important attempts to abolish or to weaken the assumptions on uncertainty. These will be discussed later in the book.[7]

The twelve basic assumptions are outlined above without being ranked in terms of order of importance. Actually, *basic assumptions 6, 7, and 8, the assumptions of preference ordering (optimization) and convexity, are the most important and characteristic assumptions of GE theory.*

3.3. The Conceptual Framework

I do not intend to survey the logical structure of the conceptual framework of GE theory. I am concerned only with what the sociologist would call "content analysis". Every ideology or school of thought possesses its own special

---

[7] In *Debreu* [50] there is a separate chapter attempting to relax the certainty assumption.

terminology which includes five, ten or twenty characteristic concepts employed in describing the phenomena investigated.

When reading a text, we can count the words most frequently used in the introduction, the theorems, the conclusions, etc. When the words which occur most often coincide with the "key words" of some school of thought, it may be inferred that the text in question belongs to that ideology.

GE theory has envolved its own special terminology. Without any pretense of completeness, let us list the concepts which may be regarded as most characteristic:

Preference, utility, optimum.
Demand, supply.[8]
Price, profit.
Equilibrium.

The above eight concepts are particularly characteristic of the works belonging to GE theory.

3.4. The Questions to Be Answered

The questions which a theory attempts to answer constitute the main distinguishing feature of that theory. As a matter of fact, all other characteristics—the basic assumptions as well as the main concepts—are largely determined by the questions to which answers are sought.

GE theory aims at providing answers to two main types of questions.

The first type of question concerns the existence and stability of equilibrium. GE theory, as indicated by its name, addresses itself to the problem of determining the conditions which must obtain for an equilibrium to exist and be stable. It asks what processes bring about an approach to equilibrium, what the necessary and sufficient conditions of equilibrium are and whether or not the equilibrium satisfies one or another of a number of criteria for stability.

Equilibrium is the central problem which is treated. Let me quote some general definitions of equilibrium in the natural sciences:[9] "Equilibrium is a macroscopic state of changeable (mobile) material systems (bodies), brought about by the interactions of external and internal forces, which remains un-

---

[8] *Samuelson*'s famous textbook, which includes a humorous motto at the head of each chapter, places the following statement before Chapter 4: "You can make even a parrot into a learned political economist—all he must learn are the two words "Supply" and "Demand"." See [213].

[9] Természettudományi Lexikon (Encyclopedia of Natural Sciences) [57a], Vol. 2, p. 212.

changed over time. If these interactions—or, rather, the indicators of state expressing their magnitude (such as pressure, temperature, concentration, etc.) undergo a change, then the equilibrium will also change. In the state of equilibrium, the effects tending to change this state (the forces originating from the interactions) macroscopically neutralize each other and their resultant effect is zero." The general definition of mechanical equilibrium, a special case of equilibrium, is the following. "Mechanical equilibrium is a state of a mechanical system or a body in which every point of the system is forever at rest (in relation to the system of reference taken as a basis) or performing rectilinear uniform motion with constant velocity. Although rest and rectilinear uniform motion are equivalent to each other, a system will usually be considered to be in equilibrium only if the system is at rest at a definite point in time and also forever at a standstill."

The concept of equilibrium is seen to be closely connected with that of *rest*. Accordingly, when examining the conditions of equilibrium for an economic system, we actually wish to determine the conditions which ensure that the system comes to rest, that it assumes a state where it is in the interest of no participant in economic life to change his behaviour and thereby, disturb the equilibrium.

There can be no doubt that in society and, in the economy, forces exist which drive the system toward a state of equilibrium just as in natural systems. The investigation of these forces is both important and interesting. It is, however, no less important to seek an answer to the following questions:

What types of disequilibrium exist?

Does the economy oscillate around equilibrium? Or is disequilibrium a lasting or even constant trend?

Is equilibrium desirable as a "norm", as the ideal state to be reached? Is disequilibrium, or some definite type of disequilibrium, not preferable from the point of view of economic progress?

The second type of question concerns the optimality of the state of the economic system. The representatives of modern mathematical GE theory (above all, *Arrow* and *Debreu*) must limit themselves with a rather weak criterion of optimality, the so-called *Pareto-optimality* criterion. The economy is said to be in a Pareto-optimal state if no other state exists which satisfies at least one consumer to a higher degree and no consumers to a lower degree.

The Pareto-optimum concept presupposes the acceptance of basic assumption 8, i.e. the existence of an ordering of consumer preferences; it does not, however, require any commensurability or additivity of the latter.

In my opinion, the acceptance of basic assumption 8 is unwarranted (this question will be dealt with in detail below), but we shall disregard this fact for the time being. At a given moment, and under different conditions, the

economy may assume a variety of Pareto-optimal states. True, that the enforcement of Pareto-optimality, the exclusion of those programs which violate
Pareto-optimality, is an elementary requirement of rationality, and therefore,
analysis of Pareto-optimality is important. But even if this requirement of
Pareto-optimality is complied with, the question remains open as to whether
the economy is in other respects functioning satisfactorily or not.

Here, we will not deal with the question of what would happen if we evaluated
the state of the economy on the basis of some macro-economic preference
function and not according to Pareto-optimality. This question will be discussed
at the end of this book when treating the different types of GE theory.

Finally, the GE theory endeavours to determine what relationship exists
between equilibrium and the Pareto optimum; under what conditions does
equilibrium ensure Pareto-optimality and *vice versa?*

In this context it is important to mention that the theorem "equilibrium
implies Pareto optimum" is independent of the convexity assumptions, listed
in Section 3.2.

3.5. What Should Be Considered a Related Theory?

I intend to criticize not only those works which fully accept the twelve basic
assumptions listed in Section 3.2, employ the concepts described in Section 3.3
and answer exclusively the questions outlined in Section 3.4. I wish to criticize a
much wider range of practical and theoretical economic thought, and especially
those theories related to GE theory.

Definition 3.1. We will consider the schools of economic thought which satisfy
criteria A), B) and C) below as r e l a t e d t o g e n e r a l e q u i l i b r i u m
t h e o r y.

A) Their *assumptions* are largely identical with the twelve basic economic
assumptions of modern mathematical GE theory. Even if they deal with a
narrower sphere (e.g. with the firm or the consumer alone), the assumptions
employed are identical with or similar to the corresponding assumptions of GE
theory. Other works may treat the economic system as a whole, employing,
if not identical, at least similar assumptions.

The closeness of the relationship to GE theory depends primarily on the
assertion of the following basic assumptions:

a) The assumption of convexity;
b) the assumption of optimality;
c) the assumption concerning the exclusivity of price information.

The severing of ties with GE theory begins essentially with giving up these
assumptions.

B) Their *conceptual framework* is largely identical to that of modern mathematical GE theory and employs those terms listed in Section 3.3.

A characteristic criterion for the existence of a relationship with GE theory
is that the various schools of thought employ the above concepts in the same
sense (and leave them, to the same degree, open to criticism) as the GE theory.

C) The *questions posed* by them are essentially identical to those of GE
theory, i.e. they are concerned with two types of question:

— with the conditions for the existence and stability of equilibrium;
— with the conditions of optimality.

From the above it is clear that the level of closeness of the relationship may
vary considerably.

Definition 3.2. The set of models of general (i.e. economy-wide) equilibrium
in the strict sense of the word, and those closely related to GE theory according
to Definition 3.1 but covering a narrower sphere, I will call the g e n e r a l
e q u i l i b r i u m   s c h o o l.

Accordingly, the term "GE school" denotes a definite school of economic
thought. The boundaries of the school are, admittedly, somewhat vague, with
several works occupying a marginal position, so that the closeness of their
relationship with the original GE theory is open to question. However, for
the majority of these works it is possible to conclude almost definitely whether
they belong to the GE school or not. My work is concerned with the criticism
of the *whole* GE school and not only with the models of GE theory in the
strict sense of the word.

3.6. First Summary Evaluation

Anyone wishing to understand the operation of the market, be it that of a
capitalist or socialist country, is bound to become acquainted with the GE
school. The work of the GE school has thrown light on numerous important
aspects of the operation of the market, particularly on the interaction between
demand, supply and price under conditions where producers maximize their
profits and price formation takes place without central interference.

The theories belonging to the GE school have brought to the fore highly
significant notions concerning adaptation to consumer requirements and allocation of scarce resources.

With due acknowledgement of the contributions of this school from the
point of view of the history of theory, it is now time to shift the emphasis to its
negative aspects since, in my opinion, *the GE school has become a brake on the
development of economic thought*. We include in this body of theoretical litera-

ture most of the work which is attributed to the "neo-classical school," or called "neo-classical price theory" in Western countries.

It is not my intention to criticize or refute one by one, in the order described in Sections 3.2 to 3.4, the basic assumptions, the main concepts, or the major questions posed. To accomplish this, I would be obliged to follow the scheme of thought of the GE school when my aim is its criticism. Instead, I shall proceed with my own analysis, submitting to criticism *in the process* one or another basic assumption, concept or method of the GE school.

To help orient the reader, *Table 3.1* presents a survey of the various places in the book where the assumptions, concepts and statements of the GE school will be subjected to criticism. Here, I will give only a preliminary outline of the leitmotivs of my criticism in order to facilitate the development of my ideas.

1. As has been pointed out above, in connection with its historical importance, the GE school has suggested two important and correct ideas: First, scarce resources should be used economically, and second, production should be adapted to needs in order to give the greatest possible satisfaction to the consumer. However, these conclusions stem from an analysis based on an unrealistic vision of the world. In reality, there are mammoth corporations and the role of the government is great. GE theory assumes atomized markets and "perfect" competition. In reality, there exist sharp conflicts of interest. GE theory sees peaceful harmony in the market. In reality there is concentration and rapid technical progress. GE theory "disregards" increasing returns to scale, one of the most significant aspects of technical progress and one of the fundamental explanations of concentration. In reality, the information structure is highly intricate and complex. GE theory describes a system governed in an entirely reliable manner by a single signal, namely prices.

Thus, the GE school is disorienting, and diverts our attention from the most important task of economic science, namely, the realistic description, explanation, and formal modelling of the actual operation of the socialist and capitalist economic systems of the present era.

2. The use of the attribute "general" in connection with Walrasian models is warranted in so far as they always describe a *whole* system (e.g. the economy of a whole country) and not only some part of it. Thus, the "general" equilibrium models may be contrasted with the "partial" equilibrium models of the GE school, which describe only a single firm or household, or the market for a single product.

It would be entirely unwarranted, however, to interpret the attribute "general" (and this is exactly what many economists would do) as implying a real-science theory of *general validity*, valid, to a certain extent, for every age, country and system.

TABLE 3.1

References to the critique of general equilibrium theory

	The critique may be found mainly in the following chapters:
Assumptions of the GE theory	
1. Static or stationary character	8, 9, 11, 12, 13 14, 20, 21, 22
2. Constancy of the set of organization	14, 17, 21
3. Exclusive existence of producers and consumers	4, 5, 6, 7
4. Constancy of the set of products	11, 20, 21
5. Simultaneous operation	5, 17, 18, 23
6. Convexity of the production sets	9, 14
7. Profit maximization	7, 10, 11
8. Maximization of consumer utility	10, 11, 14
9. Constancy of production and consumption sets and of preference ordering	10, 11, 20
10. Exclusivity of price information flows	4, 5, 6, 17, 18
11. Anonymity of market relations	5, 17
12. Lack of uncertainty	11, 14, 23
Major concepts of the GE theory	
Preference, utility, optimum	10, 11
Demand, supply	17, 18
Price, profits	5, 7, 23
Equilibrium	14, 18, 19, 21, 22
Main questions which the GE theory intends to answer	
Equilibrium	14, 16, 18, 19, 21, 22
Optimality	11, 14, 16, 21

In actuality, we are dealing with a highly special theory, valid only within a highly restricted sphere.

The question of whether the individual assumptions are "true" if taken one by one is hardly relevant. It may be said of almost all of them that there exists a restricted category of phenomena for which the assumption is acceptable, at least as an approximation. Yet, it is clear that none of these assumptions describes even roughly certain crucial phenomena of the real world.

For example, we observe in reality, diminishing, constant and increasing returns to scale. A general theory should hold for all three, but the GE school is concerned almost exclusively with the second (or, at best, with the first two).

In reality, we observe some decisions characterized by optimization. However, a variety of other decision processes exist. The GE school deals nearly exclusively with the former.

In reality the consequences of decisions are sometimes predictable with certainty and at other times, only with uncertainty. Yet, the GE school is mainly concerned with the case of certainty.

Further enumeration of these restrictions is unnecessary; these will be dealt with extensively in the sequel. Each individual assumption automatically excludes from consideration many phenomena present in reality. The joint use of the assumptions *multiplies* this effect. *The category of phenomena which can be even approximately described by the set of twelve basic assumptions is extremely restricted.*

The conceptual apparatus is similarly narrow. A great number of economic phenomena and processes cannot be described simply with the aid of the characteristic concepts listed above and the additional concepts derived from them.

Finally, *the range of questions to be answered is also extremely narrow.* A series of fundamentally important questions concerning the assessment of the operation of economic systems is left unconsidered.

3. The GE school *makes* the description of economic systems entirely *too dull;* it *over-schematizes* and *impoverishes* it. Recognizing only one type of consumer behavior, one type of motive force for the firm, and one type of information, it diverts attention from the study of complex and intricate structures, from classifying types of behaviour, motive forces and information, and from investigating the causes of differences among these types.

4. Mainly as a result of its static (or stationary) approach, the GE school offers *little explanation of the real motion of the economy.*

Some basic assumptions might be given either a more restricted or a wider interpretation. The difficulty consists in the fact that under a more restricted interpretation, the assumption might be *erroneous* and under a wider interpretation, *empty* or almost meaningless. Let us take, for example, assumption 4, concerning the constancy of the set of products. In the case of a more narrow interpretation, there would be no new products and the economy would always have to choose from among and produce the same ones. This interpretation of the assumption would obviously provide an erroneous point of departure for economic systems theory, since many important problems concern the changes in and development of new products. A wider interpretation of basic assumption 4 might assert that the list of products includes not only those presently produced, but also those which potentially can be produced in the future. Thus, the model could allow for the appearance of new products. First, the economy would turn out products with serial numbers from 1 to 1000 and later, the products with serial numbers from 1001 to 2000. This would expand the framework of the GE model. It would permit the inclusion of new products *without, however, giving any explanation* of the motive force or the factors furthering or hampering the development of new products. The expanded theory is, therefore not erroneous, but rather, poor.

The case is similar with several other assumptions.

5. Finally, a remark of seemingly technical character: the formalism of the GE school is rather *inconvenient*. To explain this remark, we must refer to the basic assumptions.

The basic assumptions include some "voluntarily" (and wrongly) chosen assumptions, as well as some erroneous or overly restrictive ones which are forced on the theorist by virtue of his "voluntary" assumptions and necessary for the solution of the model. For example, assumption 6 concerning convexity is a restriction of the latter type. Every adherent of the GE school well knows the significance of non-convex production sets, but the mathematical apparatus employed forces him to apply the assumption of convexity. Non-convex sets are much more difficult to treat mathematically. Every economist would be glad if this difficulty could be overcome. As mentioned before, there have been attempts in this direction.

With the "voluntary" assumptions, the case is entirely different. The most important from this point of view are assumptions 7 and 8 concerning the strict rationality of the economic agents. This involves the formalization of economic systems theory in terms of an extreme value problem. This assumption does not simplify the mathematical treatment of the problem; on the contrary, it makes it more difficult in contrast to a more simple causal description, according to which certain impulses produce definite reactions with some regularity. Yet, the GE school voluntarily and deliberately accepts the difficulties involved in order to abide by the postulate of strict rationality.

To put in even more clearly, the problem is not simply that the GE school chooses "voluntarily" and "deliberately" inconvenient assumptions, independently of each other. The "original sin," contained in basic assumptions 7 and 8, is the supposition that optimality characterizes the actual behaviour of economic organizations. Once this supposition has been accepted, the GE theorist is forced to assume convexity, to disregard uncertainty, and so forth. This is necessary for the determination of the points of extreme value which characterize the equilibrium state and optimal condition of the economic system as a whole. If, on the other hand, we do not insist on presenting the economic system as a set of optimizing elements, we could describe the regularities and laws relating to the operation of the economy in a considerably more convenient form.

In summary, I have listed five main objectives in anticipation of further criticism; I will have to substantiate my remarks with additional arguments.

In my critique I do not strive at originality; I will repeat many arguments also made by others. My aim is to attempt a *synthesis* of known and new critical remarks.

4*

PART II

CONCEPTS AND QUESTIONS

What seems to be exceedingly difficult in economics is the definition of categories... it is always in the conceptual area that the lack of exactness lies.

J. von Neumann: The Impact of Recent Developments in Science on the Economy and on Economics

In the course of coming into contact with empirical material, physicists have gradually learned how to pose a question properly. Now, proper questioning often means that one is more than half the way towards solving the problem.

W. Heisenberg: Physik und Philosophie

4. THE GENERAL MODEL OF THE ECONOMIC SYSTEM

4.1. A "Language" to Describe the Economic Systems Theory

In Chapters 4 to 16 an attempt will be made to develop a "language" suitable for the description, comparison and analysis of economic systems. Like the real spoken languages, this one consists, on the one hand, of a basic vocabulary, i.e., a conceptual framework. I will endeavor to suggest and define the basic concepts necessary for the description of economic systems. On the other hand, in analogy with the spoken languages, there are basic grammatical rules. This means, in our case, that in order to define the concepts, we have to describe definite relationships among the various concepts (or among the real phenomena and processes referred to by the concepts). These are, however, only *definitional relationships* and not real-science theorems or propositions.

From the above it follows that, in connection with the proposed concepts and the relationships existing among them, one cannot ask whether or not they are "true." All one can ask is whether they are suited to the purpose at hand, whether they are workable.

This undertaking, the clarification of the concepts, may seem academic to many people. In fact, however, in the development of the sciences, concepts play an important part, both in a positive, promoting sense, and in a negative, retarding one as well. The general use of a wrong, vague or ambiguous concept, which evokes erroneous associations, may impede the recognition of essential relationships. Sometimes a concept will seem entirely clear until science subjects it to thorough investigation and discovers its vagueness, its lack of precision.

In developing a "language" for economic systems theory, I have attempted to satisfy three requirements:

1. The language should be *more general* than that of traditional economics. It should contain concepts suited to the description of economic systems differing from one another. These concepts should enable one to characterize the numerous types of socialist and capitalist structure which exist in different countries.

2. The language should be suitable for a *more concrete* and penetrating description of the economic system than the traditional language of economics.

3. My intention is to further the development of operative and *workable* concepts. In many cases, the definition of a concept may be dispensed with if the ways of observing and measuring the phenomenon reflected by it are indicated instead. In general, the formation of correct concepts encourages observation and measurement. The elaboration of a language, of a conceptual framework, will thus lead to the posing of *questions*, and defining of problems of scientific research, observation and analysis.

It would be desirable to impose an additional, fourth requirement, namely that of a strictly logical structure for the conceptual framework. The various groups of derived concepts ought to be defined with complete consistency on the basis of a few basic concepts. This, however, I cannot undertake. All that I can do here is outline some elements of a new language for economic systems theory, the development of which I consider both desirable and necessary.

Although the main task of Part II of the book is, as expressed by its title, the introduction of concepts and formulation of questions, in some places I will go further than this and make *statements* concerning the structure and functioning of economic systems.

The course of this analysis will resemble that of an airplane in flight. From a high altitude we are often able to survey an entire country. Then, losing altitude, we perceive the network connecting the various points of the country—roads, railroads, transmission lines. Descending further, we can distinguish ever smaller units—towns, blocks, houses. Ascending once more, we are again able to view ever larger units—entire mountain ranges, regions, or countries.

Passing now from the analogy to the arrangement of the chapters which follow:

In Chapter 4, we will deal with the economy as a *whole* in its most general form.

Chapters 5 and 6 describe the *network* connecting the elements of the economic system and the *flows* taking place among them.

Chapter 7 presents the *"interior" of the institutions* which form the system; the organizations within the system differ from each other in the nature of their activity.

Chapters 8 to 12 penetrate even deeper, dealing exclusively with the *decision processes* within the organizations.

Thereafter, we will "ascend" again and continue the investigation on a broader scale. The subject matter of Chapters 13 and 14 is the collective behavior of and interrelations among the organizations and the related problems of the system's *autonomous operation, adaptation and selection*.

Chapter 15 deals with the problem of forming an aggregate from the ensemble of different organizations or processes, and *classifying* the organizations and processes.

Finally, Chapter 16 again analyzes the system as a whole, its subject being the evaluation of the operation of economic systems.

Now, we can proceed to clarify the most general concepts.[1]

4.2. Organization and Unit

The term "economic system"[2] is used here in a broad sense. It designates, in a majority of cases, the economy of a country. However, the concept may be used more narrowly to refer, for example, to a single branch of production or a single territorially defined area, a county or town, within a country. Conversely, a set of several countries like the countries belonging to CMEA (Council of Mutual Economic Aid) or all of the countries of Africa may also be considered as a system.

Let E denote the economic system under investigation.

Breaking down the economic system into its components, elements on three different levels can be distinguished: the *institutions*, the *organizations* within the individual institutions and the *units* within the individual organizations.

An institution is usually an intricate and complex social and economic formation, such as, a big modern corporation or a ministry. In this chapter we will not deal with the institutions in detail; this subject will be discussed in Chapter 7.

Within an institution there are organizations. Examples of organizations are the productive plant or the sales department within the firm, or the planning section or the technical development section within a ministry.

The household constitutes a special organization. In this case "institution" and "organization" are identical.

To use an analogy, society is considered here as matter is considered in physics. The molecule is the institution and the atom, the organization. In this chapter we will not deal with the molecular structure but only with the atoms.

Definition 4.1. An o r g a n i z a t i o n is a social formation consisting of persons who associate in order to perform some definite social-economic functions.

Usually, it is possible to state unequivocally what social-economic function (e.g. production, planning, etc.) the organization performs. It should also be possible to enumerate individuals who comprise it at a given moment. One per-

---

[1] In the formulation of Chapter 4 I have drawn many ideas from *Hurwicz*'s essay [97].

[2] The concept of the economic system will be defined in Definition 4.20.

son may, naturally, belong to several organizations at the same time; as a worker, he belongs to a firm and, as a family member, to a household.

In economic system E a total of m organizations function. Let these be denoted by $o_1, o_2, \ldots, o_m$. Let O denote the set of organizations. $O = \{o_1, o_2, \ldots, o_m\}$.[3]

The organization is composed of units. A general definition of the unit will be given below;[4] here, it is only provisionally defined.

Definition 4.2.' A u n i t is a nondivisible element of the economic system which behaves with definite regularity and responds to outside stimuli in a regular manner.

The difference between the organization and the unit must be clearly recognized. The organization is a real social formation. It consists of actual, living persons; its sphere of activity, in addition, is usually defined by law. The unit, on the other hand, is a mere abstraction; it serves the purpose of generating in the model the activities which take place within the organization.

The unit is a part of the organization; as will be indicated below, two units belong to each organization. A unit can, however, belong only to one organization and cannot be shared by two or more.

To expand the analogy above, if the organization is the atom, then the unit is the elementary particle within the atom.

4.3. Input, Output, State

The economic system operates over the *course of time*. In order to simplify the treatment of the subject, a discrete time scale will be assumed.[5]

Definition 4.3. Time intervals $t = 1, 2, \ldots$ of equal length, adjoining each other, will be called p e r i o d s. Several periods following upon each other constitute a t e r m.

Depending on the specification of the general model, a period may be one year, one day, one minute.

Definition 4.4'. The operation of the unit consists in receiving in every period an i n p u t, releasing an o u t p u t, and changing its internal s t a t e in

---

[3] I use set theoretical concepts in many parts of this book. References on set theory include the following works: *Kemeny et al.* [116], *Kalmár* [110], *Berge* [33], and *Debreu* [50].

[4] For a complete definition of 4.2, see Section 4.9. The prime after the number indicates the provisional character of the definition.

[5] It is, of course, arbitrary to assume that time is discrete. We may also formulate the model in continuous time. But the explanation of some notions and interrelations and the expansion of the analogy to the theory of automata are facilitated by this assumption.

the process. Inputs and outputs are called, collectively, f l o w s. The process which takes place within the unit which results in the transformation of input entering the unit into an output leaving the unit, producing at the same time, a change in the internal state of the unit, will be called an i n t e r n a l p r o-c e s s.

The basic concepts introduced above are rather general in character. The input of the unit might be some raw material to be processed, but it might also be an instruction. The output might be a product released, but it might also be a production report. The state of the unit might be characterized by the stock of goods on hand or by the technical condition of the machinery, but it might also be characterized by the documents accumulated on the office desks.

If we were dealing simply with a group of units operating independently of each other, it would not be possible to speak of a system. The ensemble of units forms a system only because they are connected to one another by the input and output flows.

Definition 4.5. The outputs of the a d d r e s s e r units are the inputs of the a d d r e s s e e units.

Every output becomes the input of some addressee unit, and conversely, every input was the output of some addresser unit. In other words, every flow has an addresser and an addressee.

4.4. Real Sphere and Control Sphere

The processes which take place within the economic system may be classified into two categories:

*Definition 4.6.** The r e a l p r o c e s s e s of the economic system are mate-rial, physical processes. These are production (including transportation, ware-housing, material services, etc.), consumption and trade. The c o n t r o l p r o c e s s e s of the economic system are intellectual processes. These include observation, information transmission, information processing, decision pre-paration, and decision-marking. Real processes are described by *real variables*, control processes by *control variables*. The unequivocal separation of the real processes from the control processes is called the d u a l i s t i c description of the economic system.

For the sake of conciseness, the following abbreviations will be used: R denotes real and C denotes control.

This sharp separation is obviously an abstraction. Actually the two types of activity are closely intertwined and interrelated. On the one hand, there are no real processes without control. On the other hand, every control process assum-es some form which is physically perceptible, whether it is writing on paper, or conveying, vocally, a message by telephone. This fact, notwithstanding, we will consistently apply in our further investigations the dualistic description of the operation of the economic system. This is one of the most essential characteristics of the "language" proposed in this book and one that distinguishes it from other modes of description.

Naturally, the classification is, sometimes, necessarily arbitrary. Should education be considered an R-process or a C-process? In marginal cases the decision may be left to those who are using the proposed "language". For the classification of marginal cases there are, presumably, no generally valid rules; it all depends on the concrete aims of the scientific analysis. On the abstract level of the present discussion it is permissible to assume that the economic activities lend themselves to unequivocal classification into one of the two categories.

It follows from the explanations above that an individual usually takes part in both control and real processes. The director of a factory both gives commands and uses paper, at home he plans the family budget and takes his supper. We do not place an individual exclusively in one or in the other sphere, but on an abstract level we distinguish two main classes of economic processes.

Let us now remind the reader of the distinction made in Definition 4.4, between flows and internal processes. On the basis of this distinction, two subcategories of the processes should be noted:

Definition 4.7. Production and consumption constitute the i n t e r n a l r e a l p r o c e s s e s. Information processing, decision preparation and decision making constitute the i n t e r n a l c o n t r o l p r o c e s s e s.[6]

The units fall into two categories:

Definition 4.8. Within the r e a l u n i t s, only internal real processes, and within the c o n t r o l u n i t s, only internal control processes take place.

Let us now return to the organizations defined in 4.1. To every organization, o_i, there belongs a real unit, r_i, and a control unit, c_i, $(i-1, \ldots, m)$. The coordinated units $[r_i, c_i]$ are called *elementary unit pairs.*

To resort to a simile of a somewhat religious tone, an elementary unit pair belongs together like spirit and body, with the control unit governing the real unit.

The relationship between the organization and the elementary unit pair is presented in *Figure 4.1*. The rectangle indicated by a thick solid line represents the economic system. The rectangles inside it, indicated by thin lines are the organizations. (In the figure only two organizations are presented.) Within an organization, there are two circles; one of these is the R-unit, the other the C-unit. Figure 4.1 will be repeatedly referred to in the following discussion.

---

[6] The other two categories—real *flows* and control *flows*—will be defined later.

Let $\mathcal{R}$ denote the set of real units $r_1, r_2, \ldots, r_m : \mathcal{R} = \{r_1, r_2, \ldots, r_m\}$. Let $\mathcal{C}$ denote the set of control units $c_1, c_2, \ldots, c_m : \mathcal{C} = \{c_1, c_2, \ldots, c_m\}$.

Definition 4.9'.[7] The economic system E is composed of two subs-ystems, the set of interconnected real units, r e a l s p h e r e. $\mathcal{R}$, and the set of interconnected control units, c o n t r o l s p h e r e $\mathcal{C}$.

The two spheres are distinguished from each other on the abstract level; in reality, however, they are closely intertwined through interaction with each other.

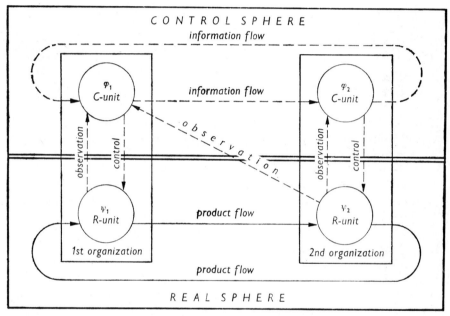

Figure 4.1
Scheme of the economic system

In Figure 4.1 the two sub-systems are separated from each other by a double horizontal line, with the C-sphere above and the R-sphere below.

[7] Attention is called to the following:

For the concepts of real sphere and control sphere, only provisional definitions are given in the first approximation. (This is indicated by the prime after the number of the definition.) A complete definition can be given only later when the economic system has been precisely defined. Sub-system $\mathcal{R}$ is not actually identical with R, the simple *enumeration* of the real units, but includes their operational *characteristics*, as well. The same applies to $\mathcal{C}$ and C.

In accordance with what has been said above, every organization includes a unit which belongs to the R-sphere and a unit which belongs to the C-sphere. Thus, the organization itself belongs to both spheres. Let us illustrate this point with some practical examples:

The R-unit of the productive plant includes the productive workshops which belong to the real sphere. Management, however, belongs to the C-sphere.

The intellectual processes connected with planning take place in the C-unit of the long-term planning section of the National Planning Office. However, even here, there appear material inputs; paper, office machines, means of conveyance are used. The latter are represented by the R-unit assigned to the section.

An attempt has been made to give an estimate of the relative proportions of the R- and C-spheres in the Hungarian economy. For the sake of simplicity, only a single unit of measurement was applied, namely, employment in organizations *mainly* concerned with real activities, and in organizations *mainly* concerned with control activities as a percentage of total employment. In the delimitation of the spheres, arbitrariness is inevitable, but as an approximation, the following statement appears acceptable: about 83 to 85 per cent of those employed in the economy act in the real sphere and about 15 to 17 per cent in the control sphere.[8] It can be seen that the control sphere is substantial even if exmployment alone is taken into consideration. However, on the basis of qualification and of renumeration, which reflects the former to some extent, the proportion would be even larger.

4.5. Products

Let us now turn to the variables which describe the operation of the economic system. Here, too, the dualistic approach worked out in the preceding section will be exmployed. Let us first survey the R-processes.

Definition 4.10. The types of real output turned out by the real units are called p r o d u c t s; these are used as inputs by real units.

The products are distinguished from each other according to all of their characteristics (similar to the procedure applied in the case of general equilibrium theory models). Let these be denoted $g_1, g_2, \ldots, g_n$. Let $\mathcal{G}$ denote the set of all products: $\mathcal{G} = \{g_1, g_2, \ldots, g_n\}$.

Let us assume that the quantities of all products can be measured in some unequivocal and additive unit.

[8] The estimation was performed by *Zsuzsa Mausecz* (National Planning Office). In accounting we did not consider the households as organizations.

Definition 4.11. The real flow which takes place in the economic system in period t is described by a p r o d u c t f l o w vector $x(t)$. Its components are distingusihed by three indices: the first index gives the serial number of the product, the second that of the addresser unit and the third that of the addressee unit. The numerical value of the component expresses the quantity of the product flow as measured in the unit of the product in question.[9]

Definition 4.12*. The state of the real units of the economic system in period t is described by a p r o d u c t s t o c k vector $y(t)$. Its components are distinguished by two indices: the first gives the serial number of the product, the second, the unit of measurement. The numerical value of the component expresses the amounts of real stocks as measured in the unit of the product in question.

4.6. Information

Let us now describe the C-processes.

Definition 4.13*. The information stored in the units of the economic system and flowing among the units can be classified according to various criteria. The finest element of classification, which cannot be further subdivided, will be called in the i n f o r m a t i o n t y p e .

The information types are given serial numbers. In the system E there are q information types. Let these be denoted $s_1, s_2, \ldots, s_q$. Let $\mathscr{S}$ denote the set of all information types: $\mathscr{S} = \{s_1, s_2, \ldots, s_q\}$.

Definition 4.14*. The information flow which takes place in the economic system in period t is described by an i n f o r m a t i o n f l o w vector $u(t)$. Its components are distinguished by three indices: the first gives the serial number of the information type, the second that of the addresser and the third that of the addressed unit.

Definition 4.15*. The state of the control units of the economic system is described by a m e m o r y c o n t e n t vector $v(t)$. Its components are distinguished by two indices: the first gives the serial number of the information type, the second the number of the unit which stores the information.

Definition 4.16. The i n f o r m a t i o n v a r i a b l e is a collective concept which includes the vectors of both the information flows and the memory contents.

---

[9] Instead of vector notation, array notation could also be employed; in this instance, the real flows would be represented by a three-dimensional array arranged according to the three indices mentioned. However, in expounding the conceptual framework, the vector notation is simpler and one to which economists are more accustomed. Therefore, this notation will also be employed when defining the other flow and state variables.

In the four definitions above, we have introduced an entire series of concepts. Let us now discuss these in some detail.

It is crucial to understand the relationship between information types and information variables. The set of information types, set $\mathscr{S}$, is a *constant* characteristic of the system E. The actual values of information variables $u(t)$ and $v(t)$ will *change*, however, from period to period.

When defining an information type from the economic point of view, it is necessary to state to which of the economic system's elements, components, processes and phenomena it relates, what properties, states and characteristics of the latter it describes, to what time period it refers, and what temporal relation between information and action exists. An information type may be, for example, the wholesale price fixed by the Hungarian National Price Board for the product marked with serial number 127. Pertinent to the economic definition of the information type is the product's name (No. 127) the price type (wholesale, fixed by central authority) and the unit of measurement. To this information type there may belong, in period t, a whole range of information *variables*. For example, the Prices Board may have announced a new price in period t of 155 Forint. In this case, we are dealing with a component of $u(t)$, the vector of information flows, the addresser of which is the Price Board and the addressees of which are the firms concerned. Or perhaps, the price was announced earlier but now it can be established from the records of the Price Board and the firms that in 1967 it was 142 Forint and in 1968 it became 155 Forint. In this case, it is a component of the $v(t)$ vector of memory contents that contains this information type.

From this it follows that to every information type there corresponds a measure which can be used to indicate the actual value of the information variable. In the case of some information types this can obviously be expressed by a real number. Thus, in the above example, the measure of the wholesale price is Forint; the actual value of the information variable is expressed by a real number, which is, in our example, 142 or 155.

In other cases, the information variable may refer to various possible discrete states, qualitative valuations, such as "yes" or "no," "small," "medium-size," or "large," "red," or "green," etc. But even in such cases, the various possible values of the information variable may be represented by some index number as, for example, in the case of the trio "small"-"medium"-"large," the numbers 1, 2, and 3 respectively.

In the definition of the information type it is, thus, necessary to state, together with the unit of measurement, also the range of values which the information variable may assume.

To illustrate the concepts "information type" and "information variable", et me outline another idea.

Let us imagine a printed form. On this form, every piece of information which flows among the units of the economic system or is stored in the units in some given period of time can be entered. The form has printed boxes with serial numbers from 1 to q. To each box we assign an economic interpretation, a definition, a unit of measurement and filling instructions. In the corresponding empty places of the box, figures can be entered.

Now, the boxes of the form represent the information *types*. The figures to be entered in the empty places of the form are the *actual values* of the information variables. (Also to be entered here are, in the case of a flow, the serial number of the addresser and addressee unit, and in the case of storage, the serial number of the memory unit owner.) In the former example, one of the form's boxes is the wholesale price of product No. 127; the figure to be entered is 155.

One more remark concerning the concept of memory contents is in order. Although this seems to be a rather abstract concept, in fact it represents well-known things. In thinking of the memory of a firm, we must not envision archives with dusty old records and statistical reports no longer used by anyone. The memory contents include only the "living" experiences which continue to exert an influence on the firm's operations; in one sphere, the information relevant to decision-making includes the basic experiences of the past months, and in another sphere maybe those of the past years, condensed in figures and accumulated information.

4.7. The Response Function: An Introductory Example

In the provisional definition 4.2' of a unit, it was stated that the unit behaves according to definite regularities and responds to outside stimuli in a predictable manner. With the aid of the concepts introduced thereafter, we can advance a step further. The relationship between the input entering the unit and the unit's initial state, on the one hand, and the output leaving the unit and the change in the unit's state will be called the *response function*.

Before giving an exact definition, let us elucidate this concept by means of a simple example.

Consider as a unit the pedestrian on the sidewalk at an intersection with a traffic light, who wishes to cross the street.

The main input of the unit is the color of the traffic light. The input may assume three discrete states: "red", "yellow" or "green". In order to simplify the discussion, let us disregard the intervening yellow light and consider the red and green lights only.

The main output of the unit consists of two types of action: "waiting" and "crossing".

Now, there is a definite relationship between input and output. When the input is red the output is waiting; when the input is green the output is crossing.

This correspondence between inputs and outputs is called, in the first approximation, the "response function" of the unit which, in the example above, is the pedestrian. The pedestrian's action is a *function* of the colour of the traffic light.

It may be necessary to explain why the pedestrian's behaviour is treated as a "function". The reader acquainted with mathematial economic models will accept unquestioningly as a function, for example, the Cobb–Douglas function, $Y = AK^{\alpha}L^{1-\alpha}$. But can the behaviour of the pedestrian be treated as a function?

Although the above example is of an illustrative nature, we will continue to use the function concept in a broad sense to include the pedestrian's response function. This fully corresponds to the *general* function concept used in mathematics[10] which regards as a function the relationship or correspondence between the elements of two sets. In our example one set is composed of the elements "red" and "green", while the other set is made up of the elements "waiting" and "crossing". It is between these elements that correspondence exists.

Thus far, we have idealized the pedestrian's behaviour. The truth of the matter is that a pedestrian will sometimes risk crossing *notwithstanding the red light*. His behaviour will be influenced by various factors:

1. How heavy is the traffic? When there is no vehicle nearby, the temptation will be greater than when automobiles and buses crowd the road.

2. Is there a policeman around? If there is one standing on the other side of the road, the pedestrian will think twice before crossing.

3. The pedestrian's behaviour will be influenced by past experiences. An adult observes the rules more instinctively than a child. A man who has already had an accident or who was fined just the day before by a policeman will have greater respect for the traffic rules than one who has never had such unpleasant experiences.

4. The pedestrian's behaviour may be influenced by his momentary mood, his state of mind. If he happens to be nervous because he was just reproached by his superior at the office, he may not pay attention to the traffic light. Or, he may flaunt the traffic rules simply because he is in high spirits and does not care.

Thus, the independent variable of the response function is not merely the sign of the traffic light. There are other information types, too, such as impressions about the momentary situation on the road. Another independent variable is the unit's "state", its "memory contents", in our example, the pedestrian's experiences and state of mind.

---

[10] See, e.g. *Kalmár* [110]. p. 7 and 9, *Szép* [248], and *Debreu* [50], Chapter 1.

Likewise, the response function's dependent variables are not only "waiting" or "crossing". The internal state may also change; for example, the pedestrian may become upset at having to wait a long time.

The general form of the response function is, accordingly, the following:

$$(\text{change of state; output}) = f(\text{initial state; input}).$$

If a given initial state and a given input determine uniquely the dependent variables, then the response function is deterministic. However, the response is usually not deterministic but rather *stochastic* in nature. In determining the response, i.e., the change of state and the output, random factors also play a part. Initial state and input do not uniquely determine the value of the change of state and of the output; the latter are random variables.

4.8. The General Form of the Response Functions

With this illustrative example in mind, let us return to our discussion of the economic system.

In the following discussion, we will, at first, deal exclusively with *deterministic* response functions. Only at the end of this section will some remarks concerning *stochastic* response functions be made.

First of all, we need to introduce some notation.

In the definitions above, operation variables $x(t)$, $y(t)$, $u(t)$ and $v(t)$ were assumed to be single vectors for the economic system as a whole. Now, however, we want to describe the relationship between input, state and output for *each unit separately*. For this purpose, we must, first of all, partition the state vectors $y(t)$ and $v(t)$ by units. Then, we must partition the flow vectors $x(t)$ and $u(t)$ by units, including each component twice, corresponding to the two types of unit, the addresser and the addressee.

The operation of the i-th unit ($r_i \in \mathcal{R}$ real unit or $c_i \in \mathcal{C}$ control unit) is characterised by the following variables:

$\overset{\leftarrow}{x}_i(t)$ = the product input entering the i-th unit as addressee;
$\vec{x}_i(t)$ = the product output leaving the i-th unit as addresser;
$y_i(t)$ = stock of products of the i-th unit;
$\overset{\leftarrow}{u}_i(t)$ = the information input entering the i-th unit as addressee;
$\vec{u}_i(t)$ = the information output leaving the i-th unit as addresser;
$v_i(t)$ = the memory contents of the i-th unit.

On the basis of our earlier definitions, the following statements may be made about the six types of variable of the units described above:

The vectors $\vec{x}_i(t)$, $\overset{\leftarrow}{x}_i(t)$ and $y_i(t)$ are not defined for C-units because no ye processes take place.

The vector $v_i(t)$ is not defined for R-units because R-units have no memor$R$ Vector $\vec{u}_i(t)$ contains the guidance information received and vector $\overset{\leftarrow}{u}_i(t)$ tht· observations reported.

It will be assumed that in each phase the unit receives inputs only once, changes its state only once and releases outputs only once. (We shall make use of this assumption later on.)

The response functions will be given separately for the C-units and the R-units.

Definition 4.17*. The r e s p o n s e f u n c t i o n of the i-th control unit:

$$(v_i, \overset{\leftarrow}{u}_i) = \varphi_i(v_i, \vec{u}_i), \tag{4.1}$$

i.e. the memory content and the information output depend on the memory content and the information input.

In response function (4.1) the relationship between input, state and output has been described independently of time. The operation of the control unit, however, takes place over the course of time and we must, therefore, take into account also the temporal relationship between the variables. This is done in the following manner:

$$(v_i(t), \overset{\leftarrow}{u}_i(t)) = \varphi_i(v_i(t-1), \vec{u}_i(t)), \tag{4.2-}$$

i.e. the present memory content and information output of the control uni depend on the earlier memory content and on the information input just received

The response function of the real unit is also defined in two steps:

Definition 4.18*. The r e s p o n s e f u n c t i o n of the i-th real unit:

$$(y_i, \vec{x}_i, \overset{\leftarrow}{u}_i) = \psi_i(y_i, u_i, \overset{\leftarrow}{x}_i), \tag{4.3}$$

or, taking into account the temporal relationships:

$$(y_i(t), \vec{x}_i(t), \overset{\leftarrow}{u}_i(t)) = \psi_i(y_i(t-1), \vec{u}_i(t), \overset{\leftarrow}{x}_i(t)), \tag{4.4}$$

i.e. the present stock of products, product output and information output of the real unit depend on the past stock of products, and on the information and product input currently received. The meaning of the $\vec{u}$ information input: the control unite *guides* the real unit. The meaning of the $\overset{\leftarrow}{u}$ information output: the real unit is *observed* by its own control unit or by some other control unit.

Both response functions describe a *transformation* which takes place inside the C-unit or the R-unit. The input and the unit's initial state are transformed into the output and the unit's terminal state.

Definition 4.19*. The transformations which take place in the economic system as a whole, the set of response functions of all interconnected units, are described by the following two response function systems: the r e s p o n s e f u n c-t i o n s y s t e m of the control sphere, which is denoted Φ and the r e-s p o n s e f u n c t i o n s y s t e m of the real sphere, which is denoted Ψ.

The response function systems Φ and Ψ constitute the "skeleton," the forms of the economic system's general laws of motion. They can receive cotent only if suitably specified.

Although the response functions have been described above in a very general form, this form, nevertheless, involves certain restrictions to which attention must be drawn:

First restriction. As has been pointed out, we have assumed a definite temporal relationship between input, state and output, namely, simultaneity of input, state and output.

This does not involve an essential restriction. It does not exclude, for example, the representation of time lags and the modelling of the interrelations of units connected through flows, etc.

At any rate, it is not even necessary to insist on the temporal relationship between input, state and output which has been formalized in functions (4.2) and (4.4). Equally, and in certain cases even more advantageous, may be the assumption of some other relationship, such as the lagging of the relation between output and input by one period.[11]

Second restriction. The response function is constant over time. This restriction is more essential than the former one. The recommended formalism is dynamic, describing the operation of the economy over time. The changes which take place over the course of time, however, are expressed exclusively by the changes in the actual values of vectors x, y, u and v from period to period. In contrast, the behavioral regularities of the units, and as a result, of the system, are not modified.

This may cause some difficulties when describing definite processes; it may become expedient to work out a formalism where not only the operation variable but also the response function itself changes over time.

Apart from the two restrictions mentioned above, the response functions (4.1) to (4.4) are entirely general. There may be investigations which can be carried out employing only the most general form. For a more profound

[11] The abstract theory of automata asks under what conditions the various automata types which describe the relationships between input, state and output in different ways, e.g. with different time lags, are *equivalent* to each other. In other words, it tries to determine when they are able to represent the same series of events in an identical manner.

On this question see e.g. *Starke* [240], [241], and *Gluschkow* [71].

description of economic systems, however, it is usually necessary to *specify* the response functions. On this general level of treatment of the subject, however, we will not deal with the mathematical form of the response function.

Here, at the end of the general discussion of response functions, I must make an essential remark:

Up to this point—thus also in formulae (4.1)-(4.4)—deterministic response functions have been described. In fact, as was pointed out in connection with the example of the pedestrian, random factors play a part, too.

Here, I do not consider it my task to formalize the effects of random factors. Formalization would require numerous further definitions and explanations; on the present highly general and "language" clarifying level of discussion, this is unnecessary. I only want to state emphatically that the *flows and changes of state are in reality the outcomes of stochastic processes*.

The operation of the units of the economic system is characterized by *stochastic response functions*. The unit's output and changes in state are influenced not only by the input and the initial state but also by other, random factors.

Hereafter, when using the term "response function" without qualification I shall always mean a stochastic response function. The deterministic response function constitutes a special case of the latter. To formalize the stochastic response function is a task for further research.

4.9. Comprehensive Definitions

Armed with the necessary concepts, we are now able to give some definitions of fundamental importance.

First of all, we can provide a final definition of the unit.

Definition 4.2. The u n i t is an indivisible element of the economic system the behavior of which can be described by means of a response function.

We can also give a definition of the economic system:

Definition 4.20*. The e c o n o m i c s y s t e m $E(O, \mathcal{G}, \mathcal{S}, \Phi, \Psi)$ is composed of organizations and units making up these organizations which operate over time. The organizations and the units are connected to each other by product and information flows. The operation of the economic system is determined by the units of response functions.[12]

---

[12] According to the definition the system is *closed*. Within a given examination the sphere that is not closely studied i.e. the "external world" can be considered—in the specific model describing the problem—as an n-th organization of the system which is connected to the first, second ...$(n-1)$th organizations by outputs and inputs. In such cases we can suppose that the output of the external world does not depend on the input received, only on the time and chance. The "resources" of the system are considered as the product outputs of the n-th organization, i.e. the external world.

Having defined the concept of the economic system, we can now replace our provisional definition of the two spheres, with a complete one.

Definition 4.9*. C o n t r o l s p h e r e C ($\mathcal{C}$, $\mathcal{S}$, $\varPhi$) and r e a l s p h e r e R($\mathcal{R}$, $\mathcal{G}$, $\varPsi$) constitute two sub-systems of economic system E defined in 4.20.

Definition 4.21. The set of organizations, O, the set of products, $\mathcal{G}$, the set of information types, $\mathcal{S}$, the response function system of the control sphere, $\varPhi$, and the response function system of the real sphere, $\varPsi$, are the c h a r a c - t e r i s t i c s of the economic system E. The characteristics jointly give the structure of the economic system.

It follows from the above that in the *five* characteristics defined in 4.28 alone, the *structural* features of an economic system can be summarized in the most concise form possible. It is, of course, also necessary to indicate the values of the state variables—i.e. the product stocks $y(t)$ and the memory contents $v(t)$ for a starting point in time, let us say for $t = 0$, in order to forecast the future operation of the economic system.

Definition 4.22*. E c o n o m i c s y s t e m s t h e o r y is a real science.[13] Its subject matter is the description, classification and comparison of the characteristics of economic systems. It focuses primarily on the arrangement of the economic system into organizations and units, and on the flows among the organizations and units. It investigates both the real and the control spheres, and especially the interconnections between the two spheres, i.e., the control exercised over the real sphere by the control sphere.

4.10. The Unit and the Economic System as "Automata"

The method of describing the economic system employed in this Chapter fits closely into the general conceptual framework of cybernetics, mathematical systems theory and the theory of automata.[14]

---

[13] The concept of "real science" has been clarified in Part I, Definition 2.2. In this chapter, however, the adjective "real" has appeared in a different sense in numerous concepts; real process, real sphere, and so forth, is always contrasted with the control flow, control sphere, etc.

I trust that this second use of the word "real" will not confuse the reader. The economic systems theory defined in 4.22 as a real science, deals, naturally, not only with the real processes and the real sphere but also with the control processes and the control sphere.

[14] In this connection I rely heavily on the studies of automata theory, works by *Ashby, Kleene, MacCarthy, Shannon, Neumann, Uttley* and others, which appeared in the "Automata Studies" [226], *Beer*'s book [29], *Gluschkow*'s manual on automata theory. *Starke*'s articles [240], [241], *Toda-Sluford*'s [257], *Van Court Hare*'s [264] and *Mesarovic*'s [178] works on systems theory.

Several books are available in Hungarian also, among them, some books by *J. Neumann*, e.g. [190] and [191], a selection [247] of classics in cybernetics, *O. Lange*'s book [148], works by *R. Tarján* [251] and *Wintgen*'s article [277].

The economic system, as defined in 4.20, is a special case of the system as described by mathematical systems theory.

The unit defined according to 4.2 can be considered an abstract automaton. At the same time, the economic system, consisting of an interconnected network of units, may be regarded as a complex abstract automaton.

This relationship opens up many important scientific possibilities:

1. The *general theorems* developed and proven by cybernetics, systems theory and the theory of automata can be used in describing and analyzing the economic system, organizations and units.[15] Some of these theorems touch upon problems which are important from the point of view of economic systems. These include, among others:

—The similarity and identity of systems and automata.
—The breakdown and composition, the complexity of systems and automata.
—The reliability of automata (especially of stochastic automata).
—The control and regulation of systems.

2. There is a close connection between the theory of automata and the development of computers. The theory has mainly served to provide a theoretical basis for the development of computers.

It is well-known that if the behaviour of a system, a series of events, can be described by means of the formalism of automation theory, then it can also be *simulated* on the computer.[16] The theory of automata bears a close relationship to the computer's simulation language. With the conceptual framework outlined above, this relationship can be extended to the language of economic systems theory.

We would like to set at ease the minds of those who are probably irritated by the analogy with automata. "Man is, after all, not a machine..." Indeed not.

---

[15] The systems theory description of the economy bears a certain affinity to the so-called "structuralist" models which have been used in other social sciences. Thus, in anthropology, philology and literature science we encounter analytical works which regard the subject of their investigations (for instance, a language) as a *system* composed of elements, and endeavor to characterize the relationships among the elements, the structure of the system. For a short survey of the structuralist schools see, for example *Miklós*'s article [179], in the special issue of the periodical "Helikon" treating structuralism. In the same issue are published *Levi-Strauss*'s, *Goldman*'s, *Jakobson*'s and others articles on structuralism.

[16] It should be mentioned here that the ideas expressed in Chapter 4 took shape when we were engaged in simulating economic processes on the computer. (On this question see the study written by *B. Dömölki* and the author of the present book [134], and *J. Tankó*'s report.) Together with my mathematical co-workers, we tried to develop a uniform conceptual framework, a common "language" as it were, for the series of independent simulation experiments conducted. This is exactly what led to the basic concepts and definitional relations.

When speaking of an abstract stochastic automata, we are not thinking of an automatic pressing machine in a factory or of the coffee-dispenser of a snack-bar. The question here is simply whether or not it is possible to observe any regularities (usually only statistical or probabalistic regularities) in the economic system and in its elements. If so, then these regularities can be formulated more or less precisely in the language of economic systems theory as well as in that of the theory of automata. If not, then that language cannot be used; nor can any other scientific language. Scientific description is only possible where regularities can be observed.

Finally, one last remark.

While emphasizing the analogy with cybernetics, systems theory and the theory of automata, I do not wish to "commit" myself too strongly to these new and rather "fashionable" branches of mathematics. The *response* function systems Φ and Ψ are meant to express *the economy's general laws of motion. The mode of description in terms of the theory of automata is only one of the possible formalizations* suitable for the description of behavioural laws. In some definite research works it may be more expedient to use some other formalism. Mathematicians have given much attention to the problem of the equivalence of formalisms, that is, of "conversion" between one mode of description and another. All the economist can do is attempt to employ the formalism which promises to be best suited to his purposes.

4.11. A "Dictionary" of Some Current Economic Concepts

Let us now turn our attention to the question of the relationship between the concepts described in Definitions 4.1 to 4.22 and other basic concepts generally employed in economic theory.

"*Economic mechanism.*" The term has been employed since the mid-fifties by Hungarian economists to denote the set of methods of economic administration and management. No uniform and generally accepted definition of the concept has yet been provided, however.[17]

According to the conceptual framework outlined in this book the general interpretation of the term "economic mechanism" is broadly covered by two characteristics of the economic system: O, the set of organizations, the organizational structure of the economic system, and Φ, the response function system of the control sphere, which includes economic management on every level.

The overlapping, however, is not complete; the set of characteristics $[O, \Phi]$ is not exactly identical to any of the standard definitions of mechanism. There-

---

[17] See *Csapó* [46]., *Hegedüs* [87] and *Kornai* [126].

fore, in order to avoid conceptual debates and confusion, I shall not use the term "economic mechanism" in my book.

"Model." Polish economists, discussing the methods of economic administration and management, have used the term "model" in the same sense as their Hungarian colleagues used the term "economic mechanism".[18] (For example, they spoke of the "centralized model" and the "decentralized model" of the economy.) The term is probably even clumsier than its synonym in Hungarian. All other sciences, including economics, use the term "model" in an entirely different sense. Accordingly, the term will be disregarded here (in the sense employed in the Polish debates).

"Control". In the terminology of the debates on the reform of the Hungarian system of economic administration and management, the term is used exclusively to refer to the directing activities of *central* government organizations which regulate the country's economic life.

The terminology is loose and contrary to the connotations associated with the term "control" in most branches of science in all parts of the world.

Here, we will follow the terminology of cybernetics. Control may be exercised by *every* control unit, whether it belongs to an organization on the lowest level (a firm or a household) or to one on the highest level, e.g. the government. Whenever we want to emphasize the fact that we are dealing with a control activity of a central-government organization the adjective "central" will always be added.

"Production forces and production relations". The concept of "real sphere" and "control sphere" used in this book are, undoubtedly, reminiscent of the terms "production forces" and "production relations" introduced by *Karl Marx.*[19] Marx did not provide comprehensive definitions of these concepts; rather, through the repeated *use* of these concepts he endeavored to evoke associations in the mind of the reader. This has made it possible for his followers to interpret the two concepts in different ways. For a long time the views of *Stalin* on the subject were generally accepted,[20] but recently, many Marxists have criticized the Stalinian interpretation.[21] I do not intend to take a stand in this debate. Due to the lack of unambiguously formulated definitions, I am not even in a position to explain precisely the relationship between the concept pairs "production forces—production relations" and "real sphere—control

---

[18] See e.g. *W. Brus* [39]. Although the Polish terminology was also taken over by some Hungarian economists, it did not find any widespread application.

[19] Some places of major importance in *Marx*'s work where the concept pair appears are the following: [169], pp. 406–407, [170], pp. 6–7, and [167], p. 940.

[20] See *Stalin* [249], pp. 649–671.

[21] See e.g. the book by *F. Tőkei* [258] and [259].

sphere". Instead, I must remain content with pointing out the affinity between the two pairs. Only one more remark should be made on the political aspects of the question.

Marx obviously included in the term "production relations", the political and power relations, ownership relations, exploitation relations existing between men and classes. The "language", the method of describing the economy which is proposed in the present and the following chapters, enables us to describe this "political" aspect of production as well. It permits a description of the behavior of groups, strata and classes in conflict with each other; we must only ensure that they are represented in separate organizations and units, and characterized by their own specific behavioral responses. This method makes it possible to describe the distribution of power among them. As a matter of fact, it is necessary to state, for each organization and unit, which activities it controls, what products and resources it has at its disposal, what relations of subordination and superordination exist among the units, and how the responsibilities of decision-making are distributed among them. These are exactly the most important characteristics of "ownership relations", surely more significant than the external legal forms.

"*Socio-economic formation*" *or* "*social-political system.*" These terms are mainly used in the political and economic literature of the socialist countries, to distinguish between the "socialist economic system" and the "capitalist economic system." According to the accepted usage, the word "system" is reserved for formations basically differing from each other, such as socialism or capitalism.

In my book, I will not employ this special interpretation of the "system" concept. Most branches of science consider it a much broader concept and mathematical systems theory has furnished an exact definition of the concept; we cannot, therefore, reserve the term for use in distinguishing between the two formations mentioned above. Nor can the social sciences of the socialist countries dispense with the term "system" in the description of numerous other types of phenomena.

Accordingly, the "socialist system" is a general class of concrete economic systems, and the "capitalist system" is another class of concrete economic systems. Economic systems E_1, E_2, and E_3 may belong to the class of socialist systems, and may nevertheless differ from each other, for example, in the behavior of their respective control spheres. This is the case, with the Polish, Hungarian and Rumanian economic systems; all three differ in their systems of control response functions and yet, they all belong to the common class, socialist economic systems.

4.12. Comparison

A comparison of the "general model" described in this chapter with the models of GE theory would not be entirely "fair". The latter are *specific* models; the authors characterize mathematically the functions, orderings and sets which enter into the models. The general model described in Sections 4.2 to 4.9, on the other hand, is not specified at all. As mentioned in the previous sections, it constitutes, rather, a framework to be filled in, a skeleton of concepts lacking the flesh and blood of specification. Thus, it cannot at this point be considered a virtue that it applies fewer restrictions, since, so far, it fails to derive theorems and theoretical statements from its more general assumptions.

The above notwithstanding, it is still necessary to compare the two approaches in order to point out the characteristic differences in the *mode of thought* between the general model described here and the models of the GE school.

Actually, Walras and his followers also employed a *dualistic description method*. In their works, there is a real sphere (production, the set of feasible productions, products, resources, consumption, the set of feasible consumptions) and a specific control sphere. However, the control sphere is highly specific, manifesting itself exclusively in prices. The "releaser" of prices is an anonymous process, namely, the "market".

It is from this fact that all additional essential characteristics follow.

1. Let $\overline{\mathcal{I}}$ be a subset of $\mathcal{I}$, containing all information-types which appear in the information-*flow* variables. Information-types appearing only in the memory are not in $\overline{\mathcal{I}}$. The number of elements in $\overline{\mathcal{I}}$ is $\bar{q}$.

According to *GE* theory, each product has a single uniform price. Since, as has been mentioned above, information flows exclusively in the form of price, the set $\overline{\mathcal{I}}$ is quite simple: to each product there corresponds a price: $\bar{q} = n$.

In reality, however, to each product there corresponds not a single price but a variety of prices: price-forecasts, preliminary demand and supply prices, differentiated, real selling prices, etc. Moreover, there are many other information flows besides prices. Accordingly, the set contains considerably more information types than the number of products: $\bar{q} > n$. If we want to characterize the system, we must describe $\mathcal{I}$ the set of information types or must at least survey the main groups and classes of information types which play an essential part in the system. This will be the subject of the next two chapters.

2. In the system of GE theory, the set of organizations, O, is also easy to define; there are only producers and consumers. In fact, however, specialized control organizations also operate (and are of increasing importance). If we want to characterize the system, we must describe the set $\mathcal{O}$, or must at least

survey the major groups of organizations specialized for control processes. This question will also be dealt with in the following chapters.

3. According to GE theory each control response function, φ_i, has a special form; the response of the real unit is given by the solution of a conditional extreme value problem.

On the producer's side the only input of the response function is the price of the products and resources released and used. The output is the production program. The nature of dependence is the following: the producer chooses the production program which serves to maximize his profit.

On the consumer's side the only inputs of the response function are the prices of the products he desires to consume and his income. The output is the consumption program. The nature of dependence is the following: the consumer chooses the consumption program which serves to maximize his utility.

The real response function ψ_i ensures the complete realization of the optimal production or consumption plan.

As can be seen, in GE theory, every organization *optimizes*. The general model described in this chapter does not, however, assume optimization, strictly rational and consistent behaviour. It is based only on the assumption that, in the economic system, *causal relationships* prevail. The model of description, according to which the units of the economic system respond to the inputs received by producing outputs and experiencing the changes in the internal state indicated by their stochastic response functions, is called the *stochastic-causal description* of the economic system.

Impulse is followed by reaction, cause by effect. The cause-effect relationship is usually stochastic in character (input and initial state do not uniquely determine output and change of state).

This question will be discussed in detail later on, in the chapter dealing with preference ordering. Here, only the axiomatic aspect of the problem should be pointed out. *The most essential axiom of the GE school is that every organization of the system has a preference ordering. The axiom of the general model proposed here is less restricting: the behaviour of every organization of the economic system can be described by a response function.*

5. INFORMATION STRUCTURE

In the next two chapters, we shall examine in detail the relationships which exist between organizations and units, and the flows which take place between them. We will not focus our attention on the activities carried out within the individual organizations and units belonging to them, but will investigate these matters in later chapters.

In Chapter 4, the flows among the organizations were divided into two main categories, product flows and information flows.

As far as the description and analysis of product flows is concerned, our discipline is reasonably far advanced. The *Leontief* models, the input-output tables (or, rather, the tables of real inputs and outputs) and the programming models describing the real processes have provided us with an easily manageable tool for investigating the problem.

Less satisfactory are the results in the description and analysis of information flows. Economics has an enormous debt which remains to be paid.

5.1. Three Main Classes of Information Flows

The description of information flows and of the memory contents of the organizations in general requires, first of all, a classification of information types. In every economic system, an enormous amount of information flows continously among units, and the number of information types is immense. In order to survey them, it is necessary to introduce various classification criteria.

Although we shall introduce strict classifications and a number of new, related notions, we wish, by no means, to convey the impression that this is the only conceivable classification. The problem requires further empirical as well as theoretical research. Therefore, Sections 5.1–5.5 should be considered a preliminary approximation to a more exact classification.

In the first place, information may be divided into three major categories:
1. Money flows,
2. Price-type information,
3. Non-price-type information.

An organic integration of the theory of money into the conceptual framework of this book is an immense task and extends beyond the scope of my work. I will not attempt to provide more than a single definition in order to facilitate further references in the book.

What does it mean if I have 100 dollars in my pocket? I have command over as many products as can be bought for 100 dollars at prevailing prices. The possession of money implies command over goods and services; if I hand over 100 dollars to a friend, I transfer to him, in the process, my right to purchase 100 dollars worth of goods and services.

Definition 5.1. The f l o w o f m o n e y is a special class of information flow. A transfer of money means a transfer of purchasing power.

It is not only the flow of money which transmits command over goods and services. If a manager goes on leave and tells the deputy-manager that, in his absence, he should act on his behalf, the manager has also handed over a right of command in his absence; the deputy manager is entitled to make decisions about matters concerning the firm. Thus, the flow of money is a special and an important kind of information which serves to transmit power. It pertains only to the real products which, under the concrete circumstances of the given economic system, can be bought for money, i.e., in terms of Marxian political economy, the commodities. Within the realm of possibilities, the actual purchases which will take place are indeterminate.[1] For 100 dollars we may buy either 100 pairs of socks or 20 train tickets. The quantity which can be bought depends exclusively on prevailing prices.

The scope of money flows covers not only the movement of cash but also the movement of means of payment of monetary value including remittances and certain credit transactions.

In the definition and comments above we have dealt only with money *flows;* disregarding the problems of money *stocks*, liquidity, etc.

In addition to money flows, by which we mean the transfer of purchasing power, there are two other major categories of information flows, the main criterion of classification being whether money is or is not used as the unit of measurement.

Definition 5.2.* Classification of information other than money flows: in the case of p r i c e - t y p e i n f o r m a t i o n the numerical value of the information variable is measured in monetary units. All other types of information will be called i n f o r m a t i o n o f n o n - p r i c e c h a r a c t e r.

The price of some product, service or resource constitutes price-type information. A figure indicating the magnitude of some real process may be of

---

[1] This indeterminate character is indicated by *P. Erdős*, in his definition of contemporary capitalist money. Besides its other criteria, money is "a thing (a piece of paper or a mere token), which can be directly exchanged for any commodity, because society immediately acknowledges it to represent—without any separate check prior to the exchange—socially necessary labour, in both senses of the word..." (See [58]).

price character if measured in money terms, for example, production value in domestic prices, or in terms of some foreign currency.

The description of some product, resource, technology or economic action by means of technical-scientific concepts, but without the use of money and price concepts, constitutes information of non-price character. Cases in point are the technical description of a product, the description of a patent or a series of activities connected with some investment project. The numerical description of the extensive and intensive characteristics of some product, resource or process in physical units of measurement also constitutes information of non-price character.

5.2. Further Classification of Information Flows

Having categorized information types into the three categories listed in Section 5.1, and together with these the information flows and the memory contents of organizations, we introduce some further classification criteria.

Classification criterion: Direct vs. indirect reflection. Each piece of information reflects some event, phenomenon or process. From this point of view, two main information groups may be distinguished.

Definition 5.3. D i r e c t   r e f l e c t i o n describes some event or process of the real sphere. I n d i r e c t   (o r   t r a n s m i t t e d)   r e f l e c t i o n describes some event or process of the control sphere.

Let us give a few examples. When a factory sends a report on its production to the statistical office, this is a case of direct reflection. However, when it reports its profits to the revenue office, this is only indirect reflection. It is true that, in the final analysis, the profit report also provides information about real processes, about the relation between real outputs and inputs, but only after several transformations.

In the case of indirect reflection the number of transmissions is important. Let us assume that firm "A" received from its bank information on the credit rating of firm "B" with which it maintains business relations. The credit worthiness of firm "B" is, in the last resort, connected with its real activities, with facts about what it produces, how much it produces of each product, of what quality, for whom, and so on. The real processes are, however, embodied, in the first transmission, in the returns and expenditures; in the second transmission they are embodied in various credit operations, in accounts payable and receivable, in payment on the payable and collections on the receivable accounts. On the basis of all this, some "opinions" will be formed, in the third transmission, about whether the firm in question is trustworthy or unreliable, and to what extent the granting of further credit is warranted. These opinions reach, in the fourth transmission, the bank of firm "A". The bank will perhaps sift and correct, perhaps even distort, the information received, and forward it, in the fifth transmission, to firm "A".

Classification criterion: The addresser. Which organization is the addresser of the information? It could be claimed that it is a matter of indifference who sends off a message; it is only the content that is essential. This view, however, is incorrect. A socialist enterprise distinguishes between information from the planning office and from the enterprise next door.

In the connection, a special kind of information should be defined: Anonymous information is information, the addresser of which is unknown to the addressee. In economic life, anonymous information like the "usual price" or the "fair rate of profit" plays an important role.

A crucial characteristic of a piece of information is whether it is sent to the addressee by only one addresser or by several addressers simultaneously. For example, the government may receive information concerning the prospect of a decline in production from the statistical office, the ministry of finance, the institute for market research, some large firms and even the press, all at the same time.

Definition 5.4. The information flow is s i n g l e - c h a n n e l if a characteristic collection of information reaches the addressee exclusively from one addresser. It is m u l t i - c h a n n e l if this collection of information reaches an addressee from several addressers simultaneously.

Classification criterion: Time-lag. What is the temporal relationship between the information and the event it reflects? (The latter may be an event either of the R-sphere or of the C-sphere.)

Definition 5.5. The reflection is s i n g l e - p h a s e if there is only one reflection emanating from an event. The reflection is m u l t i - p h a s e if there are a series of reflections following each other over the course of time, as a result of an event. In the latter case, a reflection process is said to occur.

Definition 5.6. Information can be a n t e r i o r, s i m u l t a n e o u s or p o s t e r i o r, depending on the time lag between the reflection and the event reflected.

Let us take, as an example, the production of a socialist firm, on, say, the 30th of March, 1969. The production may already have been reflected in anterior information, in the production targets of the five-year and the annual plan, and later on in the monthly and daily production program of the factory. Production is also accompanied by a great variety of simultaneous information. Then, the production data appears in numerous posterior reflections, in the daily production report which is sent to the managing director, in the quarterly report submitted to the statistical office, in the explanatory supplement to the balance sheet presented to the ministry of finance. Moreover, it may eventually

6 KORNAI: Anti-Equilibrium

appear, ten years later, in a planning model based on the mathematical-statistical analysis of long time series. Thus the reflection process is a long one with a five-year preliminary, and ten-year posterior time horizon, in the course of which the information is stored in the memory for a long time.

Classification criterion: Fineness. In how much detail does the information reflect the event? A production plan may be drawn up in minute detail, containing output estimates for all individual products. It may, however, prescribe only that products equal in value to 10 million dollars must be produced, regardless of the actual product mix.

Definition 5.7. Given two pieces of information reflecting the same event, we will call f i n e r the one which describes the event more completely, in a more disaggregated manner.[2]

In the terminology of socialist planning, the term "more disaggregated" means data supplied with a "greater breakdown".

5.3. The Complexity of the Information Flow Structure

The operation of every system is characterized by a definite information structure. The latter can be described with the aid of the criteria listed in the preceding paragraph.

The concept of complexity of the information structure will be defined in detail below. In order to clarify the notion of a complex structure as it exists in reality, the concept of simple structure will also be defined to provide a basis of comparison, although this is an abstract concept which has never had a counterpart in reality.

Definition 5.8*. The concepts of the s i m p l e a n d c o m p l e x i n f o r-m a t i o n s t r u c t u r e s are presented in *Table 5.1.*

The definition is presented in tabular form because it is the ensemble of a large number of criteria that differentiate the simple and complex structures.

The Table is used, at the same time, to summarize (in the last column) the major factors which serve to increase the complexity of the structure.

Summing up, the following statement can be made:

Statement 5.1. For every actually existing economic system, the information structure is complex. Historically, the complexity of the information structure increases with the expansion and development of the real processes. There is an ever greater amount of information connected with the same event which flows sequentially, over the course of time or simultaneously; in other words, the information flow is expanding.

To illustrate Table 5.1 and to substantiate Statement 5.1, let us consider

---

[2] *See Hurwicz* [97]. The originator of this idea was *J. Marschak.*

TABLE 5.1

Characteristics of information flow structures

Criterion	Simple information structure	Complex information structure	Factors tending to increase in complexity of information structure
Price vs non-price character	Information of price character only	Information of price + non-price character	
Direct vs Indirect	Direct reflections and simple-transmission indirect reflections	Direct reflections + Indirect reflections with one or more transmissions	Increasing number of transmissions
Addresser	Anonymous information Single-channel information flow	Specified addressers Multichannel information flow	Increasing number of channels
Time-lag	Single-phase reflection	Multi-phase reflection Anterior and posterior time horizon	Increasing number of elements in reflection series Increasing anterior and posterior time horizon
Fineness	Single degree of fineness	Several degrees of fineness simultaneously for the same event	Increasing amount of information with varying degrees of fineness for the same event

6*

the operations of a large modern firm, which operates either in a socialist country such as Hungary after the reform of economic administrations or in a capitalist country like France or the Netherlands where there exists a certain amount of central planning. Below, various characteristics will be listed. From the standpoint of the information structure, the differences between the firms, would manifest themselves in differences in the relative "weights" assigned to the individual characteristics, and differences in their interrelations. (At this point we shall not describe these differences.)

1. In the life of the firm, information both of price and non-price character may play a role. For the capitalist firm, the former may be relatively more important; for the socialist enterprise the opposite is true, although, with the reform of the system of economic administration and management, the significance of price-type information is increasing. In any case, it is characteristic of every economic system without exception that it receives information both of price and non-price character.

2. The firm will be affected by different types of information, such as indirect information concerning the intentions of other firms, business prospects, market abroad, technical achievements, and so on. The higher the standards of management, the more effort will be expended in analyzing the firm's operations; for this purpose, market researchers, operation researchers, system planners, scientific advisers, etc., may be consulted who will deal with multi-transmission information material. The organizations specialized in performing control process (e.g. the planning offices, government economic departments, market research institutes) transmit to the firms a large amount of information (mostly of multi-transmission type).

The times when it was still possible to manage a firm on the basis of single-transmission information obtained from a market partner are long past.

3. For the small-scale farmer, anonymous information obtained on market prices may be sufficient. But the modern large-scale firm knows its business partners; it maintains personal contacts both with those from whom it purchases and those to whom it sells. The effect of the information is highly dependent on its source.

A firm may have connections with a variety of information sources. Information may orginate from:

a) the firm's own warehouses, from the observation of the absolute volume of and changes in inventories;

b) the firm's own financial apparatus, from the reports on its financial position;

c) the individual market partners separately rather than a single homogeneous black-box market;

d) the competitors, in the form of open and direct information obtained in trade associations, joint ventures and cartels, or in that of information discovered illegally or even obtained by spying;

e) institutions professionally engaged in obtaining and distributing information such as statistical offices, trade journals, market research institutes;

f) banks or other credit institutions;

g) government control organizations such as the ministries or the planning office.

4. The firm's productive and sales activities are accompanied by a long flow of reflecting information. To give some examples:

Long before the real activity is carried out long-term plans are drawn up and prognoses are made.

· In the course of the years preceding the given activity investment projects are formulated and information relating to their realizations obtained.

When activity is first begun, short-term production plans are made and information on sales possibilities obtained.

Posterior signals include reporting quantities produced and sold, counting inventories and calculating expenditures within the firm.

After a longer period of time, say one year or five years has elapsed, statistics, balance-sheet reports and so forth give a final report on the outcome of the activity.

5. Within the economic system, for one and the same real event there exist pieces of information of different degrees of "fineness", with more or less disaggregation. The finest breakdown will be found in the workshop, more aggregate data at the management level, even more aggregate records at the office of the superior control organizations, the statistical offices, ministries, planning offices, market research institutes. (The question will be returned to in the next chapter in dealing with "levels" of control.)

Referring to facts well known from common experience, we have pointed out in Statement 5.1 that in modern economic systems the information structure is growing in complexity; multi-channel information reporting is increasing. This phenomenon of fundamental importance has deep-rooted social and economic *causes*.

With the advance in production and technology, not only the volume, but also the number and variety of products is increasing. Simultaneously (partly as a cause and partly as an effect), the social division of labour is becoming more intricate both within the individual organizations and among the organizations within the economic system. This trend, in itself, tends to make the flow of information more intricate and more complex.

As a result of technical progress, concentration is growing in many fields; in many sectors, larger and larger firms and organizations are appearing.[3]

[3] We shall return to the question of concentration in a later chapter.

In the wake of concentration, the minimum profitable plant size is increasing; as a result, the risk involved in setting up a new plant is also increasing. To diminish the uncertainty, the decision-makers strive to obtain the greatest possible amount of preliminary information.

Concentration involves a trend away from the atomization of economic processes. This implies that the success or failure of one decision-maker depends, to a considerable extent, on the decisions of other decision-makers. There is, therefore, a growing tendency on the part of the decision-makers to endeavour to obtain information about the plans of others.

In actually existing economic systems the main problem is not how to make a decision under uncertainty, although this is one of the favourite subjects of decision theory, but rather, how to reduce the uncertainty.

Statement 5.2. The complexity of decision problems, uncertainty, the comparative unreliability of information and the increasing risk involved in important decisions made under uncertain conditions lead to multi-channel information and the multiplication of information gathering. Concerning the same real event multiple information partly of price-character, partly of non-price character will be received through several channels, in various phases, at several times, in various degrees of fineness.

Statement 5.2 will be called the *principle of information multiplication*. Performance according to this principle is most reasonable and useful, since the goal is to increase the reliability of functioning of economic systems.

The factors listed above go a long way towards explaining why the *demand* for complex information flows is increasing. Similarly, the technical possibilities for meeting this demand are also growing. Consider the immense technical apparatus of modern *data processing;* this includes the punched-card systems and electronic computers, as well as of the techniques of accelerating the *flow of information*, the telephone and the telex machine. The development of the techniques of data processing and transmission reacts on the requirements, further increasing the latter. Interaction of these factors leads to an acceleration of the tendency indicated in Statement 5.1.

Increasing complexity is a general tendency in every modern economy. The concrete forms are, however, essentially different in the different economic systems. For example, the following traits are particularly characteristic of the information structure of the socialist countries:

— Information of non-price character plays a relatively more important role.

— The anterior time horizon is relatively longer; anterior information, i.e., *planning*, is of great significance. Planning embraces the economic system as a whole, and, in this case, requirement of mutual information and preliminary consideration discussed above is most completely satisfied.

5.4. Complex Structure of the Price-Type Information Flow

The place of prices in the information structure must be separately mentioned, albeit briefly. It is beyond the scope of this book to undertake the elaboration of a comprehensive price-theory particularly since very little empirical material is available.[4] I will restrict myself to a brief outline of the way in which prices can be fit into the conceptual framework introduced here.

As a matter of fact, I am dealing here with the analysis of a single phenomenon; I would like to show that the set of price-type information in an economic system constitutes a *complex* information structure itself, even if we disregard the non-price-type information and its complexity.

Let us examine a single product of a productive firm. Which price-type information relates explicitly to the product in question and influences the decision of the firm? We are *not* concerned with the price of the materials or machinery necessary to manufacture the product, but only with the prices of the product in question.

We shall discuss all information which is relevant to either capitalist or socialist firms or both. It should be determined, however, when describing an actual price system, which pieces of informations are, in fact, relevant to decision makers and what their real influence on the decision is.

There is no need to provide a taxonomic enumeration of characteristics; it will suffice to classify the price information according to various criteria. This will be similar, but not identical, to the classifications of earlier sections.

First criterion: character of the price information. The word "character" will not be defined; instead, it will be illustrated with the following list:

a) Actual price. This is the price at which the actual transaction between buyer and seller, as well as the accompanying money flow, takes place. Obviously, this has a central role in the price system. This is why it heads the list.

b) Contract price. Proceeding backwards in time, this price precedes the actual price (provided that there is a previous contract). Sometimes the actual price deviates from the contracted one.

c) Price offer. This precedes the contracted price. Offers can be made both by the seller to the buyer and vice versa and these may undergo several modifications in the course of preparing the contract.

d) Price prognosis. This may be prepared either by the buyer or seller or by another organization. Even if it is prepared by the party concerned, the prognosis may differ from the subsequent offer.

---

[4] Even today, an empirical work [81] done in the thirties by two Oxford economists, *Hall* and *Hitch*, to check, empirically the marginalist theory of prices is frequently referred to, but really convincing, comprehensive material based on facts has not been prepared to this date.

e) Prescribed price. This is an instruction issued (generally in socialist systems) by the government price authority or (in capitalist systems) by a multi-firm cartel to the contracting parties.

f) Price report. This can be submitted to many kinds of addressees: the price authority, statistical office, tax office, economic research institute, etc. This too, may differ from the actual price, either through inadvertent inaccuracy or through deliberate distortion.

From this survey it follows that the structure of the price system is characterized by a multi-channel type of information flow.

Second criterion: the partner. In some cases a given enterprise sells a given product to one or more buyers at a strictly uniform price. In other cases, however, prices are differentiated according to the buyers.

Third criterion: the date of issuance of information, and the fourth criterion: the date of transaction. To define a piece of price information uniquely, two dates must usually be given. For example, a firm may draw up its production and sales plan in October 1969. This plan is influenced by the information collected on price developments in the past, such as the price report prepared in 1968 covering sales prices in 1967. (1968: date of issuing information; 1967: date of transaction.) But expectations regarding the future influence the plan. These may appear in the price prognosis prepared in 1969 concerning expected prices in 1970. (In this example 1969 is the date of issuance of information and 1970 is the date of transaction.)

The decision-maker is influenced by the whole series of past and anticipated future prices and not only by the price prevailing at the moment. In addition, information is often issued on both past and future prices several times, with different time-lags; i.e., offers, prognoses, reports.

Thus, it follows that the structure of the price system is characterized by a *multi-stage* information flow.

Fifth criterion: Addresser of the price information. The structure of the price system is characterized by a *multi-channel* flow of information.

If we wanted to list all price information related to a single product of our firm, we would need, in accordance with the five criteria of classification, an array of five dimensions.[5] This array would contain empty entries, noninterpretable elements, but it would still be composed of a great number of elements.

[5] The notion of "array" is a generalization of the notions of vector, and matrix. If a set of data is arranged according to a single criterion, the data can be described with the aid of a vector, that is, by writing figures beside each other or under each other. The vector is a one-dimensional array. If a set of data is arranged according to two criteria, the data can be described with the aid of a matrix, that is, with a standard table consisting of several rows and columns. The matrix is a two-dimensional array. In our case, however, we have a five-dimensional array which cannot be represented geometrically.

The GE school considers only a single element of this five-dimensional array. That is, according to the second criterion, there is no distinction; it assumes a uniform price.

According to the third and the fourth criteria: there is exclusively simultaneous information relating to the transaction which has just taken place.

According to the 5th criterion the only source of information is the observation of the firm about its own transactions. (Or, what is equivalent to that, anonymous information originating from the market.)

Thus, according to the GE school (and using the terminology introduced in Definition 5.8) the information structure of the price system is *simple*.

In reality, however, the firm is influenced by the whole five-dimensional price-array.

Statement 5.3. Within the whole information structure, the price system itself has a complex structure. This structure is multi-type, multi-stage, multi-channel. To the same real event there correspond many pieces of price-type information which follow each other in time or flow together at the same time.

The above statement appears to be true as a real science statement, as a description of reality. But, beyond that, in a normative sense, it can be claimed that the economic systems "have done well" in establishing complex information structures in their price systems. Here we may recall the "principle of information multiplication". This multiplication is necessary for the reliable functioning of complex and complicated systems, and this is surely true of the price system.

5.5. Control Sub-Systems

It is worth while to survey the structure of information flows from one additional point of view. Within the whole information flow of an economic system, the information flows of various *control sub-systems* are clearly distinguishable. The most characteristic sub-systems are:

1. The market: the sub-system directly regulating sales purchases, and transactions involving products.
2. The monetary and credit sub-system.
3. The sub-system of national economic planning.
4. The sub-system of information about technical progress and science.
5. The sub-system of labour allocation.

The five sub-systems listed are not independent of one another; in several fields they overlap or are, at least, closely linked. For example, the direct exchange of information immediately prior to and following purchase and sales is complemented by the operations of money payment or credit. Or, in the framework of national economic planning, financial and manpower plans are also drawn up.

It is, nevertheless, justifiable to speak about several kinds of sub-system which are distinct from each other in a certain sense.

Definition 5.9*. In the regulation sphere of the economic system we find distinct c o n t r o l   s u b - s y s t e m s. Each sub-system specializes in definite functions of regulation. Accordingly, they are organizationally separated. The functions of a sub-system are carried out, within the complex system, by specialized control organizations with relative autonomy.[6] In the information flow within a sub-system, a definite class of information types is relevant and differs from the information types of the other sub-systems.

Let us discuss the sub-systems listed above in turn.

1. The market. (We speak of "market," always meaning the *commodity* market; our usage of this term does not include the labor market, the financial capital market, etc.)

Within the productive firm, purchases and sales are usually carried out by the component organizations; in addition, there exist trading organizations specialized in these activities.

The main information types are offers, modifications of offers, advertisements, bargaining, contracts, prices.

This sub-system of control and of information flows is dealt with extensively in the economic literature. In Part III we will return to the question of whether the statements given in the literature are acceptable. One thing, however, is certain; this is *not* a neglected domain of our discipline.

2. The money and credit system. We explained the special significance of money flows in Section 5.1.

In every productive firm, there are separate financial organizations. In addition, there exist organizations specialized in financial matters—banks, tax offices, ministries of finance, financial departments in state administrations. Only the printing press, the mint and the gold mines belong to the real sphere; all other essential functions of money are performed in the C-sphere.

In the sphere of processes connected with money we have a great variety of information types which are classified according to numerous criteria. Let us now mention only a few classification criteria:

—Is it actual money (banknotes or coins) that is transferred between the addresser and the addressee, or is it a claim to money in the form of assets, liabilities and credits that is transferred?

---

[6] The control organizations serving definite separate functions within complex organizations will be dealt with in Chapter 7.

—Is the money flow accompanied by a product flow (purchase and sale) or is it a movement that is unilateral: the granting or repayment of a gift or loan, etc.?

—If the transfer is accompanied by a product flow, what is the time relation between the real flow and the money flow? Does the product flow precede or follow the money flow, or do the two occur simultaneously?

The enormous literature on money and credit includes large numbers of both empirical and theoretical works. There is, however, almost no work which describes the sub-system of control and the information flow from the point of view of economic systems theory in a mathematically formalized manner.

In an interesting investigation, *Maria Augustinovics*, a Hungarian economist, has applied the technique of the Leontief model to the survey of money flows.[7] Her models are special input-output tables where money and credit, i.e. definite information types, flow between addressers and addressees.

Similar research work also is being undertaken in other countries.

3. National economic planning. This has been developed primarily in the socialist countries, but has been applied, albeit less comprehensively, in some capitalist countries, as well.

Wherever national economic planning makes its appearance, a special organization or apparatus also emerges, including, primarily, the central planning institutions. In the socialist countries, lower-level planning organizations have also been set up, namely, the planning departments of the ministries.

Special information types also appear which include preliminary plan figures, plan proposals, plan instructions, reports on the implementation of the plan. All of these flow within the information sub-system of national economic planning.

4. Information for technical progress and science. A significant part of the information relating to real economic processes consists of technical and scientific information, descriptions of products and technologies. The scientific-technical information on the inputs and outputs of production usually flows together with the information of price character.

The lower-level organizations of the sub-system consist of the technical, research and development departments of the firms. In addition, numerous special institutions belong to this sub-system: technical and scientific documentation centres, councils and offices of technical and scientific development, engineers' associations, etc.

5. The distribution of manpower. A highly important sub-system concerns the distribution of manpower and the flow of information on manpower. For this,

[7] See *Augustinovics* [19] and [20].

too, there are special organizations within the productive firm, the personnel and labour management departments. In the socialist countries, centralized institutions are also engaged in the selection of at least the leading personnel of the economic organizations. In the capitalist countries this takes place in a more decentralized manner, but even there central institutions such as trade associations of firms, employment agencies, and trade unions exist.

Information may vary greatly in form, ranging from newspaper advertisements to confidential personal character reports.

The importance of selection will be discussed below. It should, however, be pointed out immediately that in speaking of the information structure, the personal abilities of employees and especially of those holding managerial posts constitute a highly important factor. Such employees are not bought and sold "per unit", exclusively on the basis of information of price character, namely wages. Information of non-price character concerning the employee's personal abilities is important.

Surveying, now, the sub-systems 3, 4, and 5 discussed above, it may be recognized that systematic descriptions suitable for comparative analysis are hardly available in these areas. There are no studies to describe, for example, the route of information relating to planning, technical-scientific development or the selection of staff, although these are important characteristics of the economic system. The empirical observation of behavioural regularities is almost completely missing.

5.6. Comparison

Let us now turn our attention to the general equilibrium theory which was discussed briefly in Section 5.4.

The GE school assumes that the economic system is able to function with a simple information structure.

Let us survey the information structure of GE theory:

The information's mode of description: Exclusively of price character. Price provides sufficient information for decision-making. There is no information of non-price character. The independent role of money-flows is completely disregarded.

Mode of reflection: Direct. Every producer and consumer observes directly the actual prices; these prices reflect directly the real processes. There are no organizations specialized in transmitting information.

The addressers: The only addresser of price is an anonymous and atomized market. Single-channel information flow. There are no parallel information channels.

Time of information: Single-phase. In the original GE models, information flows simultaneously with real actions.[8]

Detailedness of information: Information has a single degree of fineness. In decentralized decision-making, every decision-maker decides on real inputs and real outputs, given his knowledge of actual market prices.

Of the five sub-systems of information flows described in Section 5.5, only one appears in the GE model, namely, the market. The other four are entirely neglected.

As a real science theory, describing and interpreting the information structure, GE theory is unacceptable.

No economy organized into a system has ever functioned with a simple information structure. The information structure was always complex, and its complexity, as has been emphasized above, has been growing historically.

It could be argued that the GE school deliberately describes reality in a simplified form in order to bring into relief the most significant characteristic traits. However, reality differs markedly and essentially from the schema drawn up by the GE school. Due to its one-sided oversimplification of the information structure and its neglect of numerous essential features of reality, GE theory is not suited as a reliable tool for understanding reality.

The alleged merit of the simple information structure advocated by the GE school is that it economizes on information to the maximum extent.

The investigations dealing with the stability of the system try to show that every individual decision-maker has to receive a single price vector only: the prices of his own real inputs and real outputs. Given this knowledge, if the decision-maker acts according to the appropriate rules, the system will sustain a state of equilibrium and even of Pareto optimality, once it is attained.

The truth is that the simple information structure of GE theory deals with information not in an economical, but, rather, in a totally miserly manner, ignoring a significant portion of the information which is *indispensable* for the operation of a real economy.

In this connection, the following problem in the theory of automata is particularly thought-provoking:

How is it possible to assemble uncertainly functioning sub-units into an automaton which will nevertheless function properly? The classic treatise by *J. von Neumann*[9] provided the first impetus to research in this direction. Von Neumann had two types of "machine" in view—the computer and the living organism with its nervous system. The solution of the problem of increasing

---

[8] In some modified GE models there are regular time lags. The information flow is thus posterior, but, again, single-phase. See Section 25.2.

[9] See *von Neumann* [190] and [191].

reliability, he saw in the multiplication of information flows. This can be characterized by way of the following schematic example:

Let us suppose that the same operation is always carried out simultaneously on three computers. The computer calculates the results of the operation. If all three computers arrive at the same conclusion then all is well. If the results of only two out of three are identical, the result of the "majority" will be accepted and computation continued on all three computers on the basis of that result. Computation halts only if the results of all three are different. The probability of this, however, is only a fraction of the probability of error which would occur if any of the three computers worked alone.

It may be assumed that in an economy which is also composed of units functioning unreliably, themselves, multiplication of information is both *necessary and useful*. Too much multiplication will obviously not be necessary. It is, however, certain that the operation of a system with "maximum information thriftiness", relying on a *single* type of information flow would soon get stuck. Since the GE school disregards this uncertainty as well, it is understandable that it considers justifiable the assumption of exclusive price information.

The GE school, as well as some related theories which will be surveyed below, seeks an answer to the question of the form in which the optimally simple information structure appears. To pose the question in this way, is, however, unjustifiable; we cannot seek any optimum alternative *outside* the set of feasible informations structures.

This leads to an important matter of principle to which we will repeatedly refer in the course of our criticism of GE theory, namely, the relationship between descriptive interpretative theory and normative theory, a system of thought which is useless as a real science, as a descriptive-interpretative theory.

It is impossible to disregard the fact that in *every* modern economic system a complex (and increasingly complex) information structure has evolved. There are numerous differences among the structures, but complexity is undoubtedly a common characteristic feature. This indicates that the complexity of the information structure is a *necessity* of the modern economic system which, even if not worked out deliberately as in the case of socialist planning, will develop more or less spontaneously.

The complex information structure can be modified by deliberate intervention. On the one hand, unnecessary information can be dispensed with. On the other hand, information flows formerly missing can be introduced. The reform of the Hungarian economy also shows how far-reaching and fundamental changes can be brought about in the information structure of an economic system on the basis of a suitably prepared program. There is, however, a limit to the changes, that can be realistically made.

It requires further research to ascertain the *lower and upper bounds of the complexity of an information system;* which is the comparatively least and the comparatively most complex structure that is still workable? which factors determine the bounds? do these include the degree of concentration of uncertainty, the secondary effects of erroneous decisions, etc.? It appears that this question can be investigated not only empirically, but also theoretically, by a formal model. Studying the problem, we must employ the dualistic description of economic system theory. There is dualistic correspondence between the current properties of the real sphere (degree of concentration, complexity of real connections, etc.), and between the complexity of the control sphere.

Although this requires further investigation, on the basis of our experiences both the lower and the upper bounds may be approximately discerned. Those working out the Hungarian reform soberly took into account these bounds when they refrained from suggesting an information structure as simple and incomplete as the one which would have been suggested by the GE school.

The economic system cannot be deliberately fashioned like a toy, produced from the components of an erector set. *The economic system is a living organism, the functioning of which obeys certain laws. The complexity of the information structure is a phenomenon that conforms to these laws and will enforce their assertion itself.*

6. MULTI-LEVEL CONTROL

6.1. The Types of Subordination and Superordination

In the preceding Chapter we described the relationships among the organizations and units of the economic system in two dimensional terms. However, *three-dimensional* investigation is also necessary. The economic system consists not only of coordinated organizations but also of organizations among which relationships of subordination and superordination exist. In other words, both horizontal and vertical relationships exist among organizations.[1]

A more exact treatment of the subject requires the clarification of several concepts.

We wish to determine which organizations are "higher up" and which are "lower down" within the economic system. In mathematical terms, we must introduce a partial preordering[2] on the O set, the set of organizations. This is carried out in *several steps* and in Definition 6.6 the general concepts of subordination and superordination are defined.

As a first step, we will distinguish between two types of organization: the real organizations *specialize* in real processes and the control organizations in control processes. Although information flows to and from the real organizations, their *main* activity consists in releasing real outputs. The real organizations include, above all, production, investment, technical development, marketing and purchasing organizations. Households are also real organizations.

The control organizations may use real inputs and even release real outputs but the main output is the release of information characteristic of their special sphere of activity. Characteristic control organizations are the directorates and finance departments of the productive firms, the institutions specialized in economic information processing, information transmission, decision preparation and control banks, social organizations, research institutes, and documentation centers.

---

[1] I first used the concepts of "horizontal" and "vertical" relationships in my book [126] published in 1957. The aim of this differentiation was to call attention to the proliferation of vertical relationships and the withering away of horizontal relationships under the over-centralized conditions of economic administration in Hungary at the time. The definition of the two concepts follows later.

[2] See *Debreu* [50].

SUBORDINATION AND SUPERORDINATION

Definition 6.1. The sub-system of real organizations ($O^{(R)}$) forms the l o w e r l e v e l of the economic system; the sub-system of control organizations ($O^{(C)}$) forms the u p p e r l e v e l of the economic system.

The terminology is to some extent arbitrary. It states that every organization specialized in *C*-activities is *ipso facto* "higher up" than the institutions engaged mainly in production and consumption. This terminology, however, involves no social value judgements and is not meant to reflect the prestige scale or power relationships which in society. It is, rather, the cybernetic approach that lends justification to the proposed conceptual apparatus. The regulator that exerts control is always "higher up". All *C*-organizations serve a regulatory function.

Let us now turn our attention to the upper level. Here, too, relationships of subordination and superordination obviously exist. Two types of subordinate and superordinate relationship can be distinguished:

—Subordination and superordination based on the right to command and subject to legal sanctions.

—Subordination and superordination based on a monopoly of indispensable information.

Let us consider first the type connected with the right to command. These commands take the form of directives.

Definition 6.2. D i r e c t i v e s constitute a special class of information type. The addresser of a directive is a control organization; the addressee may be any organization. The set of information types of a directive character will be denoted $\mathcal{M}$; directive $\mathcal{M} \subset \mathcal{S}$ will control the addressee's operational variables. Failure to obey a directive entails legal sanctions.

Let us examine the definition in some detail.

First of all, I wish to state my reasons for choosing this term. Instruction, order, and command are frequently employed synonyms for the word "directive". The command concept has become particularly popular in Hungary, for example, to describe the methods of economic administration in the period of overcentralization. The term "instruction" is used in a more neutral and general sense in the language of cybernetics, control theory and computer programming.[3] Since my terminology conforms to the greatest extent possible to the conceptual framework of these fields, in order to avoid confusion I have chosen the word "directive".

The most important part of the definition is the last two sentences. The directive is issued in order to influence the activities of the addressee; does not

---

[3] One would say that the *C*-unit "instructs" the coordinated *R*-unit, whereas we are thinking here of an abstract instruction transmission within the same organization. To use the former analogy, within the "one body—one soul" organism, the "soul" instructs the "body".

7 KORNAI: Anti-Equilibrium

simply convey information, but rather, transmits information entailing *legal sanctions*.

We use the term "legal sanction" in a broad sense to mean not only processes involving police, state prosecutors, courts, but also disciplinary procedures, applied according the own rules of an institution. For example the dismissal of an employee for disciplinary reasons is regarded as a "legal sanction" in the present context.

What differentiates the directive from other types of information is not its degree of effectiveness. A directive may, in the last analysis, have less effect on the addressee's activities than other information of non-directive character. (For example, the news of a slump on the stockmarket may lead to panicky sales without anybody directing the owners of stock to sell. And, conversely, there may be directives which are frequently ignored.) The essential distinctive characteristic of the directive lies in the fact that its command is backed by legal sanctions.

Definition 6.3.[4] Control organization o_1 is a n i m m e d i a t e d i r e c t i v e s u p e r o r d i n a t e of control organization o_2 if o_1 is the addresser and o_2 the addressee of some directive information type m^5 ($m \in \mathcal{M}$). In the same relationship o_2 is a n i m m e d i a t e d i r e c t i v e s u b o r d i n a t e of o_1.

Control organization o_1 is an i n d i r e c t d i r e c t i v e s u p e r o r d i n a t e of control organization o_2 if o_1 is an immediate directive superordinate of another organization, which is an immediate directive superordinate of a third organization, and so on. The concept of i n d i r e c t d i r e c t i v e s u b o r d i n a t i o n is to be interpreted similarly (o_1, $o_2 \in \mathcal{O}^{(C)}$).

In the Hungarian economy before the reform, a minister was an immediate directive superordinate of the chief director of a branch, and the chief director of a branch was an immediate directive superordinate of an enterprise manager; between the minister and the enterprise manager, however, only an indirect directive relationship existed. According to the rules which then prevailed, the minister was not supposed to bypass the chief director of a branch; he could issue indirect instructions to the enterprise only through the intermediary of the chief director.

Definition 6.4. On the set of control organization, $\mathcal{O}^{(C)}$, a partial ordering

---

[4] Numerous ideas concerning the definitions relating to directives and subordination/superordination are drawn from the pioneering paper by *Koopmans–Montias* [125]. Since that paper started from a different conceptual outlook than did my book the conceptual framework there and here is different. However, they are compatible; the concepts used there can be "translated" to the present conceptual framework and vice versa.

[5] The symbol m denotes here an element of set $\mathcal{M}$. In Chapter 4, m denote the number of organizations. Hopefully, this twofold use of the symbol m will not lead to any misunderstanding.

is introduced which is denoted by the symbol $\overset{\text{dir}}{>}$ and is called d i r e c t i v e
o r d e r i n g. Under this ordering $o_1 \overset{\text{dir}}{>} o_2$ if there exists a directive m' for
which o_1 is an immediate or indirect directive superordinate of o_2, and there
exists no other directive m'' for which o_2 is an immediate or indirect directive
superordinate of o_1.

According to Definition 6.4. there is *no* directive relationship between two
organizations if, in one matter, the first organization directs the second but,
in another matter, the second organization directs the first. The "command-
obey" relationship must be unidirectional or we cannot speak of a directive
relationship which is unequivocally one of subordination and superordination.

In the case of a directive relationship, it is a non-economic criterion, *legal*
regulation, that enables us to determine unequivocally who is "above" and
who is "below". He who is entitled, by the provisions of law, to command, is
above. However, there are many relationships which are not regulated by legal
statutes which one feels, nevertheless, represent relationships of subordination
and superordination.

For example, the central state bank controls the entire credit system in most
countries although it possesses formal directive authority, enforced by legal
sanctions, in only a limited number of areas. Its superordination is based on
the fact that is has a monopoly on the issuance of a special information type,
money. The planning office (even when it is not entitled to give directives) has
the monopoly on issuing central plan information which is indispensable for
the lower-grade planning organizations specialized in drawing up economic
plans.

Definition 6.5. On the set of control organizations, on the $O^{(C)}$ set a partial
preordering is introduced which is denoted by the symbol $\overset{\text{mon}}{>}$ and is called
o r d e r i n g b a s e d o n i n f o r m a t i o n — m o n o p o l y. In this ordering $o_1 \overset{\text{mon}}{>} o_2$, i.e. o_1 is a superordinate based on information monopoly of o_2,
if there exists some piece of information of which o_1 is the only addresser that
is an indispensable input of o_2.[6]

---

[6] It is assumed that the converse relationship cannot prevail at the same time. If
$o_1 \overset{\text{mon}}{>} o_2$ then $o_2 \overset{\text{mon}}{>} o_1$ is excluded.

7*

6.2. The General Concepts of the Vertical and Horizontal Relationship

Given Definitions 6.1 to 6.5, we can now provide a general interpretation of the concepts "above" and "below", "vertical" and "horizontal".

Definition 6.6. On the set of organizations O a partial ordering is defined which is denoted and called v e r t i c a l o r d e r i n g. Under this ordering $o_1 \overset{\text{vert}}{>} o_2$, i.e. o_1 is a s u p e r o r d i n a t e of o_2 (and o_2 a s u b o r d i n a t e of o_1) if at least one of conditions A), B) and C) below is fulfilled:

A) There is an information relationship between o_1 and o_2 and $o_1 \in O^{(C)}$, $o_2 \in O^{(R)}$

B) $o_1 \overset{\text{dir}}{>} o_2 \; (o_1, o_2 \in O^{(C)})$

C) $o_1 \overset{\text{mon}}{>} o_2 \; (o_1, o_2 \in O^{(C)})$

Organization o_1 is not a superordinate of o_2 if $o_1 \overset{\text{dir}}{>} o_2$ but $o_1 \overset{\text{mon}}{<} o_2$.

It is not clear whether or not the above definition is exhaustive. It requires further research to establish whether other conditions for the existence of subordination and superordination relations can be proposed.

From Definition 6.6 it is clear that we cannot introduce a *complete* ordering on set O, the set of organizations. If we consider any two organizations from the set, we may be able to determine which of them is "higher up" and which "lower down", provided that one of the above conditions A), B), or C) obtains. On the other hand, it may be the case that none of these conditions apply to the pair of organizations in question and we are, therefore, unable to determine their relative places on the vertical scale.[7]

The next problem is to determine the number of levels of the economic system.

Let us begin with a simple example. In the early 1950's, there was four-level administration in the Hungarian socialist industry: 1. government, 2. ministry, 3. chief directorate of branch, 4. firm management. In socialist cooperative agriculture, on the other hand, administration was five-level: 1. government, 2. ministry, 3. county agricultural administration, 4. district agricultural administration, 5. farmer's cooperative management. Had the Hungarian economic system consisted of these two sectors alone, one could state that the system was a *five*-level one.[8] In other words, the determination of the number of levels

---

[7] In describing subordination and superordination relations, a further difficulty is caused by so-called double subordination. For example, the financial section of a town council is subordinated to both the chairman of the town council and the financial section of the county council. We will neglect discussion of the question.

[8] In the next chapter we will discuss the fact that an intricate and complex institution such as a model large-scale industrial or agricultural plant must, in itself, be considered a sub-system of at least two levels. In this case, the number of levels in our example above is increased by one.

is based on the hierarchy containing the greatest number of vertical deg-rees.

Definition 6.7. Let us call a v e r t i c a l   c h a i n a subset of the set of organi-zations, O, in which every element (except the last one) stands in an immediate superordinate relationship to the subsequent one. The number of levels of the economic system is determined by the number of elements in the longest chain contained in O. The economic system having more than one level is called a m u l t i - l e v e l   e c o n o m i c   s y s t e m.

The definition is in harmony with the description of "lower" and "upper" levels in Definition 6.1. That definition is now complemented with a method of calculating the number of levels in the entire system. (The number of upper levels is one less than the total number of levels, the remaining one being the lower level.)

This concept is illustrated in *Figure 6.1*. Proceeding downward, we see a number of chains: [1, 2, 4, 13, 19], [1, 2, 5, 14], [1, 3, 9] and so forth. The following three chains are the longest ones: [1, 2, 4, 13, 19], [1, 2, 4, 13, 20] and [1, 2, 4, 13, 21]. This set of chains implies a *five-level* system.

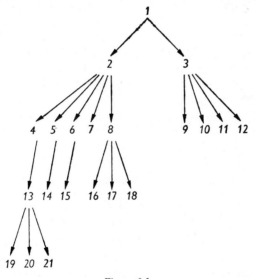

Figure 6.1
Vertical ordering

Having clarified the notion of "vertical" let us now turn to that of "hori-zontal". As a matter of fact, this is a special *equivalence* concept which refers to relationships existing among the organizations of the vertical ordering. The term applies exclusively to organizations which are unequivocally at the same level. For example, we may consider the levels of two ministers who are members of the same government or two firms which are both at the lowest level identical. The chief director of a branch and the president of the county council are not at the same level because each of them belongs to a different vertical chain, and within their own particular chain, they are separated by different numbers of elements from their common superordinate, the prime minister.

Definition 6.8*. In the vertical ordering introduced on the set of organizations, O, the equivalence between organizations o_1 and o_2 is denoted by the symbol $\overset{\text{vert}}{\sim}$ and called a h o r i z o n t a l   r e l a t i o n s h i p. We may also say that o_1 and o_2 are on the i d e n t i c a l   l e v e l. This is the case when at least one of the two conditions below is fulfilled:

A) $o_1 \in O^{(R)}$ and $o_2 \in O^{(R)}$

B) $o_1 \in O^{(C)}$ and $o_2 \in O^{(C)}$,   and

there exists an organization $o_3 \in O^{(C)}$ where $o_1 \overset{\text{vert}}{<} o_3$ and $o_2 \overset{\text{vert}}{<} o_3$, and o_1 and o_3 can be connected by a chain of the same length as that connecting o_2 and o_3.

In practice, Definition 6.8 means that:

A) All real organizations are on the identical level, and the relationships between them are, necessarily, horizontal in character.

B) Two control organizations are on the identical level, and the relationship between them is horizontal, if they are at the same distance from their common superordinate and separated from the latter by the same number of immediate and indirect superordinates.

This concept is illustrated in *Figure 6.1*. Organizations 4, 5, 6, 7, 8 have no immediate superordinate in common with organizations 9, 10, 11, 12. They are nevertheless on the identical level because both are separated from their common indirect superordinate, organization 1, by one immediate superordinate.

From what has been said above it follows that we can speak of a single-level economic system only if the system does not contain any C-organizations specialized in control activities.

The economic system has more than two levels if 1) certain C-organizations have a monopoly over some information inputs which are indispensable for other C-organizations, or

2) there are directive relationships among some C-organizations. (Each of the two conditions is sufficient.)

6.3. Once More about the Information Flow

In Chapter 5, information flows were classified according to several criteria. This classification can now be completed by the addition of an important criterion.

Definition 6.9*. An information flow is v e r t i c a l and directed downward if the addresser is a superordinate of the addressee. (In the converse case it is vertical but directed upward.) An information flow is h o r i z o n t a l if the addresser and the addressee are on the identical level.

The most frequent and most characteristic horizontal information flows take place among the real organizations. The majority of what the traditional conceptual system of economics would call "market relations" are information flows of this type.

There exist numerous information flows which are "diagonal", i.e., neither vertical nor horizontal.

6.4. Actually Existing Systems Are Multi-Level

Several branches of science have been investigating the multi-level character of control for quite some time now.

The importance of multi-level control in the operation of complicated technical equipment is well known.

In the control of unmanned space-ships, for example, there are at least three, but possibly more, "levels" of control. Some of the spaceship's actions are controlled by automatons on board; others are controlled from earth by automatic equipment but with the intervention of the research staff; some orders are issued to the space-ship solely on the basis of individual decisions taken by those controlling the space-ship from earth.

Biologists and physiologists have also dealt extensively with the problems of multi-level control. Consider another simple example.

We know that the mechanical equilibrium of living organisms is primarily, and automatically, controlled by the organ of balance situated in the inner ear. Secondarily, it is controlled by the organisms visual and kinaesthetic impressions about the body's position and by the conditioned reflexes which act in response to these impressions. On the "highest level", the organism can intentionally regulate its position.

In both examples, multi-level control serves the purpose of reciprocal correction of the separate control systems. The "minor tasks" of governing, i.e., the provision of constant and continuous control, are carried out in both cases,

by automatic devices, whereas the more complicated interventions involve
deliberate centrally-made decisions.

The multi-level phenomena of social life and especially the hierarchy of
administration, control and bureaucracy have engaged the attention of a
whole series of sociologists since the time of *Max Weber*.[9]

It is surprising that economics should have neglected for so long the problem
of multi-level phenomena, expecially given the analogies provided by other
branches of science.

In the early stages of economic theory this was hardly surprising. It is true,
that the English economy of the early and mid-nineteenth century was not a
single-level system, in the sense of the definitions given above. Central bank,
stock exchange, revenue office, customs, and so forth all existed in England at
that time. It is, therefore, possible to speak of a two-level, non-directive economic system in which the influence of the upper level was comparatively slight.
But vertical relationships were clearly insignificant, in comparison with horizontal relationships.

Since the thirties, as a result of the tremendous upheaval caused by the great
depression and also the suggestions of the Keynesian school, government
monetary and fiscal intervention in economic life has gained considerably
in intensity.

Participation in two World Wars, especially in World War II, greatly increased the government bureaucracy involved in the control of the economy.
In many countries the special apparatus set up during the war was not fully
abolished after the war and has survived in modified form but with a diminished
sphere of authority.

In several capitalist countries, certain industries were nationalized. In the
control of the nationalized industries, multi-level verticality can be observed.

The role of government in the control of investment has increased, partly
as a result of Keynesian employment policies and partly as a result of the
requirements of the armed forces, on the one hand, and the nationalized industries on the other.

Local government organs have come to play more important economic role.

In several capitalist countries, economy-wide planning has been undertaken.
In some countries such as France, planning was multi-level from the start;
the planning secretariat, controlled by the government, directs the plan commissions of the individual industrial branches and these, in turn, influence the
planning of firms.

The banking system has become increasingly centralized. In several countries,
the financial and credit system is operated on a multi-level basis.

---

[9] See e.g. *Weber* [272], *Blau–Scott* [34], and *Evan* [59].

Large-scale and mammoth enterprises, corporations, and trusts have evolved in which multi-level management is the rule.

After World War II, highly influential international economic organizations were set up. These constitute a new supra-national level of the control.

Statement 6.1. The present-day capitalist economy is, in most countries, a system of more than two levels. Among the levels there exist even if not predominantly, relationships of directive subordination and superordination.

The above proposition, although based on generally known observations, is still rather superficial in this form. It needs more detailed and more precise formulation as well as factual substantiation based on suitable empirical investigations.

So far we have spoken only of the capitalist countries. In the socialist countries, the multi-level character of the system and the significant role of directive relationships has been conspicuous from the very beginning. The reforms of economic planning which took place in recent years have, in several countries, served to diminish the role of directive relationships and have, in some cases, even brought about a decrease in the number of levels.

It appears, nevertheless, that the following proposition holds for present-day socialist countries:

Statement 6.2. All socialist economies are systems of more than two levels. In their operation, directive relationships play a considerable role, although to an extent that varies from country to country.

The above proposition, too, could be enriched and rendered more precise by a methodical comparison and dynamic investigation of the economies of the socialist countries.

Statement 6.3. The share of vertical information flows in all information flows has been increasing historically.

The truth of this proposition is beyond dispute. However, it would be worth-while to obtain more precise observations and to assess the share of vertical information flows and their relative importance. The concepts proposed in these pages serve to render such investigations more precise.

6.5. Comparison

A large number of purely descriptive monographs did not commit the error of failing to recognize the multi-level character of economic systems. Dozens of works could be quoted which have, albeit in some other terminology, pointed out the multi-level character of the system.[10] It is most regrettable that this

---

[10] For the capitalist economy this is shown, for example, by *Galbraith* [69] and [70].

recognition has never been incorporated into a formal theory of multi-level control of the economic system.

Modern mathematical equilibrium theory never advanced beyond the phenomena of the mid-nineteenth century. *The world of Walras is a strictly single-level economic system.* This fact is reflected in the second basic assumption of the GE school, according to which, the economic system consists exclusively of real organizations: producers and consumers. This basic assumption makes any further study of the multi-level control phenomenon impossible.

It is only in recent years that mathematical models have appeared which represent multi-level economic systems even if only partially and mainly in connection with planning.

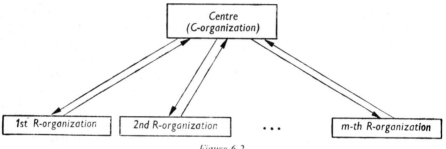

Figure 6.2
Scheme of two-level planning

Originally, these models owed their existence to technical computing, considerations; in order to facilitate the solution of large-scale linear programming problems, so-called decomposition methods were worked out.[11] Later on, it turned out that any decomposition algorithm could be interpreted as an abstract description of a partly centralized, partly decentralized process of decision preparation. In other words, the mathematical decomposition algorithms constitute a model of a multi-level planning process. Each iteration of the algorithm represents a particular phase of decision *preparation*, and the program obtained in the last iteration represents the *decision* itself.[12]

In the two-level planning models there appears a centre, i.e. a *C*-organization, and lower-level organizations which may be considered the control units

[11] The first, and still most widely employed decomposition method, was developed by *Dantzig* and *Wolfe* (see [48], 1960 and [49], 1961). Since then, several other procedures have been worked out. (see, among others, [136] by *T. Lipták* and the author, the articles of *Abadie–Williams* [1], *Rosen* [211], and *Weitzman* [273], etc. See also the comprehensive surveys by *Künzi–Tan* [145], and *Ligeti–Sivák* [152].

[12] For an economic interpretation of the decomposition methods see *Malinvaud* [157], pp. 170–210, and the author's works [127] and [128].

of the *R*-organizations. Between the two levels, information flows. This is illustrated in *Figure 6.2*.

Without describing and comparing in detail the various multi-level planning procedures and decomposition algorithms, let me point out some of their common features which obtain significance from the point of view of the information flow problems dealt with in Chapters 5 and 6.

The following properties are characteristic of all multi-level planning algorithms:

1. In the information flow there appears information both of price and non-price character. Generally, price information flows in one direction and information of non-price character in the other direction. Thus, in the *Dantzig— Wolfe* algorithm, information of price character (the shadow prices of the central constraints) flows downward, and information of non-price character (the aggregate real input requirements and real output obligations of the lower-level organizations) flows upward. In the algorithm worked out by *Tamás Lipták* and the author, the direction of the flows is just the opposite; information of non-price character (the input quotas and output obligations suggested for the lower-level organizations) flows downward, and information of price character (the marginal valuations, shadow prices of the central constraints) is sent upward.

2. Two grades of aggregation are contained in the information flow; on the lower level of the *R*-organizations, finer magnitudes are used, whereas on the level of the central *C*-organizations less detailed and more aggregated magnitudes are employed.

The multi-level planning models generally employ basic assumptions (convexity, optimization, absence of uncertainty) identical with the basic assumptions of the GE theory. As a result, their information structure is also rather simple, resembling that of the GE models. They lack the multiplication of information characteristic of the real world primarily because they are deterministic in character and assume away the existence of uncertainty. Although vertical information flows appear in these models, horizontal information flows are missing.[13] In general, the decomposition methods must be regarded as models of decision preparation and planning, and *not* as a simulation of the *whole functioning* of the economy (including the actual real processes and the control of those processes).

In relation to our subject, they are nonetheless, significant. By taking into account the multi-level character of actual economic systems, they come closer to approximating the actual information structure. *In the multi-level models*

[13] Lately, *György Simon* has worked out a decomposition method where there are both vertical and horizontal information flows. (See [232] and [233].)

we do not deal with simple information structures but rather with a complex (although not highly complex) ones.

It will be useful to determine whether or not the multi-level *planning* models can be further developed into models describing multi-level *control* of economic systems, involving uncertainty as well as the formalization of the multiplication of information. This is one of the most important problems of research on the economic system theory in the near future.

7. CONFLICT AND COMPROMISE WITHIN THE INSTITUTION

7.1. Notion of the Institution

In describing the structure of Part Two of this book, I indicated that first the whole system would be surveyed "from above", so to speak, and then the networks connecting the elements of the system (the organizations and units), that is, the relations and flows would be reviewed. Now our "plane" is descending; we are going to scrutinize the small *sub*-systems of the economic system, the institutions.

Definition 7.1*. An i n s t i t u t i o n is a sub-system of the economic system, and is differentiated by its organizational and legal characteristics. The s i m-p l e institution coincides with an organization; it consists of a single elementary unit-pair (e.g., a household). The c o m p l e x institution consists of several organizations; each of the organizations contains an elementary unit-pair, (e.g., an enterprise, an office). The organizations within a complex institution, distinguished by purpose or scope of activity, will be called f u n c t i o n a l o r g a n i z a t i o n s.[1]

It follows from Definition 7.1 that each organization belongs unequivocally to a single, specific institution, and constitutes a part of that institution.

7.2. Functional Organizations of the Productive Firm

A modern productive firm is a complex institution.[2] Because its real and control processes are quite complicated, the various functions are performed more or less separately. This separation is, as a rule, apparent also in the organization.

Within the *R*-sphere of a major enterprise, autonomous organizations: factories, units, plants and workshops can be distinguished. We will not deal, however, with the classification of the organizations of the real sphere.

---

[1] Within complex institutions organizations separated according to other criteria (e.g. territorial competence) may be distinguished. I will not deal with this type of organizational separation in my book.

[2] I have used extensively the ideas and concepts of *H. Simon, J. G. March, R. M. Cyert* and others of their school. The work [45] may be regarded as a compendium of their views on the productive firm. See also [159], [236], [237], and [238].

In the C-sphere autonomous organizations: various departments, sections, groups also exist. There is, of course, no generally valid organizational scheme that can characterize all large modern firms, but the following separate functional organizations can generally be found;

—Organizations which directly control production.
—Organizations concerned with research and technical development.
—Organizations which control and execute investment plans (perhaps linked with technical development).
—Organizations responsible for selling the firm's products.
—Organizations in charge of buying the firm's inputs (perhaps organizationally linked with selling).
—Organizations involved in the selection of personnel.
—Organizations which deal with the monetary and credit transactions of the firm.

The relationships among the institution, organization and unit in productive firms is shown in *Figure 7.1*. The rectangles inscribed within it in continuous lines are the organizations. The two circles within the organization are the units.

The relative separation of the functional organizations manifests itself in various ways. One inportant fact is that they have separate information chan-

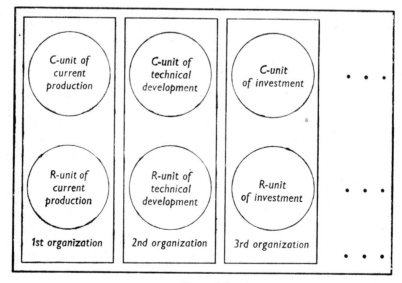

Figure 7.1

Scheme of functional organization in a productive firm

nels which connect them to the world outside the institution. Perhaps only those organizations directly in control of production are exceptions to this rule; they obtain the bulk of information "from within", from the other sections of the firm, and above all from the people engaged in procuring and selling. On the other hand, information passes through separate channels into and out of the organizations engaged in research and technical development, selling and buying, personnel management and in finance. These organizations might be regarded as the "nerve-endings" of the various control and information sub-systems reviewed in the preceding chapters. The information sub-system of market relations is in direct contact not with the whole firm, but only with that portion of the firm responsible for selling and buying. Similarly, the information sub-system of monetary movements has immediate contact only with the firm's financial apparatus.

Statement 7.1. The complexity of the information structure coincides with the complexity of institutions and leads to the separation of functional organizations called upon to perform definite information and control activities within the institution.

7.3. Manyfold Character of Motivation

The behaviour of the functional organizations within the productive firm is overwhelmingly determined by their roles relative to the division of labour within the firm. Their behaviour may also be influenced by special material incentives such as bonuses. But one of the outstanding motives which explains their behaviour derives from the propensity of man to *identify himself with his task and role*, to become the "representative of an interest". It is most conspicuous in socialist relations. Those working in the industrial departments of the Planning Office consider it only natural to "fight" for the "interests" of their industry, to obtain more investment funds for the branch in question, although they are influenced by no material incentives and by considerations of prestige only remotely.

Let us investigate the influence of this motive on the behaviour of the functional organizations within the firm. In the course of the analysis, both the modern capitalist corporation and the socialist firm interested in profits will be considered, assuming with respect to the latter, that the changes envisaged by the reform have been implemented.

Although motivation is not quite identical in all countries I should like to stress a few, fairly general tendencies.

Let us consider in turn the various functional organizations:

1. *Those directly in charge of production (e.g. plant managers, engineers*

directly responsible for production, etc.). Their desire is to insure that production is continuous and smooth; there should always be work, material and labour; there should always be proper maintenance. They are unhappy if production is suddenly cut back or stepped up. They prefer relative stability of production.

2. *Those engaged in technical development, product development, investment and research.*[3] They wish to promote the introduction of technical innovation, to commerce the production of new products, to expand production.

3. *The selling apparatus.* Those in charge of sales would like to sell as much as possible and as easily as possible. They expect those in charge of production to help them in this activity to the largest extent possible; thus, if sales possibilities suddenly increase, they would like production to expand immediately. The wishes of buyers (e.g. of foreign buyers) should be met, as far as possible, with regard to quality, terms of delivery and even price.

4. *The financial apparatus.* Those responsible for financial arrangements of the firm wish to attain the maximum profits possible. They want the firm to possess monetary reserves and to be credit-worthy. The firm should not incur overly large debts, should not immobilize inventories superfluously or raise unnecessary credit; it should not have an unnecessarily large interest burden.

7.4. Conflicts and Compromises

The interests of the functional organizations within the firm are in constant conflict with each other, even if the conflict is not always articulated.[4]

There are always conflicts of interest among all of the functional organizations and interest groups, regarding the allocation of investments, bonuses, and the raising of remunerations.

There is a conflict between the desire for stability of those in charge of production and the desire of the selling apparatus for flexibility in production.

There is a conflict between the efforts of the financial apparatus to reduce inventories and those of the production and the selling apparatus to increase them.

There is a conflict between the innovation efforts of those engaged in research and technical development and the claims of those in charge of production that production should be continuous and smooth.

An endless list of similar conflicts could be produced; for the sake of illustration, these will suffice.

So far we have spoken only of the motivation of functionally separate

---

[3] In the book by *Cyert–March* already quoted [45] this interest group is omitted.
[4] *McGuire*'s book [174] can be consulted on the subject of the motivation of firms.

organizations within the firm, and the conflicts stemming from differences in motivation. There exists, however, another type of conflict of fundamental importance within the firm, related to the question of *ownership, power and income*.

While the first type of conflict, based on the division of labour between the control processes in the firm, appears in essentially the same form in all modern economies, the latter obviously depends essentially on the political institutions, the forms of ownership, of the society.

Within a capitalist firm at least three interest groups should be distinguished: the owners, the managers, and the workers in subordinate positions. I do not intend to discuss here the extent to which a manager identifies with the owners, subordinates his actions to the interests of the owners, or acts in accordance with his own, separate interests and motives for prestige, career, security, higher personal income.[5] It may be taken for granted that his behaviour is *not exclusively* motivated by the interests of the owner. He may attempt in asserting his own point of view when a decision is made about what portion of gross profits should be paid out to the owners as dividend and what portion should be allocated to retained earnings for the development of the firm. The interests of the manager may conflict with those of a minor or major group of shareholders. Some shareholders are interested mainly in current dividends rather than in total profits. The masses of atomized small shareholders typically display such behaviour. The large shareholders are interested, of course, not only in current dividends but also in the entire operation, life, stability, development and expansion of the firm.

Minor employees and workers in capitalist firms identify less with the so-called "interest of the firm" although it is a characteristic feature of the modern style of management to promote such identification, by encouraging workers and employees to buy the shares of the firm and so on. In spite of this, there remains a fundamental conflict of interest between the employees, the managers and the owners. In everyday life this conflict of interest becomes manifest in disputes about wages, income and social benefits which are rooted in the deep class conflicts revolving around the question of power and ownership.

In the state-owned socialist firm the owner is "invisible". (I will neglect, in this discussion, the characteristics of Yugoslav firms.) The authority and the attitude of the "owner" is partly represented by the higher authorities: the ministry supervising the production of the firm, the ministry of finance which collects the major part of the net income of the firm. But an "ownership" attitude also develops in the firm itself, mainly in its leaders, particularly if they are interested both in profits and in the entire development of the firm.

[5] For further literature on the subject see e.g.: *McGuire* [174], *Williamson* [275] and *Cyert–March* [45].

As a result of the development of an "ownership" attitude among the firm's managers, there exist conflicts in socialist firms also between the workers, the minor employees and the senior executives of the firm as well as between the managers of the firm and the central institutions representing the global interests of the state. One of the aims of the reform of the system of economic administration and management was to provide more desirable or healthier mechanisms for resolving these conflicts in order to eliminate, as far as possible, behaviour which retards the development of the economy.

The firm as an institution, as a living organism, is capable of operating, in spite of its internal conflicts, because *there emerge compromises among the conflicting interests*. The activity of the manager, the representative of so-called "firm interests", involves mediation to a significant extent; he smoothes conflicts and harmonizes in his decisions the contradicting proposals. That is, the manager formulates compromises.

A suitable compromise can be reached because each member of the institution desires to see it *survive*, and even develop and grow. *The motives of survival and expansion lead to the emergence of compromise.* In what follows we will frequently return to this idea, namely, to the importance of the survival motive. If the internal conflict in an institution is so strong that survival as a *common* motive can no longer assert itself with sufficient force, the institution will become incapable of maintaining itself, it will fail, fall into ruin and be dissolved.

Undoubtedly, the profit motive has an outstanding role in the capitalist firm. One of the major ideas underlying compromises in a capitalist firm is the belief that the separate activities of the sub-organizations should serve to increase the total profits of the firm. Stable and, if possible, growing profits are an indispensable condition for the survival and expansion of the firm.[6]

Since the reforms, the interest in profits has grown in socialist firms as well.

I do not wish to get involved in the debate over whether profits are the ultimate motive in capitalist firms or only a tool employed to achieve the goals of survival and expansion. In my opinion, debates about "ultimate" motives lead to philosophizing rather than genuine analysis of reality. The real task which confronts us is to describe the internal conflicts which exist within institutions including the capitalist firms, the variety of interests which collide and produce conflict and the mechanisms through which compromises are agreed upon.

Thus far we have described only the productive firm and its internal conflicts and compromises. Obviously, analogous phenomena can be found in all other institutions (public offices, banks, universities, etc.).

[6] Though I rely mostly on literature, I feel that the role of the profit motive is stronger in the actual operation of the capitalist system than assumed by the "behaviourist" school of thought.

To sum up, we may, on the basis of well-known empirical facts, establish the following statement:

Statement 7.2. In all complex institutions including productive firms, internal conflicts are a regular occurrence. The functionally separated organizations, and interest groups, divided over questions of power, ownership and income, act on the basis of particular motivations, which differ among organizations. Survival and expansion of the institution is made possible through compromises. Therefore the behaviour of the institution as a whole is characterized by a motivation emerging from complex compromises.

7.5. Acceptable Compromise

Formation of a compromise within the institution is an important type of *control process.* In the later chapters the C-processes will be repeatedly discussed. Here, however, the features of those processes related to the problems of conflict and compromise will be stressed.

At this point we will neglect the fact that the preparation of decisions takes place over time; the problems stemming from this assumption will be discussed in the next chapter.

Let us denote the set of all possible decision alternatives of the institution by the symbol $\mathcal{A}$.

The set of decision alternatives which are *acceptable* to the i-th organization, D_i, consists of all decision alternatives, the realizations of which are acceptable from the point of view of the interest of the organization. Information about the set of acceptable decision alternatives is given by the i-th organization in the form of *bounds on acceptance.*[7]

In the next chapter we will deal more extensively with the bounds on acceptance, but the notion will be illustrated here by means of a few examples. In the preparation of an investment decision, for example, the knowledge that the financial section accepts only proposals which return the investment outlay in three years and cost at most 100 million dollars, constitutes a bound on acceptance. Another bound on acceptance results from a statement by those in charge of production that investment activity must not start in the plant before August 1st, because it would interfere with production and orders could not be met. The people engaged in technical development may put a bound on acceptance by stating that they expect the new workshop resulting from the investment to reach a certain critical value in terms of various technical characteristics.

[7] The bounds on acceptance of the organizations belonging to the same institution are, of course, not independent of each other. This will be treated later, in Sections 8.3 and 9.1.

8*

Some of the bounds on acceptance may individually restrict definite characteristics of the set of possible alternatives. Often the common bounds on acceptance of several, simultaneous decisions can be given.

The intersection of the sets of decision alternatives acceptable to each organization of the institution is the set of decision alternatives acceptable to the institution as a whole, in short: the set of *acceptable compromises*.

$$\mathcal{D} = \bigcap_{i=1}^{M} \mathcal{D}_i \subset \mathcal{A} \tag{7.1}$$

The concepts of acceptability are illustrated in *Figure 7.2*.[8] Assume that an institution is to carry out two actions (say, two kinds of investment). Each action is described completely by a single indicator; the first is represented by the non-negative element of variable Y, the second by those of variable Z. (E.g., Y is the number of machines of type "A" to be purchased, and Z those of type B. The problem of indivisibility will be neglected.)

Accordingly, $\mathcal{A}$, the set of possible decision alternatives, corresponds in the figure to the whole positive quadrant in the plane.

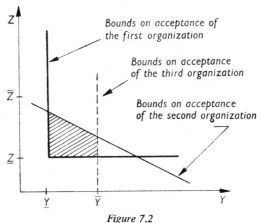

Figure 7.2

Set of acceptable compromises

There are three functional organizations which comprise the firm. The first organization sets two lower bounds on acceptance, $Y \geqq \underline{Y}$ and $Z \geqq \underline{Z}$. Accordingly, the area above the thick lines is $\mathcal{D}_1$, the set of decision alternatives acceptable to the first organization.

[8] Figure 7.2 (as well as Figures 8.2A, 8.2B, 8.2C, 8.3A, and 8.3B), illustrates the sets of decision alternatives by *convex* sets. However, the figures serve as illustrations only and from our point of view it is of no importance that these sets be convex.

The second organization sets an upper bound on acceptance, namely $Y \leq \bar{Y}$. The area to the left of the broken line represents $\mathcal{D}_2$.

Finally, the third organization prescribes that $Z + 1/2Y \leq \bar{Z}$. The area below the thin line represents $\mathcal{D}_3$.

The intersection of the three sets is the shaded polygon. This is the set of acceptable compromises for the firm, $\mathcal{D}$.

It is possible that $\mathcal{D}$ will be an empty set. But the decision process takes place over time as will be indicated in the next chapter. If in some phase, for example, at the beginning of the preparation of a decision, a compromise acceptable to all functional organizations cannot be found, so that $\mathcal{D} = \varnothing$, some of the bounds on acceptance will be modified. Modification of the bounds continues until an acceptable compromise can be found.

7.6. Comparison

The GE school regards each producing and consuming institution as a "black box". It abstracts from its internal conflicts and describes only the final compromise.

This is, in itself, a permissible abstraction. In certain investigations the biologist takes the cell as a primary indivisible unit and does not examine what goes on within it. But biology too would arrive at false conclusions if it neglected entirely the processes which take place within the cell. Unfortunately, the GE school has been content with a single approximation and its abstraction has led to distortion in its descriptions. "Idealizing" the behaviour of institutions, it overestimates the harmony and rationality in their actions; it senses only a single homogeneous enterprise interest where in reality, the interest of the firm is composed of many motives and is born out of compromises among conflicting desires.

Neglect of the functional division within the institution leads to an oversimplified description of the motivations underlying behaviour, just as neglect of the complexity of the information structure, as indicated in the preceding chapter, fails to do justice to reality.

In real economic systems the various aspects of the system's operation (such as current production, purchases and sales, investment, monetary movements, the supply of labour) appear to be separated from each other to a certain extent. Investigations based on actual observations—whether they are econometric models, simulation experiments, or case studies—describe these processes separately. In fact, the global activity of the firm cannot be directly observed, but only its more or less distinct functions.

The GE school performs only an apparent synthesis when it describes the "producing unit" as a single, undivided whole. In fact, it models only that which characterizes current production and the related purchases and sales; all other functions get lost.

7.7. Recapitulation: Microstructure

Before describing in detail the decision processes which take place *within* the organizations, it will be useful to review, for the sake of repetition, a portion of the conceptual system introduced in the foregoing. It seems especially expedient to do because the concepts are explained in different chapters.

The survey will be presented in *Table 7.1*. For the sake of illustration, analogies are presented from other disciplines.

TABLE 7.1

Division of microstructure

Economic system theory	Physics	Biology
Economic system	Matter	Living organism
Institution	Molecule	Organ
Organization	Atom	Cell
Unit	Particle	Parts of the cell (cell-wall, nucleus, etc.)

The economic system, the institution and the organization are collectives existing in reality. The unit only serves to distinguish two characteristics scopes of activity of the organization, namely, the control processes and the real processes.

The division of economic systems into institutions, organizations, and units is called the *microstructure* of the system.

The microstructure is shown in *Figure 7.3*. The outer rectangle, outlined by the continuous thick line is the system. In our example, the system consists of two institutions; the institution is indicated by a continuous thin line. In each institution there are three kinds of organizations, "A", "B" and "C". (E.g., production; sales and purchases; finance.) Within the organizations, the elementary unit-pair consists of a C-unit and an R-unit represented by circles.

Within a system with a given microstructure, *sub-systems* of varying composition may exist. For example, all upper circles, the set of C-units, constitute the sub-system of the control sphere and all lower circles, the set of R-units the sub-system of the real sphere. Or similarly, all "B" organizations (e.g., all purchasing and selling departments) constitute the sub-system "B", which deals with realization and procurement, buying and selling.

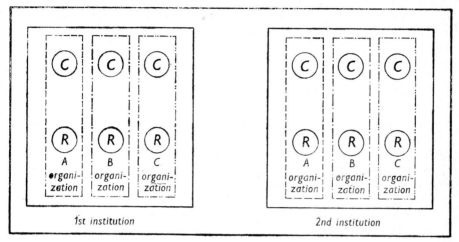

Figure 7.3
Microstructure of the system

In describing the economic system, sub-systems may be defined according to various criteria; that is, well-defined parts of an entire system may be distinguished which consist, themselves, of elements (units, perhaps organizations or even institutions).

8. THE DECISION PROCESS

8.1. Decision: Elementary Decision Process

In the preceding chapter we investigated the internal life of an institution composed of several organizations in conflict. In Chapters 8–12, we shall penetrate even deeper into the problem of compromise formation and analyze the decision processes which operate *within the organizations*. These processes make up one of the most important components of the control processes.

According to Definitions 4.7 and 4.8, the preparation of the decision and the decision itself are made in the C-unit of the organization. Therefore, we shall focus our attention on the C-unit.

The functioning of the C-unit will frequently be "personified" in the course of discussion. Thus, the following three expressions will be used in an equivalent sense: the decision of the "organization", of the "control unit" within the organization (C-unit) and of the "decision maker".

Although the following analysis holds for all economic systems that can be described with the aid of the general model of Chapter 4, the examples given refer to some functional organizations of large modern firms, such as the production, investment and technical development sections.

The notion of decision will be defined in two stages. The provisional character of the first definition is indicated by a prime after the serial number.

Definition 8.1'. The d e c i s i o n is a special kind of information output of the decision-making organization. Its function is to control the processes of other units. There exist two major groups of decisions. The *internal decisions* of an organization are instructions, the adresser of which is the control unit and the addressee of which is the real unit of the same organization. The *external decisions* of the organization are directives or other types of information, the adresser of which is the control unit of an organization and the addressee of which is the control unit of another organization.

For example, the decision of the production section of the enterprise on the production program for the next day is an internal decision. If the production section orders a definite quantity of material from the purchasing section, however, this is an external decision.

In the case of internal decisions it is merely an abstraction to speak about

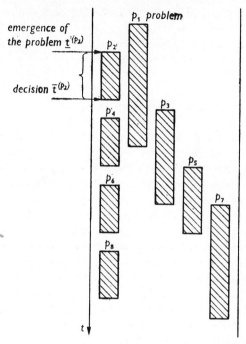

Figure 8.1
Complex decision process

"adresser" and "addressee" since this is a matter internal to the organization which is carried out by the R-unit with more or less accuracy.

Every C-unit deals, as a rule, with many kinds of problems simultaneously, although it makes decisions concerning these problems at different points in time.

The process of decision-making and decision preparation are illustrated, as taking place over time in *Figure 8.1*.

Definition 8.2*. $\mathcal{P}(t)$ is the set of d e c i s i o n p r o b l e m s in period t. Its elements are the problems dealt with by the C-unit in period t. The unit begins to deal with the problem $p \in \mathcal{P}(t)$ in period $t^{(p)}$; this is the date of emergence of the problem. The problem is concluded with the decision in period $\bar{t}^{(p)}$. The period $[\underline{t}^{(p)}, \bar{t}^{(p)-1}]$ is the time of d e c i s i o n p r e p a r a t i o n. The preparation of and making of the decision relating to problem p is called an e l e m e n t a r y d e c i s i o n p r o c e s s. The set of all elementary decision processes taking place in the C-unit is called a *complex decision process*.

The shaded columns in Figure 8.1 symbolize the elementary decision processes related to problems p_1, p_2, ..., p_8.

The height of the column indicates the length of time necessary for the elementary decision process to terminate in a decision.

8.2. The Possible Decision Alternatives

In the course of preparing the decision on problem p, the decision maker will weigh several decision alternatives. For an exhaustive description of an individual decision alternative, perhaps a hundred or a thousand characteristics should be given. However, only a few outstanding characteristics of the alternatives are deemed especially important by the decision makers. In the preparation of an investment decision in an industrial enterprise these might include the amount of additional output which will result from the investment, the type of output which will be produced, the cost of investment, and starting and completion dates of the investment project. Thus, the decision alternative is characterized by at most a few dozen indicators instead of a hundred or a thousand.

Definition 8.3*. Let us call the types of information used to describe the major characteristics of the decision alternatives for problem p, the i n d i c a t o r t y p e s belonging to problem p. To problem p there belong K_p indicator types. The set of indicator types $\mathscr{S}^{(p)} = \{s_1^{(p)}, s_2^{(p)}, \ldots, s_{K_p}^{(p)}\}$ is a subset of the set of information types. The decision alternatives can be described with a vector of K_p components (the information variables belonging to the indicator types listed), the indicator vector[1].

In the above example the components of the indicator vector, (capacity, cost, dates of completion, etc.) are the most characteristic data of the investment alternatives.

In the following discussion we will deal with a single elementary decision process related to the solution of problem p.[2] For the sake of simplifying the notation, we will drop the subscript p. (Thus, instead of K_p, we will speak about an indicator vector with K components, etc.)

---

[1] The Definitions 4.13–4.16 of the type of information and the information variable have been formulated in such a way that the value of the information variables can always be expressed with real numbers. The information types are either quantities to be measured with real numbers, or qualitative statements indicated by index numbers. Therefore, it is justifiable to speak of *an indicator vector*.

[2] The interrelations among the individual elementary decision processes within a complex decision process (actually, one important type of interrelation) will be treated in Chapter 12.

Definition 8.4*. Let $\mathcal{A}$ denote the set of all decision alternatives connected with some elementary decision process which, given the economic indicator types characterizing the alternatives, could conceivably be taken into account. The set $\mathcal{A}$, the set of p o s s i b l e   d e c i s i o n   a l t e r n a t i v e s, is a subset of the $\mathcal{L}^K$ vector space of K dimensions. Its general element, $a \in \mathcal{A}$, is the decision alternative.

In the following discussion we shall deal with many types of sets, all of which are subsets of the set $\mathcal{A}$.

It may be necessary to explain what it means to say that the elements of set $\mathcal{A}$ are all the alternatives which, given the economic indicator types, may be taken into account. In preparing an investment decision one indicator may be output which can be described by a non-negative real number. Another indicator may be the proportion of two kinds of products in percentages, this indicator is situated in the interval [0, 100]. A third indicator is connected with whether the investment employs technology "A", "B" or "C"; in this case the indicator may assume the values 1, 2, or 3.

In defining the set $\mathcal{A}$, we do not consider whether or not the alternative can be implemented, but rather if the indicator describing the alternative can be interpreted.

The decision alternative, a, frequently means not a single real action to be taken in a single period, but rather a series of actions taking place over a long period of time. It is not, however, necessary to dwell on this point. We assume that the indicator vector (and, the economic definitions of the indicator types) give the major characteristics of the time requirements envisaged in the decision. Accordingly, there is no separate subscript attached to a or $\mathcal{A}$ indicating time necessary for implementation of $a \in \mathcal{A}$.

The following discussion is illustrated by *Figures 8.2A–B–C*. It is assumed that the decision alternatives can be described by means of two indicators. Suppose that the decision process concerns the preparation of a production program. Let Y denote the value of the first indicator and Z the value of the second one where Y refers to the output of the first product and Z, the output of the second product. In the figures, set $\mathcal{A}$ is the entire positive quadrant; both variables may assume any non-negative values.

The essence of the elementary decision process is that the decision maker must choose an element from the set of possible decision alternatives. Thus, the decision can be described in the indicator space of dimension K. We may now provide a more complete definition of "decision".

Definition 8.1*. The d e c i s i o n $a^* \in \mathcal{A}$, is an element of the set of possible decision alternatives and can be described by means of an indicator vector. It is a special part of the information output of the C-unit and its function is to control the processes of other units.

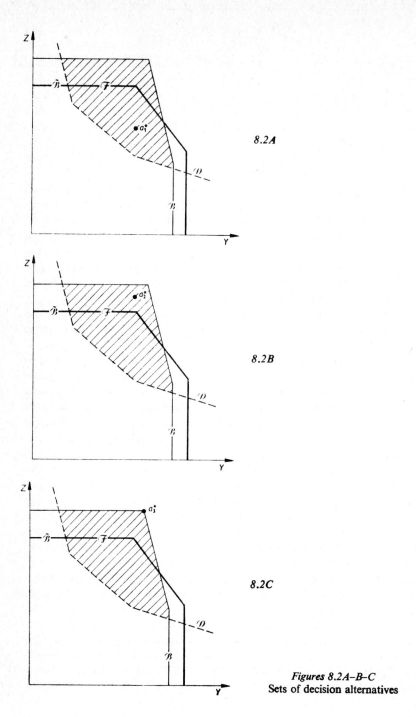

8.2A

8.2B

8.2C

Figures 8.2A–B–C
Sets of decision alternatives

The elementary decision process can be described with the notions hitherto introduced by explaining what the decision-maker *learns* about the set $\mathcal{A}$, and how he selects from the subset learned, a^*, the decision. In Figures 8.2. A–B–C we have plotted three decisions: a_1^*, a_2^* and a_3^*.

8.3. The Sets of Alternatives

Definition 8.5*. The set of i m p l e m e n t a b l e decision alternatives, $\mathcal{B}$, is a subset of the set of possible decision alternatives: $\mathcal{B} \subset \mathcal{A}$. Its elements are all the decision alternatives that actually can be implemented.[3]

In the case of an internal decision, the set $\mathcal{B}$ expresses the possibilities of the real sphere; every possible instruction which the real unit associated with the control unit is capable of fully executing is included in $\mathcal{B}$. Thus, if the decision is the compilation of the program for the next day, $\mathcal{B}$ will comprise all production plans for which the necessary material and technical conditions are available, that is, all outputs which can be produced with the existing machinery of the plant, with the existing knowledge of the workers, and from the material available. In the case of an internal decision, the communication a^* is a *direct* anterior reflection of an event in the real sphere. In contrast, external decisions are *indirect* anterior reflections. For example, in several socialist countries with a multi-level economic administration where no reform has been carried out, the ministry can decide what and how much should be produced in a firm. The adresser of the decision is the ministry and the addressee is the production section of the firm (in our model, the C-unit representing the administration of the section), which again passes the production directive received to its own R-unit. In the final analysis, "implementability" reflects, even here, the possibilities of the real sphere, only a transmission link intervenes.

The polygon delimited by a heavy solid line in Figures 8.2A–B–C represents the set of implementable decision alternatives, $\mathcal{B}$. It indicates the individual upper bounds on capacity of the variables Y and Z, respectively, and the common bound on the capacity of both variables. (For the sake of simplicity as mentioned before, only convex sets are drawn, although in reality this need not obtain.)

The set $\mathcal{B}$ is defined by actual, genuine limits on the implementability of the decision. The decision maker, however, has, as a rule, no precise knowledge about the real limits.

Definition 8.6*. The set of e x p l o r e d decision alternatives $\tilde{\mathcal{B}}(t)$, $(\underline{t} \leq t \leq \bar{t})$ is a subset of the set of possible decision alternatives: $\tilde{\mathcal{B}}(t) \subset \mathcal{A}$. Its elements

---

[3] Definitions 8.4 and 8.5 will be complemented by further explanation in Section 11.3.

are all the decision alternatives deemed implementable by the decision maker at the time of decision preparation.

If the knowledge of the decision maker about the problem is complete, then $\tilde{\mathcal{B}}(t) = \mathcal{B}$ for all $t(\underline{t} \leq t \leq \overline{t})$. As a rule, however, his information is not quite accurate, and the sets $\tilde{\mathcal{B}}(t)$ and $\mathcal{B}$ are not identical. This case is illustrated in Figure 8.2. The decision maker underestimates the production possibilities of the first and overestimates those of the second product: he also judges inaccurately the rate of transformation on the common capacity boundary. Accordingly, the frontiers of the set $\tilde{\mathcal{B}}(t)$ delimited by the thin solid line and of $\mathcal{B}(t)$ delimited by the heavy line are not identical.

One of the most important aspects of the decision process is the formation of the set $\tilde{\mathcal{B}}(t)$. How are the indicators for the decision alternative chosen? Are the requirements for implementation assessed correctly or incorrectly? What portion of set $\mathcal{B}$ is, in fact, considered? Are the alternatives considered only those in the neighbourhood of erlier decisions or are alternatives farther from previous decisions also weighed? The set $\tilde{\mathcal{B}}(t)$ also changes in the course of the decision process, as a consequence of increasing knowledge.

The set $\tilde{\mathcal{B}}(t)$ consists of alternatives which *can be* implemented (at least according to the knowledge and information of the decision makers). Which alternative it is *worth while* to implement is another question entirely.

Definition 8.7*. The set of a c c e p t a b l e decision alternatives, $\mathcal{D}(t)$ $(\underline{t} \leq t \leq \overline{t})$ is a subset of the feasible decision alternatives: $\mathcal{D}(t) \subset \mathcal{A}$. Its elements are all the decision alternatives deemed acceptable by the decision maker, given his own interests, and the communicated desires, proposals and directives of the other units.

In the preceding chapter, we spoke about the bounds on acceptance which emerge from mediation among conflicting interests within the institution. Now the problem will be dealt with in a more general form. The many possible kinds of bounds on the acceptance of a decision alternative will be classified here into two main groups.

One group of bounds is based on the *internal expectations* of the organization making the decision; it expresses the self interest, motives, attitudes and established habits of the decision makers.

A second group of bounds on acceptance expresses the *external expectations* imposed on the organization making the decision. Included here, in the case of multi-level directive control, are the directives issued by higher organizations within the framework of vertical flow of information. Also in this category are requests by other institutions which the decision maker regards as important to satisfy, even if they are not expressed in a legally sanctioned form. In this group of bounds on acceptance belong the strong wishes of the functional organizations within the same institution which were mentioned in the preceding chapter.

The broken line in Figure 8.2 indicates the lower boundary of the set of acceptable decision alternatives. For simplicity only lower bounds are assumed although, in reality, many other cases are possible.

In addition to formulating $\tilde{\mathcal{B}}(t)$, it is also necessary to define the set $\mathcal{D}(t)$. What criteria of acceptance should the decision maker employ? To what extent should he consider the wishes and expectations of others? How should he harmonize the possible contradictions among them? As indicated by argument t, $D(t)$ also changes over time, in the course of the decision process.

The set $\mathcal{D}(t)$ is delimited by bounds on acceptance. The bound on acceptance is a simple formal expression which captures the essential features of the decision process. In my book on national economic planning I emphasized that according to my experience, the highest political and economic leadership formulates economic policy mainly in the form of quantitative requirements or targets to be met, that is, in the terminology just introduced, in the form of bounds on acceptance.[4] "The growth rate of national income should be at least 4 per cent." "Industrial output should reach X billion dollars by 1980."

In lower-level institutions, such as offices or firms, the situation is similar. Motives, interests, attitudes and expectations appear mostly in the form of bounds on acceptance. For example, in an industrial firm, typical bounds are: "The rate of profits should be at least 8 per cent." "Our share in the market should not fall below our share last year." "At least 10,000 units should be produced."

This type of behaviour is called "satisficing" by *H. Simon*.[5]

The bounds on acceptance are empirically observable. Most of them appear in the information flow between the organizations. Sometimes they appear in written documents, in office or firm memoranda. Often, they may be discovered by questioning the decision makers. At any rate, there is much more opportunity to observe the bounds on acceptance than the "elusive" utility functions. But more will be said about this later.

It is necessary to stress emphatically the difference between the sets $\tilde{\mathcal{B}}(t)$ and $\mathcal{D}(t)$. The former reflects the possibilities of the real sphere, although the reflection may be inaccurate and $\tilde{\mathcal{B}}(t)$ may deviate from $\mathcal{B}(t)$. But accurate or inaccurate, it is supposed to summarize the *physical, material* bounds on activities in the real sphere. In contrast, $\mathcal{D}(t)$ is a phenomenon of the *control sphere;* it represents the *intellectual* limits established by the interests, motives and expectations of those making the decision.

Having defined the sets $\tilde{\mathcal{B}}(t)$ and $\mathcal{D}(t)$, we may turn to an explanation of the next concept:

[4] See [128], Chapter 27. Similar ideas may be found in the book by *Tinbergen–Bos* [256].
[5] See *H. Simon* [236] and [237].

Definition 8.8*. The set of e l i g i b l e decision alternatives $\tilde{\mathcal{F}}(t)$ ($\underline{t} \leqq t \leqq \bar{t}$) is a subset of the set of feasible decision alternatives. It is the intersection of sets of explored and acceptable decision alternatives.

$$\tilde{\mathcal{F}}(t) = \tilde{\mathcal{B}}(t) \cap \mathcal{D}(t) \subset \mathcal{A} \tag{8.1}$$

The actual decision eventually selected by the decision maker is chosen from among the alternatives regarded as both acceptable and implementable.

In Figures 8.2. A—B—C, the set $\tilde{\mathcal{F}}(t)$ is represented by the crosshatched polygon. As can be seen, decisions a_1^* and a_2^* are within the polygon $\tilde{\mathcal{F}}(t)$. One of them, namely a_1^*, is really implementable, but the other, a_2^*, is not. (See Figures 8.2A and 8.2B.)

The reader accustomed to the conceptual system of the mathematical theory of the firm, or of mathematical programming and operations research may ask why we describe the model of the decision process in this way? Should it not be guaranteed in advance that the decision maker will choose exclusively from among implementable alternatives?

Recall that the C- and the R-spheres have been strictly separated from each other. Implementability is a characteristic of the R-sphere; it is based on real limits on the production, turnover, and consumption processes. In contrast, the decision is made in the C-sphere.

Consider, for example, the production section of a firm. Let us assume that the C-unit makes decision a_2^*, which cannot be implemented. In this case the information transmitted by the C-unit to the R-unit, the instruction, will be the decision a_2^*. But, obviously, the R-unit cannot realize a decision which cannot be implemented. The ψ_i response function of the R-unit must describe what will happen in the real unit when the instruction received is unrealistic. (e.g., in the case of an exaggerated production plan, production will reach the technically feasible upper limit, but perhaps quality will deteriorate or machinery will break down.)

This treatment of implementability is related to the *dualistic* manner of describing the economic system. Decision a^* is like an event of the "soul"; implementation concerns the "body". The fact that the "body" does not obey an incorrect command of the soul, must find expression in the model of the real sphere, in an appropriate description of the system of response functions ψ, and not in the elimination of wrong decisions from the model at the outset.

8.4. Choice of the Decision

The final step of the elementary decision process occurs in period t; from the set of eligible decision alternatives a single element, $a^* \in \tilde{\mathcal{F}}(t)$, must be chosen.

It is a general assumption of this book that choices are random. The intellectual efforts of the decision maker are concentrated partly in exploring which alternatives *can* be implemented, $\check{\mathcal{B}}$, and partly in determining which ones are *worth* implementing, $\mathcal{D}$. If, however, the set of eligible alternatives $\mathcal{F} = \check{\mathcal{B}} \geqq \mathcal{D}$ has been sufficiently narrowed down, there are no unequivocal or deterministic rules which indicate which element of the set will become the actual decision.

In the course of further discussion, the indicators of the decision $a^* = (a_1^*, a_2^*, \ldots, a_K^*)$ will be considered random variables.

Let us call the probability distribution which gives the probability that the decision will fall in a given subset $\mathcal{H}$ of set $\mathcal{F}$ the *decision distribution* and denote it by $\xi(\mathcal{H})$.

The decision distribution $\xi(\mathcal{H})$ is highly characteristic of the functioning of an organization. Let us list only a few possibilities.

One possibility is that the distribution $\xi(\mathcal{H})$ is uniform. This means that the acceptance of all alternatives considered is equally probable.

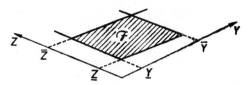

Figure 8.3A

Set of eligible decision alternatives

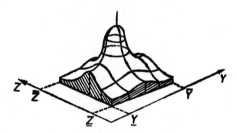

Figure 8.3B

Density function of decision distribution

More frequently the distribution has some "area of density"; a definite group of the alternatives have a greater chance of being accepted than others. For example, for conservative organizations it is more likely, in recurring problems that alternatives similar to those chosen in the past will be decided upon than radically different ones.

The "area of density" of the distribution may reflect an established decision routine, the regular enforcement of a rule of thumb. (This will be dealt with in Chapter 9.)

9 KORNAI: Anti-Equilibrium

In other cases the area of density may be centered around definite targets, desired performance and "aspiration levels". (The notion of aspiration level will be dealt with in a separate chapter, and is only casually mentioned here.)

The density function of such an uneven decision distribution is illustrated in *Figures 8.3A* and *8.3B*. In Figure 8.3A again for a two-indicator case the set $(\mathcal{F}(\tilde{\imath})$ is shown in perspective. For the sake of simplicity it is assumed that the two indicators Y and Z have only lower and upper bounds.

$$\underline{Y} \leq Y \leq \bar{Y}, \quad \underline{Z} \leq Z \leq \bar{Z} \qquad (8.2)$$

Accordingly, $(\mathcal{F}(\tilde{\imath})$ is a rectangle.

In Figure 8.3B the density function of the decision distribution is shown. Again, for the sake of ease in illustration, it is assumed that (8.2) is a truncated normal distribution. As can be seen from the illustration the probability of choosing the alternatives near the "middle" is greater than that of choosing the alternatives near the edge of the set.

8.5. The Example of National Economic Planning

The conceptual framework of Chapters 7 and 8 has been illustrated mainly with examples taken from the operation of the firm. It would be worth while to ask if we can describe national economic planning with the same conceptual framework. The problem needs further research; a few ideas are discussed, however, for the sake of illustration.

Literature on national economic planning generally focuses on a single-moment in the decision process, namely, the decision itself, although, the preparation of a final decision is usually a lengthy process. In five-year planning this may last 1, 2, or 3 years.

Preparation of the decision involves two kinds of cognition process. The first is the search for possibilities, that is, formulation of the set $\hat{\mathcal{B}}$. This is performed mainly by planning specialists, working in the Planning Office and other economic institutions. The second process is the definition of political bounds on acceptance, that is, the formulation of set $\mathcal{D}$. This involves, primarily, the organizations responsible for making political decisions.

This process includes the formulation of political compromises, that is, the setting of bounds which are acceptable to all who participate in decision-making.

The two processes can, of course, be strictly separated only in theory; in practice, they are intertwined and involve mutual interaction. The "planners" calculate in advance, when submitting their proposals, how they will be "viewed" by the political bodies and whether or not there is a chance of their being accepted. Nor do the political decision-making organs live in a vacuum; the formulation of their expectations conforms, to a certain extent, to the possibilities.

The two processes are linked by a continual exchange of information. The planners obtain new information (e.g., fresher data) about the possibilities; they work out new variants of the plan and communicate them, from time to time, to the political bodies. Similarly, the political organs modify their wishes and expectations, partly on the basis of information received from the planners and partly as a result of changes in the political situation. Accordingly, both $\tilde{\mathcal{B}}(t)$ and $\mathcal{D}(t)$ change over time.

By the end of the process, the intersection of the two sets, the set of eligible alternatives, $\mathcal{F}(t)$, will have been considerably narrowed down. At that point it is of no particular importance exactly which element of the set is accepted and becomes the approved plan, the decision. Within the narrow bounds of the set $\mathcal{F}(t)$ the planners themselves do not consider the figures to be exact.

The properties of the sets $\tilde{\mathcal{B}}(t)$ and $\mathcal{D}(t)$ and their development over time are highly characteristic of the planning work in a country. Do the two sets have an intersection at the beginning of the process? That is, how sound are the first expectations of the politicians? Is $\tilde{\mathcal{B}}(t)$, the set of alternatives *deemed* implementable, a good approximation of $\mathcal{B}(t)$, the set of actually implementable alternatives? Are the planners opportunists? Do they leave with their proposals $\mathcal{B}(t)$, only in order that they should be more easily accepted?

We shall return to the description of national economic planning later on, in Section 12.7.

8.6. Comparison

Let us compare the elementary decision process described in Sections 8.1–8.5 with the decision model characteristic of the GE school.

1. We have described a decision *process* which takes place over time. This usually consists of a phase of decision preparation which lasts for several periods and leads, in the last period, to a final decision. In contrast, the traditional GE models describe only the concluding moment, the decision itself; they do not examine the process of decision preparation.

2. In our description of decision making, we distinguished four kinds of sets, the sets $\mathcal{B}$, $\tilde{\mathcal{B}}(t)$, $\mathcal{D}(t)$ and $\mathcal{F}(t)$.[6] The GE theories, however, deal with

---

[6] I will not mention here the first set, $\mathcal{A}$, the set of possible alternatives, which is used for the purpose of definition.

only a single set, the set of "feasible" alternatives, corresponding essentially to our set $\mathcal{B}(t)$.

But distinguishing among the other three sets is neither superfluous nor pedantic. It is necessary if one wishes to describe the actual process of social decision making. On the one hand, $\mathcal{B}$ and $\tilde{\mathcal{B}}(t)$ may indeed differ, since it is not always possible to implement every alternative which has been explored and deemed implementable. On the other hand, decision makers refuse, in principle various, otherwise implementable, alternatives, and this can be represented by indicating the bounds on acceptance, by specifying the set $\mathcal{D}(t)$. Thus, in the final analysis, the choice is made from the eligible set, $\mathcal{F}(t)$, which may differ substantially from the set of implementable alternatives, $\mathcal{B}$. (See Figures 8.2.)

3. We have assumed that the actual choice from among the set of eligible alternatives (i.e., explored and acceptable alternatives) is made at random, even though the choice may be characterized by stochastic regularities, some group of alternatives having a greater probability of acceptance than others. There is, however, no rigid, deterministic rule of choice. For example, it may easily happen that, in the end, some interior point of set $\mathcal{F}(t)$ will be accepted. In Figure 8.2, both a_1^* and a_2^* are such interior points.

As opposed to that, the GE school assumes the existence of a strictly deterministic rule. Most important, only an element on the boundary of the set of feasible alternatives can be accepted, the one which maximizes the utility function $U(a)$.[7] (This is illustrated in Figure 8.2C by the point a_3^*.)

Accordingly, the GE school model is a narrow, special case of the decision model d escribed in Sections 8.1–8.4, and may be characterized as follows:

Characteristic No. 1.

$$\tilde{\mathcal{B}}(t) = \mathcal{D}(t) = \mathcal{F}(t) = \mathcal{B} \tag{8.3}$$

Ch aracteristic No. 2. There exists a utility function $U(a)$ interpreted over the set $\mathcal{B}$, and there is an element $a^* \in \mathcal{B}$, for which

$$U(a^*) = \max_{a \in \mathcal{B}} U(a) \tag{8.4}$$

Characteristic No. 3. The decision distribution function is a degenerate probability distribution concentrated at the element a^*:

$$\xi\,(\mathcal{H}) = \begin{cases} 1, \text{ if } a^* \in \mathcal{H} \\ 0, \text{ if } a^* \notin \mathcal{H} \end{cases} \tag{8.5}$$

In conclusion, we may sum up as follows:

---

[7] The utility functions will be treated in a later chapter in greater detail.

According to the GE school, to describe a decision process we must give the set, $\mathcal{B}$, of feasible alternatives and the utility function $U(a)$, interpreted over the set $\mathcal{B}$. In this book we tried to give a real-science description of the decision process, in other words we tried not to prescribe what it *should be* like but rather, to generalize the main characteristics of the actual decision processes. Accordingly, we suggest that, in order to describe a decision process, the regular development over time of the sets $\mathcal{B}(t)$ and $\mathcal{D}(t)$ should be given, together with $\xi(\mathcal{X})$, the decision distribution function.

The subject of the next chapter will be a description of the decision process as an *algorithm*.

9. DECISION ALGORITHMS

9.1. General Concept of the Decision Algorithm

When a particular problem first confronts him, the decision maker already possesses some information relevant to the problem. This information is stored, to use a concept described in Chapter 4, in his "memory".

During the period of decision preparation, additional information concerning the problem at hand pours in; some is obtained through purposeful search activity by the decision-maker, but some is completely unsolicited. This information may relate to what *can be done*, expanding knowledge about $\tilde{\mathcal{B}}(t)$, the set of explored alternatives, or it may concern what it *is worth while* to do, expanding knowledge about the set of acceptable alternatives, $\mathcal{D}(t)$.

In the meantime, the decision-making organization also issues information to the other organizations.

The preparation of the decision is a cognition process. Choice is not made from given alternatives, on the basis of existing preferences. The decision process essentially involves examining, on the basis of previous experience and new information, what the decision maker can do and what it is expedient for him to do.

The process will be illustrated schematically in *Figure 9.1*. The passage of time is illustrated in the figure on the vertical axis from the top. The diagonally striped column on the right symbolizes the elementary decision process associated with the solution of problem p. The problem emerges in period t. At this point, the information related to the problem becomes "activated" from the memory. For example, in the case of an investment decision, the memory would contain the previous experiences of the firm, the professional knowledge of the engineers about technical problems related to the project, the memoranda accumulated in the document files of the firm, literary information about similar foreign investments, the balance sheet figures of financial funds available, and so on.

The activation of information originating from earlier periods at the beginning of the process is symbolized by an arrow at t pointing from left to right; the memory content $v(t)$ affects the decision.

In the course of preparing the decision, information is continually flowing: received and transmitted; —through other functional organizations operating

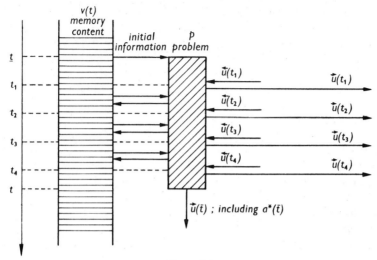

Figure 9.1

Scheme of information flow belonging to an elementary decision process

within the institution, or through other institutions—this is symbolized by arrows pointing from left to right $(\vec{u}(t_1), \vec{u}(t_2)\ldots)$ and from right to left $(\overleftarrow{u}(t_1),$ $\overleftarrow{u}(t_2)\ldots)$. For example, the investment section may have asked the selling section of the firm and a market research institute whether or not the additional output resulting from the investment can be sold. It may also have asked the financial department what profits they expect. In the case of multi-level control, directives were received in the framework of the vertical flow of information.

Furthermore, new and old information interact throughout the course of the decision process, and *interim*, partial results, are stored during this time in the memory.

At the conclusion of the elementary decision process, a special information output, namely the final decision, a^*, will be an element of the information output vector $\vec{u}(t)$ released in period $\bar{t}$.

Thus, in the final analysis, the decision may be conceived of on an abstract level, as a series of transformations. From definite *data* (initial information taken from the memory and the information received in the course of preparing the decision) a *result* is calculated, the decision.

Definition 9.1*. The ensemble of procedural rules which a control unit of a certain organization applies, in the case of problem p, to arrive at a final decision based on the information stored in its memory and received in the course of decision preparation (the data), will be called the d e c i s i o n   a l g o r i t h m and denoted $F(p)$.[1]

In the above definition, the concept of "algorithm" has been used in the same sense as it is employed in mathematical logic and the theory of algorithms. An "algorithm" is a series of moves made according to given rules and leading to the solution of a definite problem; it is a chain of logically linked instructions. "Do this; then, with the result, do that; with the new result perform the following operation. . .and so on."

A decision algorithm can be described in various ways. In some cases everyday language can be used. In other cases, it is expedient to formulate an algorithm mathematically. This mode of description can be coupled with a so-called flow-chart, illustrating the logical connections and ramifications of the algorithm. The algorithms given to computers are described in computer language and consist of a series of instructions given to the machine.[2] Which form of description is expedient depends on the concrete algorithm and on the use to which it is put.

In describing economic systems we resort to the axiomatic assumption that the control unit has a decision algorithm $F(p)$, leading to a decision for every problem p.

The above assumption does not mean that, with the aid of the algorithm $F(p)$, only a single decision, a^*, can be deduced from the given data, we do not allege that there is a unique correspondence between the data and the result, between the stored and received information and the decision. Chance plays a role. It is only assumed that the series of steps leading from the data to the result, from the information to the decision, is characterized by definite (deterministic or stochastic) regularities.

9.2. Declared Rules versus Conventions

In studying a decision algorithm it is necessary to determine what regulates the individual steps.

Declared rules are all steps of the decision algorithm which are taken in accordance with legal presciptions or other officially fixed rules. The other steps of the algorithm are based on convention, that is, on customs established in connection with decision making.

For example, it is prescribed by legal rules of the Hungarian state that before making an investment decision which involves building activity, the opinion of

---

[1] In the algorithm there may be stochastic and non-deterministic steps as well, such as, "Choose any element from a given set".

[2] For the latter, see, for example, *Ledley* [149a].

several authorities, including the building and firefighting officials, must be obtained. It is not obligatory but "in good taste" to consult the industrial ministry and the deputy minister if a major decision is involved. The former is a declared rule, the latter only a convention.

criteria established in Hungarian economic life, its own business experience, and the wishes of the government and others, this step of the algorithm is based on convention.

Convention frequently assumes the form of *"rules of thumb"*. These are simple steps within the complex algorithm. For example, one step in an algorithm for price formation may be, "Let us add at least 10 per cent to prime costs". Or a step of an algorithm designed to prepare a production program may be, "The target for the next year should be at least 5 per cent higher than the present goal".

Rules of thumb are extremely important in any economic system and in almost every institution.[3] We shall return on several occasions to this point. Custom, habit, tradition, and inertia favor a continuation of earlier behavior; all promote the continued use of rules of thumb which heavily influence actual decisions.

9.3. Standard versus Fundamental Decision Processes

Two kinds of elementary decision processes are fairly clearly distinguishable in the life of economic organizations.

Definition 9.2*. S t a n d a r d decision processes are repeated periodically or almost periodically, employ algorithms composed of a few simple steps and require little information. F u n d a m e n t a l decision processes do not occur regularly, involve algorithms composed of many and frequently complicated steps, and require a great deal of information.[4]

The "standard" and "fundamental" decision processes are two ideal types. In reality, of course, many decisions border on being one of the two types or are some mixture of the two. Yet, most decision processes can be fairly unequivocally classified into one category or the other.

Both the standard and the fundamental decision processes have algorithms. The essential difference between the two is that one algorithm is very simple, while the other is quite complex.

Statement 9.1. A majority of economic decisions are standard decision pro-

---

[3] See *Katona* [113], [114], [115].

[4] In the works quoted, *Katona* uses the concepts "habitual behavior versus genuine decisions", in a related sense.

cesses. The everyday activities of offices, enterprises and households are typically, of this nature.

Consider an industrial firm. *Standard* decisions are generally involved in the following real processes:

a) Small changes in the volume of production. The word "small" is relative; it depends on the product and on the firm, but such a change does not, in general, require investment, changes in the stock of fixed assets or major changes in material inventories; material requirements can be met from existing stocks, and at most, some adjustments must be made in the replacement of stocks at a later time.

b) Small changes in the range or quality of production. For example, with the given capacity of the firm, a variant of a product already in production is introduced, such as a cheaper desk for students in addition to the more expensive types already being produced. Or, a component of a machine already produced may be modified. Thus, a standard decision of an automobile factory may be to produce a 900 cu.cm engine instead of the 850 cu.cm type. The decision to produce cars in addition to tractors is, however, a fundamental decision.

c) Small changes in production technology. Let us return to the decision to manufacture desks with 2 drawers or with no drawers instead of only desks with a single drawer. This involves a slight change in the construction of the product, the order of operations, etc. But an essential criterion of its smallness is that the change in technology requires no new fixed assets or additional personnel.

The real changes resulting from standard decisions are mostly reversible. After a small rise in production, production can be reduced again; after a modification of the material mix, the original mix can be restored, and so on.

The real change can be arbitrarily small. Accordingly, the real changes controlled by standard decisions usually can be described quite accurately with the aid of continuous variables.

The simplicity of the standard decision process stems from the fact that, as emphasized in its definition, it requires little information. The decision-maker usually weighs only those alternatives situated "in the neighborhood" of earlier, similar standard decisions. Accordingly, $\tilde{\mathcal{B}}(t) \subset \mathcal{B}$; in most cases, that is, the explored alternatives can indeed be implemented. At the same time, $\tilde{\mathcal{B}}(t)$ is only a small subset of $\mathcal{B}$; only a small percentage of the implementable alternatives are explored. No particular effort is expended in discovering additional alternatives. As a result of similar, previous experiences, the bounds on acceptance are also well known, and thus $\mathcal{D}(t)$ can also be easily defined. In the final analysis, the set of eligible alternatives, $\mathcal{F}(t)$, is rather small.

Thus, the algorithm of the standard decision process merely involves the application of a few rules of thumb, and for this reason the decision is made in a short amount of time.

The standard decision processes enable the economic system to economize on the intellectual and material inputs of the control processes.[5] It is impossible to devote great energy to each and every problem of economic life, to explore all implementable alternatives, to predict all consequences of their acceptance, to weigh extensively all bounds on acceptance of every interested party, and so on. The standard decision imparts "automaticity" as it were, to a considerable number of control processes. It is true that the decisions which are reached may not be strictly efficient according to the efficiency criteria of mathematical decision theory; a more favorable alternative than the one chosen might be found. *But the loss in efficiency resulting from standard decision-making may be counter-balanced by the gains which are realized because the decision process is cheap for the economic system as a whole, and for the institutions making the decisions.* This is due to the fact that a decision is reached which is, perhaps, not fully optimal but not particularly inefficient either, with little information and little analytical work, that is, with small costs of decision preparation.

Standard decisions can usually be formalized in a simple manner. It is neither necessary nor possible, in general, to describe the preference ordering of the decision-maker. It suffices, instead, to give the stochastic form of the relevant "rule of thumb", since the actual decision is distributed around the decision given by the "rule".

This does not mean that systems organizers, "management consultants," operation research workers, and mathematical planners should not attempt to improve the standard decisions, to propose better and more efficient rules to replace the rules of thumb employed.

Let us consider the *fundamental* decisions. In the life of an industrial firm, the fundamental decisions are related, among other things, to the following real processes:

a) Creation of a new plant. This involves setting up either an entirely new firm or else a new plant within an existing firm. Depending on the given level of technology and the industry, the new plant cannot be smaller than a definite minimum size. A textile mill can be relatively small, a petroleum refinery cannot. The new plant requires a substantial investment in new assets and personnel.

b) Introduction of an entirely new product. For example, a plant that has been producing only radio receiver sets, may start producing TV sets in addition. For this, investment is needed, and perhaps the re-training and expansion

---

[5] In the language of our general model we formulate this as follows: the existence of the standard decision processes means that the *R-units of the control organizations* use relatively little real input.

of personnel as well. The new product cannot be produced in quantities smaller than some definite minimum because it would be unprofitable.

c) Introduction of a brand new technology or reorganization of production. An example of this is a transition to the use of the conveyor belts.

The real changes resulting from fundamental decisions are generally irreversible. The new plant cannot be partially built; it is either constructed or not. But if it has been completed, its creation cannot be undone. Nor can a new product be "half" introduced; if it has been introduced, production cannot be abandoned. If transition to an entirely new technology has taken place, if the conveyor belt has already been installed, its use cannot, as a rule, be abandoned in favor of the old methods again.

The real changes stemming from fundamental decisions cannot be arbitrarily small; they cannot be described by means of continuous variables. In some cases the variable assumes a value of either 0 or 1; the new product is or is not produced, the conveyor belt is or is not installed. In other cases the alternatives may be characterized by integer variables: one, two, or three new "towers" may constructed in a chemical plant; one, two, or three turbines may be ordered for the power plant. In some cases the variable must assume a value of either zero or a positive figure larger than some given number; there is a break between zero and the minimum size. Either no passenger cars or at least 10,000 cars annually should be produced.[6]

The complexity of the fundamental decision process is connected with the fact, emphasized in its definition, that it requires a great deal of information. The decision-maker usually attempts to weigh many alternatives, that is, to explore the set $\mathcal{B}$, the set of implementable alternatives. In the terminology of computer programming, the algorithm is usually *cyclical*. The decision-maker seeks alternatives deemed implementable and confronts them with the bounds on acceptance. It may turn out that at some times there is no alternative that has been explored and is acceptable given the bounds on acceptance; that is, $\mathcal{F}(t_1) = O$, the set of eligible alternatives is empty. Then a new cycle begins. The decision-maker seeks additional alternatives, that is, he expands the set $\mathcal{B}(t)$. He scrutinizes his own expectations and asks other interested organizations to correct their own expectations; that is, he modifies the set $\mathcal{D}(t)$. In this way the set $\mathcal{F}(t_2)$, a new set of eligible alternatives, will be formulated. If it turns out again to be empty, a third cycle will begin. This may continue until, at last $\mathcal{F}(t) \neq O$, that is, an alternative is found which is both implementable and acceptable.

---

[6] The assertion of increasing returns in production is clearly related to the phenomenon mentioned. An individual fixed input can "bear" a definite cost which is more with the given fixed input, the smaller the fixed cost per unit of product.

Thus, the algorithm of the fundamental decision process may consist of many iterations and each iteration may itself be composed of several steps. Accordingly, more time is required to reach a decision. The preparation of investment decisions of major importance may even take one or two years.

Precisely because they are of great importance, economic systems apply considerable intellectual and material inputs to the fundemantal decision processes.

This does not mean, of course, that these processes take place in the economy in the way prescribed by the models of mathematical decision theory. The decision-makers usually cannot survey the entire $\mathcal{B}$ set; they only repeat the search for new alternatives until they find an acceptable one. In the final analysis, $(\hat{\mathcal{B}})$ is only a small subset of $\mathcal{B}$, the set of all implementable alternatives. Furthermore, the formation of the set $\mathcal{D}(t)$, the establishment of bounds on acceptance, is based, even in the case of the fundamental decisions, frequently on relatively simple rules of thumb, habits or biases. Thus, the operation researchers and mathematical planners have a broad field for raising the standards of the fundamental decision processes, through the use of models and recommendations based on the models.

In Chapter 4, when the general model of the economic system was first expounded, we made the axiomatic assumption that each control unit, $c \in \mathcal{C}$, has a response function, φ, which characterizes its functioning. In Chapters 8 and 9, we assumed that each control unit, $c \in \mathcal{C}$, has a *decision* algorithm, F, characteristic of its functioning. *The two assumptions are equivalent to one another.*

The basis for the identity can be easily seen. What happens in the case of the decision algorithm? There are *data*, the initial information stored in the memory and the new information received in the course of the decision process (see Figure 9.1), which undergoes a series of transformations. *The result*, the decision is transmitted as outgoing information (like other information issued in, during, or at the end of the process, or stored in the memory). Thus, in the final analysis, the following transformation takes place over time in a sequence of steps: information received plus the contents of the memory yields outgoing information and new memory contents. The response function describes the same transformation in another form.

The usual brief comparison with the GE school will not be made in this chapter as in earlier ones, because two separate chapters will be devoted to the decision model of the GE school; one deals with preference orderings and the other with the utility function. The comparison requires a longer and more detailed argument since it is the core of the GE school, its most characteristic feature.

10. PREFERENCE, UTILITY FUNCTION, RATIONALITY: A SURVEY

In this chapter I will briefly review and clarify the basic tenets of the theory of preference orderings and utility functions. A critique of the theory is presented in Chapter 11.

10.1. The Concept of Preference Ordering

The theory of utility functions and preference orderings was formulated over a century ago and has developed a great deal since then. Originally it was described in an elementary mathematical form and was based on very strong constraining assumptions (e.g., it may be assumed that utilities are additive, that the marginal utility function is monotonically decreasing, etc.). Later on, the mathematical formulation of the theory became much more exact and, simultaneously, several constraints were removed and some strong assumptions weakened.

However, many unrealistic features of the model remain even in its more general and mathematically precise present-day form.[1]

As far as possible the notation introduced in Chapter 8 will be used to clarify the connections and the differences between the theory of utility functions and our own conceptual system.

Definition 10.1. The set $\mathcal{A}$ of possible decision alternatives is given. The elements of the set are indicator vectors with K components. The decision-maker possesses a complete preordering over the set[2] $\mathcal{A}$, according to which he is capable of stating for any pair of elements (a_1, a_2) $(a_1 \in \mathcal{A}$ and $a_2 \in \mathcal{A})$ whether he prefers a_1 to a_2 $(a_1 \succ a_2)$, or a_2 to a_1 $(a_1 \prec a_2)$ or is indifferent between them $a_1 \sim a_2$). The complete preordering of the set $\mathcal{A}$ according to these relations will be denoted by the symbol P and referred to as the p r e f e r e n c e   o r d e r i n g.

---

[1] As indicated in Chapter 3, the most important assumptions of the GE school are based on these features of the model and it is this at which my critique is aimed. (See the basic assumptions reviewed in Section 3.2 under 7 and 8.)

[2] The literature frequently distinguishes complete ordering from pre-ordering. If this distinction is applied here, we have complete preordering. See *Debreu* [50], pp. 8, 54, and 61.

Preference orderings, like all complete preorderings have the following two properties, which should be stressed separately because of their economic importance.

The ordering is antisymmetric: if the decision-maker prefers a_1 to a_2, he cannot prefer a_2 to a_1.

The ordering is transitive: If the decision-maker prefers a_1 to a_2 and a_2 to a_3 it follows that he prefers a_1 to a_3.

Definition 10.2 The preference ordering P on set $\mathcal{A}$ may be represented by a function interpreted on set $\mathcal{A}$ called the u t i l i t y   f u n c t i o n and denoted by $U(a)$. This representation derives from the fact that the relations $a_1 \succ a_2$ and $U(a_1) > U(a_2)$, as well as the relations $a_1 \sim a_2$ and $U(a_1) = U(a_2)$, are equivalent to each other.

Therefore, if we say that the decision-maker refers one alternative to another, this may also be expressed by saying that the former is characterized by a higher and the latter by a lower "utility". So, in the following discussion these two statements will be considered equivalent: "the decision-maker has a preference ordering (complete preference preordering)" and "the decision-maker has a utility function".

The preference ordering P and the utility function $U(a)$ are not specified by all of the authors of the GE school in exactly the same way. Most authors assume that the preference ordering is *convex*, perhaps even strictly *convex*.[3] This assumption is illustrated in *Figure 10.1*. For the sake of simplicity, a two-dimensional decision problem is illustrated. Let Y be the output of one product

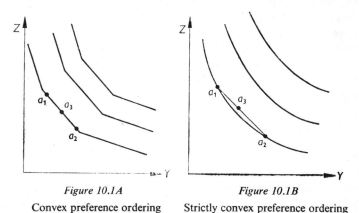

Figure 10.1A	Figure 10.1B
Convex preference ordering	Strictly convex preference ordering

[3] More precisely, the indifference hyper-surfaces representing the preference ordering are convex (strictly convex) and the corresponding utility functions, $U(a)$, are concave (strictly concave). In what follows the convexity of the indifference surfaces is always intended.

and Z that of another. All points in the positive quadrant represent plans calling for different outputs of the two products. The producer first weighs two alteratives a_1 and a_2. Both are situated on the same indifference curve, so $a_1 \sim a_2$. The assumption of convexity implies, in the first place, *continuity;* the decision-maker is not obliged to choose between a_1 and a_2. but may combine or mix them at his discretion. The straight line connecting the two alternatives illustrates their linear combinations.

There are two possible cases to consider: (Weak) convexity is shown in *Figure 10.1A* where the indifference curve consists of linear segments. Here the straight line connecting a_1 and a_2 coincides with the indifference curve. This means that the decision-maker is indifferent among a_1, a_2 and all convex linear combinations of a_1 and a_2, such as a_3.

$$a_3 = \chi a_1 + (1-\chi)a_2 \sim a_1 \sim a_2, \quad 0 \leqq \chi \leqq 1 \qquad (10.1)$$

In the case of strict convexity, a much stronger assumption is required. It is assumed that the decision-maker prefers a mixture of the two alternatives to either of the two separately. This is shown in *Figure 10.1B* where alternative a_3 lies *above* the indifference curve connecting alternatives a_1 and a_2.

$$a_1 \sim a_2, \quad a_3 = \chi a_1 + (1-\chi)a_2 \quad (0 < \chi < 1)$$

$$(10.2)$$

$$a_3 \succ a_1 \quad \text{and} \quad a_3 \succ a_2.$$

10.2. Dynamic versus Static Interpretation

Using the conceptual framework of preference orderings and utility functions, two kinds of model may be built, *dynamic* or *static* decision models.

In the case of a dynamic interpretation, a series of decisions $a^*(t_1)$, $a^*(t_2)$, $a^*(t_3)$, ... are made sequentially over time. For each decision the decision-maker may choose from the actual set of implementable alternatives available at that time, $\mathcal{B}(t_1)$, $\mathcal{B}(t_2)$, $\mathcal{B}(t_3)$, ... The actual set of implementable decisions is a subset of $\mathcal{A}$, the set of possible decisions, which remains constant over time: $\mathcal{B}(t) \subset \mathcal{A}$. The choice is made in accordance with the preference ordering P on set $\mathcal{A}$ or, equivalently to the utility function $U(a)$, which is also constant over time. Since both the set of feasible alternatives $\mathcal{A}$, and the preference ordering P (as well as the utility function $U(a)$) remain constant over time, the dynamic interpretation may also be regarded as a dynamic-*stationary* one.

In the case of a static interpretation only a single decision, occurring in a single period is examined, Here, obviously only a single set of implementable decision alternatives $\mathcal{B} \subset \mathcal{A}$, and a single preference ordering P are considered.

Much of the work on the subject of decision making fails to state whether the model should be interpreted as a static or as a dynamic-stationary one. The static interpretation is most frequent. Recently, however, particularly since the appearance of *Samuelson*'s[4] famous "revealed preference" theory, the dynamic-stationary interpretation has become more common. For this reason *Samuelson*'s theory will be discussed in detail.

10.3. Revealed Preferences

To understand the theory of revealed preference it is helpful to consider the problems connected with the empirical determination of the preference ordering.

We may learn the preferences of the decision-maker *by asking him* outright whether he prefers a_1 to a_2 or whether he is indifferent between the two alternatives. By methodically asking him about pairs of alternatives we may state the preferences of the decision-maker and may even establish whether or not his preference ordering is consistent, that is, whether or not he has violated the transitivity requirement, (e.g. preferred a_1 to a_2 and a_2 to a_3 but preferred a_3 to a_1).

Direct questioning may be replaced by more indirect methods, such as psychological *experimentation*. In a game or experiment the decision-maker is faced with decision problems and his choices indicate his preferences.

However instructive these investigations may be, they are not completely convincing. The hypothetical situation outlined for the decision-maker is unrealistic. The answer to a hypothetical question may differ from actual behavior under the given conditions. This is the reason that Samuelson proposed a model based on the actual decisions of the decision-makers, on the preferences revealed in their behavior.

Although the dynamic interpretation is generally not stressed in formulations of the theory or in comments on it, we consider it worthy of special emphasis. Its importance lies in the fact that it does not examine *simultaneous* preference *statements* made in answer to hypothetical questions, but rather, considers a series of decisions $a^*(t_1)$, $a^*(t_2)$, $a^*(t_3)$, ... *actually made over time*. Therefore, the Samuelson theory will always be interpreted dynamically.

Reviewing the theory, using Samuelson's approach, we will deal first with the consumer's decision. This method can, however, be extended to decisions in general.

Let us denote the vector consisting of the quantities bought by the consumer in period t by $x(t)$. Set $\mathcal{A}$ is the non-negative portion of the $\mathcal{L}^K$ vector-space

[4] See: *Samuelson* [214], and *Uzawa* [263].

of K components: $\mathcal{A} = \{x : x \in \mathcal{L}^K, x \geqq 0\}$. (This is an indicator vector of K components.) In *Figure 10.2*, for the sake of simplicity, a vector of two components is shown. On the horizontal axis the purchases of the first product, and on the vertical axis those of the second product, are shown.

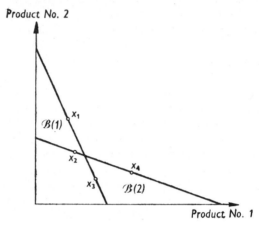

Figure 10.2

Revealed preferences

The prices of the products in period t will be denoted by $p(t)$ and the income of the consumer in period t by $r(t)$.

Accordingly, the set of implementable programs, $B(t)$, can be determined as follows :

$$\mathcal{B}(t) = \{x(t) : p(t)\,x(t) \leqq r(t)\} \subset \mathcal{A}. \qquad (10.3)$$

Let us assume that prices and income differ in two consecutive periods. In the first period, given the preveiling prices and incomes, the set of implementable consumption programs was $\mathcal{B}(1)$, which is represented in Figure 10.2 by the taller triangle. In the second period, with the new prices and incomes, the set of implementable programs was $\mathcal{B}(2)$, represented by the shorter triangle in the figure.

Let us consider four kinds of decisions, represented by the points x_1, x_2, x_3 and x_4. All four programs are efficient; they lie on the boundary of the sets $\mathcal{B}(t)$, i.e. on the budget lines.[5]

Case a) : Consistent decisions. Let us assume that the consumer chose x_1 in period one and x_2 in period two. In period one both alternatives were avail-

---

[5] In this summary only the axiom of so-called "*strong* revealed preference" is reviewed.

able: $x_1 \in \mathcal{B}(1)$, $x_2 \in \mathcal{B}(1)$. Program x_1 does not dominate program x_2 since the first yields more of product two, but less of product one than the latter. If he decided on x_1 nevertheless, he has revealed that $x_1 \succ x_2$. His subsequent decision in favor of x_2 does not contradict his observed preference for x_1 since x_1 is no longer available in period two; $x_1 \notin \mathcal{B}(2)$. In this case the decision-maker has acted consistently.

Case b): Non-comparable decisions. Let us assume that in period one the consumer chose x_1 and in period two he selected x_4. In this case we know nothing about his preferences. In the first period the alternative x_4 was not avalable, while, in the second period, the alternative x_2 was unavailable; $x_4 \notin \mathcal{B}(2)$. These two decisions are therefore not comparable.

Case c): Inconsistent decisions. Let us assume that the decision-maker chose x_3 in period 1 and x_2 in period 2. Both alternatives were available in both periods.: $x_3 \in \mathcal{B}(1)$, $x_3 \in \mathcal{B}(2)$, $x_2 \in \mathcal{B}(1)$, $x_2 \in \mathcal{B}(2)$. If he reveals first that $x_3 \succ x_2$ and second that $x_3 \prec x_2$, his preferences are inconsistent.

This analysis applies not only to consumers' decisions but to decisions in general. The three cases of consistent, non-comparable and inconsistent decisions can be clearly distinguished not only in the special case where $\mathcal{B}(t)$ refers to consumption possibilities but in every case in which $\mathcal{B}(t)$ is a convex set.

It has been theoretically demonstrated that as long as the decision-maker acts consistently (in the sense described under c) above), his behavior can be described by a convex preference ordering, P, or its equivalent, a concave utility function, $U(a)$ both of which are constant over time.

On the basis of what has been said, the requirement traditionally referred to as the "strong axiom of revealed preference" may be formulated as follows:

Definition 10.3. The requirement of c o n s i s t e n t decision-making:

Let $a_1 \in \mathcal{B}(t_1)$ and $a_1 \in \mathcal{B}(t_1)$ in period t_1. The decision-maker who chooses a_1, i.e. $a^*(t_1) = a_1$, has revealed his preference: $a_1 \succ a_2$. Let us assume that in some period, t_2, $a_2 \in \mathcal{B}(t_2)$. In the case of a consistent decision he can choose a_2, that is $a^*(t_2) = a_2$, only if $a_1 \notin \mathcal{B}(t_2)$.

The requirement of consistent behavior is not violated if the decisions made in different periods are incomparable, that is, in case b). The requirement is, however, violated in case c).

The attribute "rational" has been employed by the advocates of the GE school in too narrow a sense. They use it to characterize those decision-makers who always "optimize", that is, who always choose the alternative $a \in \mathcal{B}(t)$, which maximizes the utility function, $U(a)$. In the case of a dynamic interpretation, this is equivalent to calling decision-makers "rational" only when they never violate the requirement of consistent behaviour as defined in 10.3.

In the vernacular, the attribute "rational" is used more broadly to characterize behaviour which will be termed "clever behaviour" in Section 11.8. There-

10*

fore, to avoid any conceptual confusion, in the following discussion I will always use the word "consistent" which has a narrower meaning (in the sense of Definition 10.3) even in cases when the GE school would use the words "rational" or "optimal".

10.4. Recurring versus Non-Recurring and Comparable versus Non-Comparable Decisions

In connection with the static and dynamic interpretations of the preference ordering model it is necessary to classify decisions according to their patterns of recurrence.

Definition 10.4. If, in the same indicator space $\mathcal{A}$, the same decision-maker (individual or organization) makes the series of r e c u r r i n g decisions $a^*(t_1)$, ..., $a^*(t_Q)$, $(Q > 1$, $a^*(t_1) \in \mathcal{A}$, ..., $a^*(t_Q) \in \mathcal{A})$, more than once, this constitutes a s e r i e s o f d e c i s i o n s. Non-recurring decisions will be called s i n g l e decisions.

In the class of recurring decisions, we distinguish two sub-classes according to whether or not elements in the series of decisions are comparable. For example, in poorer countries even wealthy people buy a car only once every five or ten years. In five or ten years, however, the assortment of cars changes. If a decision is made in 1960, the available models are mainly those of 1957, 1958 and 1959; future models, from the sixties are obviously unavailable. Hower, when choosing in 1970, models from the fifties are no longer offered (at least as new cars). Accordingly, the sets of implementable decisions $\mathcal{B}(1960)$ and $\mathcal{B}(1970)$ have almost no intersection.

In contrast, the housewife, over shorter periods, say in the course of one or two years, is faced mainly with comparable, recurring decision alternatives with respect to food purchasing; she must decide in what proportions to include vegetables, fruit, meat, and so forth, in the family diet. We will disregard the seasonal changes which affect the available alternatives and preferences.

We will now present some general definitions.

Definition 10.5. Let us call c o m p a r a b l e a decision $d^*(t_i)$, a member of a series of decisions, if it can be compared from the standpoint of consistency with at least one additional member, $a^*(t_j)$, of the series of decisions, that is,

$$\exists_j, \ 1 \leq j \leq Q, \ \ j \neq i \tag{10.4}$$

$$a^*(t_j) \in \mathcal{B}(t_i) \cap \mathcal{B}(t_j). \tag{10.5}$$

Let us call e v a l u a b l e a series of decisions, every member of which is comparable. Let us call n o n - e v a l u a b l e, a series of decisions which includes non-comparable members.[6]

The substance of our definition is simple. A series of decisions is called evaluable if it can be determined whether or not individual decisions were or were not consistent in the sense of "revealed preference" theory as described in Definition 10.3. If consistency or lack thereof cannot be established, the series of decisions is non-evaluable.

Consumers' decisions obviously constitute a series of evaluable decisions (where consistency can also be empirically checked) if price changes are more frequent than changes in tastes.

For the purpose of practical empirical investigation a looser definition is sufficient; it is only necessary for a series of decisions to be "quasi-evaluable". This condition holds if most of the decisions are comparable.

10.5. Deterministic Decision versus Uncertainty

Decisions may also be classified according to whether or not they are made under uncertainty. Most models are of a *deterministic* character. A known set of alternatives, $\mathcal{A}$, is given; similarly, $\mathcal{B}(t)$, the set of implementable alternatives is also known with certainty. There is a given preference ordering P over set $\mathcal{A}$, according to which the decision-maker is capable of unequivocally stating whether $a_1 \succ a_2$, or $a_1 \prec a_2$ or $a_1 \sim a_2$; $(a_1 \in \mathcal{A}, a_2 \in \mathcal{A})$. If, in addition, strict convexity holds with respect to both the set of implementable alternatives and the preference ordering, then only one decision, $a^*(t)$, is possible.

Non-deterministic models may be classified in many ways.

The term "decision under uncertainty" has been reserved for the case where utility depends not only on the decision but also on external conditions independent of the decision-maker, on the state of "nature". The utility function has the form $U(a, \Theta)$ where Θ signifies the state of nature. The decision-maker

---

[6] The pairs of concepts introduced by Definitions 10.4 and 10.5 may remind readers of the pair of concepts defined under 9.2, namely, the "standard" and "non-standard, fundamental" decisions. There is some overlap between them, but they do not completely coincide. In Definition 9.2 the basis of distinction is whether *the decision algorithm is simple or complex*. Here, however, the distinction has formal characteristics connected with the sets of alternatives $\mathcal{A}$ and $\mathcal{B}(t)$. At any rate, there are certainly some relations between the different pairs of concepts: The evaluable series of decisions, that is, the recurring and comparable decisions, belong to the sphere of standard decisions. The converse is not always true, there may be standard decisions which, owing to the strong stipulations of Definition 10.5, cannot be evaluated.

It is always true that fundamental decisions are either single or non-recurring, or, if recurring, non-comparable.

knows the possible values of Θ; he also knows the value of the function $U(a, \Theta)$ for all possible values of Θ. ("Pay-off function.") But he has no information about the probabilities of the various values of Θ.[7]

The decision-maker has more information if, although he has no knowledge of the actual value of Θ, he knows the probability distribution of Θ. This is the type of problem usually described with the aid of various stochastic programming models.[8]

Both families of models described are similar, however, insofar as it is not the preferences of the decision-maker which are uncertain but rather his knowledge of external circumstances.

10.6. Descriptive-Explanatory versus Normative Theory

The theory of preference orderings may be interpreted in two ways from the standpoint of scientific significance.

It can be regarded as a descriptive *real-science* theory. This means that the decision-makers actually behave as hypothesized by the preference ordering models. That is, from the set of alternatives $\mathcal{B}(t)$ they always choose the alternative a^* which maximizes the utility function, $U(a)$. Thus, in the case of a dynamic interpretation their series of decisions satisfy the consistency requirement 10.3. If the theory is viewed this way, the key question to be asked by the critic is whether or not the theory is confirmed by experience.

The preference ordering model may also be viewed as a normative theory. In this case, there are two questions which are basic to a critique of the theory. The first question to ask is whether the theory is mathematically-logically correct within the framework of its own assumptions? This may be immediately answered in the affirmative; the theory needs no revision in this regard. The second question is, whether the theory is usable? It is proposed that, on the basis of the model, the decision-makers should attempt to maximize their utility functions; they should optimize; they should be consistent. The key question then is whether or not this is good advice? Is the decision-maker wise to follow this advice?

In economic literature both the real-science and the normative-theoretical interpretations can be encountered. Although the normative interpretation is more common, many authors consider the preference ordering model to be

---

[7] The starting point is the classical work on the theory of games by *von Neumann* and *Morgenstern* [192]. From the vast literature on the problem we mention the works by *Savage* [218], and *Milnor* [180]. A comprehensive survey may be found in the works by *Arrow* [9], and *Luce–Raiffa* [154].

[8] See e.g. *Headley* [85], Chapter 5.

a good, at least an approximately acceptable, model of the behaviour of individuals and economic organizations, that is, an adequate descriptive, explanatory real-science theory.

10.7. Fields of Application: The Consumer, the Firm, the Government

The concept of utility function was first used to describe the behaviour of the consumer. Its scope of application, however, has been considerably broadened. In present-day literature, the actual behaviour of the following organizations is typically assumed to be characterized by complete preference ordering:

The consumer, the households. Here the conception is still, almost exclusively, that the consumer maximizes his utility function.[9]

The capitalist productive firm. Though opinion is divided, many authors still assume that the behaviour of a productive firm may be described by a utility function. Opinion differs, to a great extent, about the nature of this utility function. According to some, including the Walras–Arrow–Debreu equilibrium model theorists, the utility function of the firm is the profit function.[10] The behaviour of the firm is characterized by its striving to maximize profits.

According to other economists it is more correct to speak about the desire of the firm to maximize its sales.[11]

Recently there has been much talk about the division between owners and managers in the capitalist firm. The consequence of this division is supposedly that the managers maximize their own utility functions.[12]

The socialist productive firm. Several authors have made attempts at describing the behaviour of the socialist firm starting from the assumption that the firm makes its decisions on the basis of complete preference ordering.

But even these studies interpret the utility function of the firm in different ways. The differences are associated with the particular socialist country or period under consideration.

For example, under conditions of directive regulation and strong centralization it is assumed that the socialist firm maximizes its output.[13]

In an article written in co-authorship with *T. Lipták* in 1962, the author assumed that after the introduction of profit-sharing, firms would maximize

---

[9] See *Houthakker*'s article [96]. It is in this manner that the general theory of equilibrium also models the behaviour of the consumer, as we have stressed in our list of basic assumptions. See *basic assumption No. 8*.

[10] See *basic assumption No. 7* of the GE theory.

[11] This view is presented in *Baumol*'s book [26].

[12] See the study by *Williamson* [275].

[13] See *Portes* [205].

either profits or profits per income.[14] In the discussions about the reform of the Hungarian economic administration some articles assumed that, after the reform, firms would maximize either profits or the profit-share falling to workers.[15]

Ward wishes to characterize the special system of incentives of Yugoslav firms by the maximization of personal income per worker.[16]

The planner, the government. Some authors use the same system of concepts to describe the behaviour of the government, or the planning office or planning collective functioning on behalf of the government, in both capitalist and socialist countries. In many works the decision-makers or planners of the country are assumed to possess a preference ordering.

The existence of a utility function or preference ordering seems completely self-evident to a large number of economists. They find it almost impossible to conceive of a model describing the economy or some sub-system of it in which the decision-makers do not possess utility functions. This, indeed, is necessary for the model to be "well-clad", just as a necktie is an essential compliment to a suit of clothes.

[14] See *Kornai–Lipták* [135].
[15] See, e.g. the studies by *E. Megyeri* [176] and [177].
[16] See *Ward* [271].

11. PREFERENCE, UTILITY FUNCTION, RATIONALITY: A CRITIQUE[1]

The group of theories reviewed in Chapter 10 may be classified according to several criteria:

—Is the model static or dynamic?
—If the model is dynamic, does it examine comparable or non-comparable decisions?
—Does the model allow for uncertainty or disregard it?
—Is the theory descriptive or normative?
—What institution—the consumer, the firm or the government—is the model meant to describe?

This critique endeavours to consider these theories from the standpoint of each criterion listed above. This is a complicated task; therefore, to facilitate the survey, in *Table 11.1*, the logical framework of the critique has been summarized. A study of Table 11.1 either before or after reading this chapter should facilitate understanding of this rather complicated analysis.

11.1. On the Static, Descriptive Interpretation

Let us consider first the static interpretation of the theory. Can it be used as a descriptive-explanatory real-science theory?

In my opinion, the theory is obviously *true* under this interpretation but it is empty, tautological. The theory reduces to the statement that in period *t* the decision-maker chooses what he prefers. Had he not preferred the choice actually made he would have chosen something else. This is a statement which cannot be refuted, but which contains no information. Regardless of the decision made by the decision-maker at a given time, one can always say that he chose the alternative which maximized his own utility function.

The task is to explain *why* he chose precisely *this* alternative rather than another one. This task cannot, however, be solved if the model is interpreted in a static sense.

[1] In my critique I have used the works of *H. Simon* [236], [237] and [238]. I have also found some inspiration in the studies of *R. Hoch* [90], and [91], mainly in his critical remarks on the changes over time in the set of alternatives and preference ordering, as well as in the "maximizing" behaviour of the decision-makers.

TABLE 11.1

Survey of the critique of the preference ordering theory

Serial number of section	Interpretation, static or dynamic	If dynamic: dealing with comparable or non-comparable decisions	Deterministic model or accounting for uncertainty	Descriptive or normative interpretation	The decision-maker is a household, a firm or government	Other viewpoints of discussion
11.1	static	—	deterministic	descriptive	all three	
11.2	dynamic	comparable	deterministic	descriptive	consumer, firm	
11.3	dynamic	non-comparable	deterministic	descriptive	consumer, firm	11.3 effect of changes in circumstances
11.4	dynamic	non-comparable	deterministic	descriptive	consumer, firms	11.4 changes in the relative position of decision-maker
11.5	dynamic	both	deterministic	descriptive	consumer, firm	11.5 other effects
11.6	dynamic	both	uncertainty	descriptive	consumer, firm	
11.7	both	both	both	descriptive	consumer, firm	summary of Sections 11.1–11.6
11.8	both	both	both	normative	consumer, firm	
11.9	both	both	both	both	government	

There is nothing more that can be said about the static, descriptive interpretation of the model; in the next sections we will deal with the dynamic, descriptive interpretation.

11.2. Consistency of the Comparable Decisions

Let us consider first the class of evaluable decision series. Recall that (see Definition 10.4) a series is evaluable if the elements of the series are comparable and, with the aid of empirical investigation, it may be stated whether or not they satisfy the (strong) axiom of revealed preference, that is, whether or not they are consistent.

Employing Definition 10.5 as a starting point, our analysis can be safely extended to quasi-evaluable series of decisions in which most, but not all, decisions are comparable with at least one other element in the series.

We must ask first whether or not many decisions are, in fact, comparable decisions? Are there really many evaluable (or quasi-evaluable) series of deicisons?

Statement 11. 1. Only a fraction, neither a negligible nor an overwhelmingly large fraction, of decisions can be considered elements of some evaluable series of decisions.

Confirmation of this statement would be a substantial research task, but one which should eventually be performed. Here, only an indirect argument in support of this statement will be offered. Every empirical investigation concerned with the consistency of decision-makers has focused on essentially the same type of decisions, namely, frequently recurring consumers' decisions. *Koo* analysed food choices, *Thurstone*, the choice of outer garments, and *Benson*, choice in a restaurant.[2] To my knowledge no consistency analysis has been performed in the context of firms or government agencies.

To characterize the consistency of comparable decisions we introduce the following terminology:

Definition 11. 1. The decision-maker is s t e a d i l y   c o n s i s t e n t if, in a long, evaluable series of decisions he never violates the consistency requirement defined in 10.3. The decision-maker is s t e a d i l y   i n c o n s i s t e n t if, in a long, evaluable series of decisions he violates the requirement of consistency in the majority of cases. The decision-maker is r e s t r i c t e d l y c o n s i s t e n t if, in a long, evaluable series of decisions he violates the requirement of consistency only occasionally.

---

[2] See *Koo* [122], *Thurstone* [254], and *Benson* [31]; a comprehensive survey is given by *Arrow* [12].

Statement 11.2. In evaluable series of decisions, the decision-makers are generally restrictedly consistent.

The statement should be proven or refuted with empirical evidence. The few empirical works available, and the investigations mentioned above, in particular, seem to support the validity of Statement 11.2.

Let us review in detail the study of *Koo*. Koo processed the food purchase data of American households using a large and representative sample of observations.

It is not necessary to discuss in detail the mathematical and statistical aspects of the analysis. What is interesting is that from Koo's data, an indicator, called the degree of consistency and denoted by γ, may be calculated.[3] The indicator may assume any value between zero and one, $0 \leq \gamma \leq 1$. If a household is steadily consistent, then $\gamma = 1$; if it is steadily inconsistent, then $\gamma = 0.5$; if, however, it is restrictedly consistent, then $0.5 < \gamma < 1$.

The numerical results of the Koo investigation may be summarized as follows:

0.93 per cent of the households are steadily consistent ($\gamma = 1$).

1.87 per cent of the households are steadily inconsistent ($\gamma < 0.5$).

86.4 per cent of the households, an overwhelming majority, are restrictedly consistent ($0.6 < \gamma < 0.8$). The average value of γ was 0.72 for the complete sample.

Koo's investigation supports Statement 11.2, but the statement is also confirmed by our everyday experience. Most people are not totally consistent but neither are they completely inconsistent. This is why it is incorrect to make the real-science assertion that decision-makers are steadily consistent. We cannot accept, even as a "first approximation", the assumption that the degree of consistency is 1, since the Koo study shows that it is not even approximately unity but rather, about half-way between 0.5 and 1.

Later in this chapter I shall endeavour to explain why γ is smaller than 1, that is, why decision-makers are only restrictedly consistent. Here we simply note this phenomenon.

---

[3] The author arranges the aggregated and adequately transformed data of the observations in a special square matrix. The order of the complete matrix is 13; within it, the order of the greatest consistent sub-matrix can be established by households. This shows a dispersion between 4 and 13. The γ indicator I have used is the order of the greatest consistent sub-matrix as established by Koo and divided by 13, the order of the complete matrix.

If the empirical investigation and its mathematical processing is carried out with a method deviating from that of Koo, the measure of consistency, which should bear some relationship to the γ indicator here used, must be, of course, defined in some other way.

11.3. Changes in the External Conditions of the Decision

Let us now turn our attention to the non-evaluable series of decisions as well as the non-recurring, single decisions.

In connection with this class of decisions, the following questions must be raised:

What factors are responsible for producing a change in set $\mathcal{B}(t)$, the set of implementable alternatives, over time? When does $\mathcal{B}(t)$ change at so fast a rate that the recurrent decisions fail to constitute an evaluable series?

What factors produce changes in the preference ordering, $P(t)$, over time?[4] Can we speak about a preference ordering, $P(t)$, that does not change, that is, that remains constant over time?

We may state at the outset that the changes in $\mathcal{B}(t)$ and $P(t)$ are closely interrelated and mutually effect each other over time. Therefore, in what follows the two processes, which can be separated only in the abstract, will not, as a rule, be sharply distinguished.

We shall consider only the deterministic case, at first; the problem of uncertainty will be introduced later. The changes over time in $\mathcal{B}(t)$ and $P(t)$ are explained by three major groups of factors:

Group 1: changes is in the *external circumstances* which are independent of the decision-maker.

Group 2: changes in the *relative position* of the decision-maker in comparison with his surroundings.

Group 3: *other factors* influencing preferences.

In the first group of factors, changes in external circumstances, the most important one is *technical progress*. Much more will be said about this in Part III of this book; we will discuss it here only in connection with the theory of preference ordering.

By technical progress one means the continuous change over time in products and services turned out by the economic system as well as the development of alternative procedures and technologies in production, consumption, turnover and in the information and control processes.

From time to time, and in recent times especially, innovations of revolutionary importance involving new products and new processes have appeared. Think only of some of the achievements of recent years: penicillin, atomic energy, plastics, supersonic aircraft, TV, and electronic computers.

---

[4] Here and in the following paragraphs we assume that a preference ordering $P(t)$ exists, since it is simpler here to debate the theory *within* its own conceptual framework. Later we will explain why it is better to use a model where the existence of $P(t)$ is not assumed.

In addition to the great "revolutionary" changes, millions of tiny changes also take place. There were motorcars and bathrooms thirty years ago, but the car and bathroom of today are quite different from those of former times.

It would be desirable to develop quantitative measures of the continuous changes in production and consumption. (This is no trifling matter since "qualitative change" must be measured in terms of quantity. The problem will be considered later.) At this point it is, however, sufficient to state the generally acknowledged empirical fact that *the process of technical progress does exist* and, though the rate of change depends on the special characteristics of the economic system, it *proceeds rapidly everywhere.*

As a Greek philosopher once wrote: "You cannot step twice into the same river." This can be said about many things, but it characterizes well the situation of the economic decision-maker; due to the incessant flow of new products and processes he almost never faces the same problem of choice twice.

Let us first consider the household. What was said above relates mainly to the purchase of durable consumer goods; the longer the lifetime of the article the less likely it is that the decision will be evaluable. Namely, by the time it is necessary to replace a durable consumer good, simply repurchasing the same article is rarely a possibility; the choice has to be made from new alternatives. The original car, refrigerator, TV set, washing machine, simply cannot be replaced, supply has been completely transformed.

But the set $\mathcal{B}(t)$, of realistic offered alternatives frequently changes drastically even in the case of recurring decisions. In a household, preferences concerning "beef or pork" will not change; this choice, therefore, belongs to an evaluable series of decisions. But, with the progress of the canning and deep-freezing industries, the problem of choice among fresh, pre-cooked or ready-made food does change.

The situation is similar in the case of the productive firm; the available alternatives are continuously modified by technical progress. But even the activity of specialized control organizations is influenced by the process of technical change. Let us recall, for example, the effect of the telephone, Xerox-copying, Telex-connection, bureau machines, punch-card data processing and electronic computers on the information and control processes.

The daily material purchases of a firm also belong to the class of comparable decisions. But the situation is quite different in the case of major investments (which are rather rare for the firm). During the time which typically elapses between two major investment decisions, the set of actually offered technological alternatives changes radically in most fields, particularly in rapidly technically developing areas.

Technical progress results in changes over time not only in the set of offered, implementable alternatives, $\mathcal{B}(t)$, but also in the preference ordering, $P(t)$.

The value judgements and tastes of the decision-makers are not independent of the available alternatives, that is, in the final analysis, of technical progress.

Let us sum up the lessons to be derived from the above:

Statement 11.3. The faster the rate of technical progress, the greater the number of regularly recurring decisions that come into the category of non-evaluable series of decisions and the faster the modification over time of the set of implementable alternatives and the preference ordering.

In connection with the discussion of technical progress, an additional remark must be made on Definitions 8.4 and 8.5, pertaining to the set, $\mathcal{A}$, of possible alternatives and the set, $\mathcal{B}(t)$, of implementable alternatives. In the case of producers' and consumers' decisions we assume that the set $\mathcal{A}$, which is independent of time, includes all technological alternatives that have arisen throughout the course of some long historical period. Set $\mathcal{B}(t)$, however, comprises only those alternatives which are actually available to the decision-maker given the level of technology existing in period t. If, for example, the decisions on aircraft-purchases of the airlines are examined, set $\mathcal{A}$ may comprise all kinds of aircraft which were commercially available through the whole history of air transport. In contrast, the set $\mathcal{B}$ (1969) includes only the types produced in the sixties; older types are, in reality, unavailable.

We have interpreted the sets $\mathcal{A}$ and $\mathcal{B}(t)$ in this sense up to now, but it is useful to emphasize this distinction in this context.

Of all the changes in the environment of consumers and producers, households and firms, technical progress is one of the few factors which displays unequivocal tendencies. But this is not the only relevant external factor. The changes in the implementable set of alternatives, $\mathcal{B}(t)$, depends for example, on the international situation of the country (whether it is at peace or war), on the general economic situation (whether there is a boom or a recession, whether growth is rapid or slow) and so on.

11.4. Changes in the Relative Position of the Decision-Maker

Now we turn to the second group of factors responsible for the changes over time in the set of implementable alternatives, $\mathcal{B}(t)$, and in the preference ordering, $P(t)$.

Firts let us consider the *household. Figure 11.1A* shows the food consumption of a poor family. Their diet consists primarily of simple foodstuffs; they purchas e luxury foods only on exceptional occasions. This situation is shown Figure 11.1A by the point a_1^*.

If we asked the members of the family what they would eat if they had much m ore money, they would list the luxury foodstuffs which they cannot afford in their impoverished condition—caviar, salmon, choice meat and cake with cream. Thus, if the income line shifted from the position $\mathcal{B}^{(2)}$, indicated by the broken line, the new choice would correspond to the point a_2^*.

But let us assume the family acutally grows rich. The new situation is illustrated in *Figure 11.1B*. The members of the family soon become satiated with the luxury foodstuffs which they desired, with the caviar and the cream If they then consume at a_3^* they consume more luxury food than when they were poor but much less than they believed they would when they were dreaming of being millionaires.

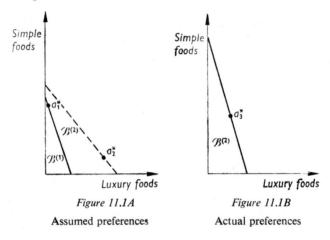

Figure 11.1A   Figure 11.1B

Assumed preferences   Actual preferences

The point of the example is simply this: We may question the decision-maker to determine his ordering over the entire set $\mathcal{A}$ of possible decision alternatives. However, statements about the set $(\mathcal{A} - (\mathcal{B})t)$ the alternatives which cannot be implemented by the decision-maker, are not reliable in reality. The preference ordering can only really be interpreted over the implementable set which depends on the personal situation of the decision-maker.[5]

The preferences of the consumer are affected not only by changes in his income but also by changes in his status, his place on the scale of social prestige, his family position, his domicile and other factors influencing his relative position.

Statement 11.4. The consumer possesses no constant preference ordering over the entire set $\mathcal{A}$ of possible decision alternatives. His preferences depend crucially on the actual set of decision alternatives which are implementable, given his circumstances, including his position in society.

[5] Though differently worded, the study by *Hoch* [90] and [91] called attention to this idea.

It is also a shortcoming of the preference ordering theory that it fails to recognize the conceptual distinction introduced in Definitions 8.5 and 8.7. In those definitions, a distinction is made between the set of implementable alternatives, $\mathcal{B}(t)$, and the set of alternatives acceptable to the decision-maker, $\mathcal{D}(t)$ considering his self-interest, motives, and expectations. Consideration of the relative position of the decision-maker in society leads to a recognition of those constraints which, in our own terminology, describe the set $\mathcal{D}(t)$.

Although we have used the consumer, the household, for the purpose of illustration, this analysis is also relevant to the firm. The preferences of the firm (as to the choice of trading partners, or productive technology) are dependent on the "surroundings" in which current decisions are made. If the relative position of the firm changes substantially (e.g., it expands or contracts), its preferences will also change.

11.5. Other Effects Influencing the Preferences

There are numerous other factors which produce changes in the preference ordering, $P(t)$ over time. Although these are related to the factors discussed under 11.3 and 11.4, they still deserve a separate treatment.

A) *The effect of "public opinion"*. Value judgements, preferences, and tastes are widely influenced by informations flowing in the economic system, by what is regarded by the decision-maker as "the general value judgement of society" or *"public opinion"*. The decision-maker's knowledge of public opinion comes mainly through the media of mass communication (the press, TV), through advertisements, through education and dissemination of scientific knowledge, through social contacts. It is through these means that people are "manipulated".

Value judgements are frequently formed by imitation. Individual consumers are affected by fashion. They identify with definite "reference-groups"; for example, many people from various social strata imitate the consumption standards of a social group higher than their own.[6] Similarly, firms frequently model their behaviour after that of the "leader".

B) *Changes of personnel in the organizations*. Up to now, we have used the term "decision-maker" as though its meaning were self-evident. In fact, the decisions are made by organizations and the individuals participating in the decision-making are changing.

This is even true for households, although the personal composition of the household, the family, may be constant over a relatively long time. But changes

---

[6] On this subject see the study of *Duesenberry* [55]. This study also presents empirical material worthy of attention. The author supports the idea that preferences move autonomously over time as a function of various social factors.

in personnel are a regular aspect of the life of productive firms, offices and control organizations. If the composition of the decision-making unit changes, the ranking of alternatives may change solely on this account.

C) *Power shifts within the institutions.* In Chapter 7 we dealt in detail with the internal conflicts of institutions composed of several organizations especially in the case of productive firms. Survival of the institution is based on the fact that, given the existing power structure, compromises emerge. The complex behaviour, and the preferences manifest in the decisions, all reflect the internal power relations of the moment. If, these power relations shift, however, the preferences and value judgements may also be modified. If, for example, the technical development section gains additional power, the spirit of innovation may gain the upper hand in the firm; the firm may try harder to introduce new technologies and new products.

It seems necessary to mention some additional factors as well. These include the experience gained in the implementation of earlier decisions, the correction of mistakes, and so on. This will be discussed, however, in the next paragraph since the treatment of these factors is closely connected with the problem of uncertainty.

11.6. Uncertainty

In examining the relationship between uncertainty and decision-making, the GE school focuses its attention on the following questions:

Suppose that in a given situation the consequences of the decision depend not only on the actual choice but also on unpredictable events of the external world. How would one rationally make a decision in such a situation?

Although, as suggested by the formulation of the problem, this branch of theory has been imbued mainly with a normative significance, it is frequently suggested as a descriptive-explanatory theory as well. Thus, some authors allege that a majority of decisions may well be described with the aid of stochastic utility functions. The utility achievable by the decision-maker depends not only on the decision but also on the state of nature; the role of the latter may be expressed by the probability distributions of the random variables describing the possible outcomes. The decision-maker maximizes the *expected* value of the stochastic utility function.

I do not wish to argue about whether or not the above may be accepted as a real-science theory. In Statements 11.3–11.4, I challenged the existence of a preference ordering that remains constant over time for a broad class of decisions. Already several arguements have suggested the impossibility of demonstrating empirically the existence of preference orderings. If these doubts are justified *in general*, for any kind of utility function, they are certainly applicable to the *special*, stochastic variants. In characteristic fashion, empirical tests of the "hypothesis of expected utility maximization" have been undertaken only in connection with a very special set of decision problems—those associated with betting and games. Here, indeed, we are concerned with uncertainties where the decision-maker can consciously estimate the probability of success.

Ignoring for the moment the literature on the problem, which employs a highly sophisticated apparatus, let us provide some simple answers to simple questions. Consider the following simple question; "What does the decision-maker do in an uncertain situation?" The answers are simple enough. He hesitates. If he is clever he tries systematically to experiment in recurring decisions and to learn from his previous experiences. If he is even cleverer, he gathers information prior to each decision to lessen the uncertainty.[7]

Let us consider each simple answer in turn.

1. *Hesitation.* Most decision-makers have no definite, unequivocal preferences. And since, in a given situation, many alternatives are available, it is rather coincidental that they choose one alternative rather than some other one, similar to the accepted one.

Most decision-makers are inclined to hesitate. This is particularly true if their desires are contradictory and conflicting, as is often the case. For example, the decision-maker might like to settle his pressing debts from a given income but, at the same time, he might also like to make new investments in order to expand. It may happen that in the case of internal conflicts of motives a rather stable compromise develops which is acted upon over a fairly long time period. It is frequently the case, however, that, in internal conflicts, one of the motives completely dominates the decision. The preferences and judgements as to the importance of the various motives may change cyclically.[8]

If the decision-maker hesitates, this indicates that a whole group of alternatives is equally acceptable to him and it is immaterial which of them will be realized. The actual choice is a matter of chance.

All of this could be described with the formal apparatus of the non-strictly

---

[7] "As a matter of fact, uncertainty is nothing but a lack of information, while information is nothing but the lessening of uncertainty (negative uncertainty). Therefore uncertainty and information mean the same thing, actually, viewed from different sides, they differ only in signs"—writes *Alfréd Rényi* who, with a mathematician's perspective, points out the same characteristics of uncertainty which are emphasized in the present discussion. See [209], p. 277.

[8] Here and in Section 8.4 we have mentioned that *decision* depends not only on the established taste of the decision-maker, on his *a priori* preferences, but also on chance. This should not be confused with be stochastic utility function mentioned in this section. Here it is the *utility* attainable with the given decision, that is, the *consequence* of the decision which depends on chance.

convex or, rather, non-convex preference orderings. Examples are shown in *Figures 11.2A and B*. In Fig. A the indifference curve is piecewise linear (convex but not strictly convex). In this case, every convex linear combination of a_1^* and a_2^* is equally acceptable to the decision-maker. In Fig. B an expressly non-convex indifference curve is illustrated; all of the tangencies with the income line are equally acceptable to the decision-maker (a_1^*, a_2^*, a_3^*, a_4^*).

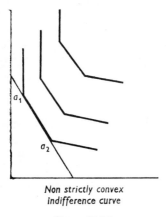

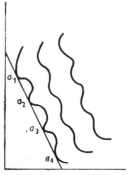

Non strictly convex
indifference curve

Non-convex indifference
curve

Figure 11.2A

Figure 11.2B

Non strictly convex indifference curve

Non-convex indifference curve

But the formal apparatus of a non-convex preference ordering is rather inconvenient; in addition, it does not incorporate what has been said about uncertainty or about the random character of the decision process. The decision model described in Chapter 8 is more relevant here. The value judgements, motives, expectations of the decision-maker delimit the set of alternatives which are acceptable to him. Then, within this set, the actual decision is selected at random, according to some probability distribution.[9]

Whether more or less hesitation on the part of most decision-makers is formalized in this or some other manner, one thing is certain: the model of a

---

[9] In relation to the strictly convex preference ordering I wish to mention a further consideration:

Let us assume that the decision-maker values a_1 (e.g. travel abroad) and a_2 (e.g. refurnishing his flat) equally $a_1 \sim a_2$. If the indifference curves are strictly convex (see Fig. 10.1B) any convex combination of a_1 and a_2 e.g. a shorter trip and the refurnishing of half the flat) is more desirable for the decision-maker than only the first or the second original alternative. Why? There is not a single economic argument that would support his choice *deterministic, strictly convex*[10] preference ordering, *P*, that is *constant over time* contradicts reality, because most decisions are actually somewhat random.

2. *Learning.* One of the major tools which the decision-maker can use to diminish uncertainty is learning, through analysis of earlier experiences. This results in a gradual improvement in the series of decisions.

In terms of the conceptual framework of the preference ordering theory, this means a successive correction of $P(t)$, a gradual reordering of value judgements.

Let us take a simple example. A consumer may repeatedly choose between two kinds of canned goods. He always chooses brand "V", in spite of the fact that it is worse and more expensive than brand "W". According to the theory of "revealed preference" he is to be praised; he has revealed that he prefers "V". His choice is completely consistent. Since he never tries brand "W" he is never guilty of the deadly sin of violating the antisymmetry commandment.

But most consumers are not so "consistent" (or rather hard-headed). Most people experiment; they buy "V" once, then "W", and decide from their own experience which brand to stick with. This, again, may involve shifts in the preference ordering.

3. *Collecting information.* The decision-maker need not, of course, learn only from his earlier experiences. In each and every decision process he can attempt to collect information in order to lessen uncertainty.

The decision-maker does not know exactly the set of implementable alternatives, $\mathcal{B}(t)$, but only, as was pointed out in Definition 8.6, the set of alternatives *deemed* implementable $\tilde{\mathcal{B}}(t)$. The two may differ from each other. On the one hand, the decision-maker is unable to survey all really available alternatives: $\mathcal{B}(t) - \tilde{\mathcal{B}}(t) \neq 0$. On the other hand, he believes that certain alternatives are implementable when this is not the case in reality: $\tilde{\mathcal{B}}(t) - \mathcal{B}(t) \neq 0$.

Thus, the key problem of decision-making under uncertainty is not what *should the decision be given the uncertainty*, but rather the relation between decision and the flow of information: what should the decision-maker do in order to lessen uncertainty? As has been emphasized in Chapters 8 and 9, the decision process is a learning process. In standard decisions little, but in fundamental ones, great intellectual and material inputs are expended on searching for new alternatives and on forecasting in the consequences of decisions. In the course of information collection, the value judgements related to the alternatives may also be modified.

[10] The *optimality* required by the GE school is stronger than the requirement of *efficiency*. An optimal decision is efficient, but the converse is not true. For example, with a given, strictly concave utility function over a strictly convex set of alternatives there is only a single optimal decision, while every point of the hyper-surface delimiting the set is efficient; that is, there are an infinite number of efficient points.

Let us sum up the major point made in connection with uncertainty:

Statement 11.5. The uncertainty which attaches to the sequences of decisions results in more or less hesitation on the part of the majority of decision-makers. The decision-makers attempt to lessen uncertainty by learning and collecting information. This results in frequent shifts in preference orderings over time.

11.7. Superfluous Link in the Explanation of Decisions

In Sections 11.1—11.6 I dealt with the descriptive-explanatory, real-science interpretation of the theory of preference ordering. Let us sum up what has been said thus far.

The preference ordering model has two main components: the preference ordering P, over the set of decision alternatives, $\mathcal{A}$, which is independent of time and a series of sets of actually implementable alternatives $\mathcal{B}(t) \subset \mathcal{A}$ over which the decisions are comparable.

Actually, the majority of decisions cannot be forced into the preference ordering model. We have listed many factors which cause both $\mathcal{B}(t)$ and $P(t)$ to change rapidly over time relative to large numbers of decisions.

These factors explain the phenomenon described in Section 11.2; even in the case of a really comparable series of deicisions we find only restricted consistency, $(0.5 < \gamma < 1)$. This is not because decision-makers are generally stupid or frequently make mistakes but rather, because many of them hesitate in the presence of uncertainty and modify their preferences in the light of earlier experiences. In addition because of technical progress, and changes in the political and economic situation, the relative social position of the decision maker, public opinion and fashion, and power relations within the organization, the preferences themselves undergo frequent changes: (cf. Sections 11.3—11.6). These factors explain why preferences occasionally deviate from the requirement of consistency with respect to a fixed preference ordering.

Statement 11.6. The existence of a fixed preference ordering, P, on the set of decision alternatives, $\mathcal{A}$, cannot be regarded as proven for many decisions.

Statement 11.6 does not contradict the idea that the decision-maker may have a *partial* (and, in addition, only *stoachastic*) preference ordering, which may prevail for longer or shorter periods. This is a much weaker assumption than the assumption of a complete, deterministic and constant preference ordering.

The logical structure of the problem can be surveyed as follows.

Let us call explanatory factors and denote by $q(t)$, $q(t-1)$, $q(t-2)$, ... the effects influencing the decision which were expounded in detail in Sections 11.3—11.6.

In the spirit of the preference ordering model, we have to learn and describe the following *indirect* function:

$$a^*(t) = f[P(t)\,(q(t),\,(q(t-1),\,q(t-2),\,\ldots)].\qquad(11.1A)$$

or schematically

$$\text{explanatory factors} \to \text{preference ordering} \to \text{decision.}\qquad(11.1B)$$

Let us trace this chain, consisting or three links, from the standpoint of *observation*. We may observe explanatory factors at one end of the chain and actual decisions at the other end. But we know very little about the middle term of the chain. We always know the actual *single* decision $a^*(t)$ and the fact that it has been preferred to other implementable decision alternatives. Therefore, we do not consider it worth while to devote much time to investigating the function $P(t)$ in a given period t, since it will change in the following period and since, in any case, we can learn very little about it.

The essential task is to know the relation between $q(t), a(t-1), q(t-2), \ldots$, the explanatory factors, and the decision $a^*(t)$. What is the characteristic form of this relation? How can one describe its regularities over time?

In other words, we simply drop the intermediate term from (11.1B) and ontent ourselves with a *direct* function:

$$a^*(t) = f(q(t),\,q(t-1),\,q(t-2),\,\ldots).\qquad(11.2A)$$

That is, schematically:

$$\text{explanatory factors} \to \text{decision.}\qquad(11.2B)$$

The relation (11.2) is the same as that which was described in Chapter 4, the stochastic response function of the unit. It is simply the relation between input, internal state and output, which characterizes the functioning of organizations. The preference ordering of the decision-maker (if it exists at all) remains *within* the "black box", that is, the behaviour of the decision-maker, the economic organization, is affected by the inputs stored in the memory and those received in the course of decision preparation.

In the final analysis, I propose the stochastic causal model defined in 4.12, describing and explaining the behaviour of decision-makers, thereby eliminating the whole framework of utility functions and preference orderings.

The presence of the $P(t)$ link in the explanatory relation would be immaterial if its inclusion did not make the problem more complicated. Unfortunately it does. It is as if an Alpine climber put heavy stones into his knapsack before embarking on his trek to the summit.

The description of the operational rules of economic systems is an immensely difficult task. And we, the mathematical economists, have *voluntarily* made it more difficult for ourselves. We always formulate our problem as one involving determination of extreme-values, as a problem in optimization. However, once we have subjected ourselves to this restriction, we become the prisoners of our own mathematical apparatus. We are also compelled to put up with other onerous restrictions: the convexity of the set of alternatives, the elimination of increasing returns, and so on.

If we grant that these efforts are superfluous, we can eliminate the convention of describing the operation of every economic system in terms of units which optimize; if we simply drop the ballast of the preference ordering model, we can formalize the laws of motion, the rules of behaviour of economic systems with greater freedom.

11.8. The Norms of Clever Behaviour

Having summarized and concluded our discussion of the descriptive-explanatory interpretation of the preference ordering model, let us turn now to its *normative* interpretation. As in the preceding paragraphs we are dealing, for the time being, only with the lower level decision-makers, that is with the households and firms.

For decades the GE school has extolled the virtues of "rational" behaviour; it is disposed to grant this distinction exclusively to those who A) maximize their utility function or, its equivalent in the case of a dynamic interpretation, and who B) are fully consistent in the sense of Definition 11.1. Such an interpretation of rationality is, however, rather narrow and sometimes even misleading.

Criterion A: It is empty advice to recommend that the decision-maker maximize his utility function. It reminds me of a popular saying in Budapest: "If you want my opinion, you should do as you please."

Criterion B: The requirement of complete consistency translated into every day language, suggests that the decision-maker should be true to himself, to his own earlier preferences. There are many situations when this is sensible, or, at least morally noble. There are, however, situations—and these occur rather frequently—when this is foolish advice. By no means can it be considered a basic principle of rationality. On the contrary, common sense most frequently demands that we modify our preferences over time.

Since, for the mathematical economist, the word "rationality", has too many connections, I will use the word "cleverness" instead.

When does the decision-maker behave in a clever way?

1. In the realm of standard decisions: he should not rely entirely on routine; from time to time (though not too frequently) he should revise his rules of thumb, his simple decision algorithms. For example, the housewife, even if she has

grown accustomed to a standard brand, to a shopping market, or to some routinized distribution of her household outlays, should occasionally revise her buying habits. This is even more necessary in the case of standard decisions established in a firm. In the language of GE theory, it is occasionally necessary for the decision-maker to revise and modify his preference ordering.

2. In the case of recurring decisions, the clever decision-maker analyses his earlier experiences, he *learns*. He systematically tries different decisions in order to obtain adequate experience for comparison. The housewife experiments with different brands and gradually develops her preferences for some of them. The material purchasing agent of a firm tries various sources of materials at different times and compares them.

3. A clever decision-maker will not spare intellectual efforts and material inputs in gathering information when preparing an important fundamental decision and for the better organization of the flow of information. He will try to learn as much as possible about the set of implementable alternatives, $\mathcal{B}(t)$, and reduce the discrepancies between the explored set $\tilde{\mathcal{B}}(t)$, and the implementable set $\mathcal{B}(t)$.

4. He will choose, as far as possible, an efficient alternative. If alternative "I" is in no respect worse than alternative "II", but is more favourable in some respect, he will obviously choose the former.

5. He will attempt to know his own mind; he will try to sort out his frequently contradictory wishes and motives.

The above list of criteria of clever behaviour are, in fact, platitudes. But they are commonplace precisely because they are trivially true. The norms listed in points 1–5, and other similar ones, constitute the normative system of "clever behaviour", but not the rationality and consistency requirements set forth by the GE school. The normative requirements of the GE school coincide with the normative system reviewed above in point 4; they do not include requirements 1, 2, 3 and 5 at all.[11]

The decision models created in the workshops of the GE school may indeed be proposed to the decision-maker; this is not incompatible with the above statement.[12] The mathematical programming models can, for instance, be used to advantage in the preparation of production, technological, inventory, investment, and other decisions of the firm.

Of course, not even the most skilfully constructed system of constraints for a programming model is capable of describing $\mathcal{B}(t)$, the set of implementable alternatives, nor does its objective function express $U(a)$, the utility function of

[11] I have already indicated this in Section 2.3.
[12] I have presented this idea in greater detail in my book on mathematical planning [128], particularly in Chapter 27.

the decision-maker. Were it able to do so, a single calculation carried out with an electronic computer would hand us down the "optimal" decision.

Rather, $\mathcal{B}(t)$ is not known precisely and $U(a)$ does not exist at all. In mathematical programming techniques we do not perform a single calculation but a *series* of calculations. This constitutes, however, nothing more than a contribution to the preparation of the decision, to the *cognitive process* preceding the decision.

We have no exact *a priori* knowledge about $\mathcal{B}(t)$, the set of implementable alternatives. *In the course* of constructing the model, through data collection and economic analysis following the individual calculations in the series, we expand our knowledge about what *can* be done. (See point 3 of the normative system of a "clever behaviour".)

The system of constraints of a programming model usually express not only the *real* possibilities of implementability, that is, the limits of the set $B(t)$, but also the limits of *acceptability*, that is, of the set $\mathcal{D}(t)$. This is the case when some constraint of the programming model requires that a definite level of profits should be attained, places a limit on credit availability, or requires that a minimum quantity be sold. The various expectations, interests and motives may also be stated in the form of an objective function instead of as constraints. Given a whole series of alternative combinations of constraints and objective functions we may "feel out" how the various expectations and interests can be reconciled with the possibilities offered by the real sphere.

Accordingly, we change the various objective functions in the series of calculations. The point in using an *objective function* at all is to select an efficient alternative which dominates the non-efficient programs. (See point 4 of the normative system of "clever behaviour".) By *alternating* the objective functions we wish to assist the process of internal cognition. (See point 5 of the normative system of clever behaviour.) Thus the decision maker will see more clearly his own desires and interests which, like the set $\mathcal{B}(t)$, he does not know exactly *a priori*, before beginning the preparation of the decision.

The sober operation researchers and mathematical planners all over the world who evaluate their research tools modestly and calmly, are aiming at this and nothing more. They wish to aid the decision-maker (who is obviously not strictly "rational" in the sense of the preference ordering model) to make more clever decisions by enlarging his knowledge of the possibilities and the consequences of his decisions.

I believe that for the purposes of *advising*, the *ex ante* use of conditional extreme value calculations, the use of models applying the mathematical techniques of "optimization", is in no way logically contradictory to the fact that I discard the same models as an *ex post description and explanation* of real economic behaviour.

11.9. Government Decisions

In the section above I have illustrated the statements only with the decisions
of lower level organizations (the household and the firm). Now, as a recapitu-
lation, we will again survey some major ideas of the chapter, but this time,
from the point of view of government decisions.

Let us examine the activity of the government of a socialist economy (e.g.
the Hungarian economy), including the council of ministers and the main
government offices controlling this economy.

*I believe that Statement 11.6 is completely relevant in this case; the largest
number of government decisions cannot be characterized by saying that the
decision-makers act on the basis of a fixed preference ordering P.*

Let us consider in turn the arguments contained in Sections 11.2–11.6.

There exist many comparable decisions in government work. Included in
this category are the routine price fixing activities of the central price office
or the authorization of major import transactions. It would require empirical
investigation to determine whether or not these evaluable series of decisions
satisfy the requirement of steady consistency. On the basis of my impressions
I have some doubts about this. I believe that they, too, qualify rather as restrictedly consistent series of decisions.

But even this cannot be stated about the really essential decisions.

*Statement 11.7. The really essential decisions, those deeply affecting the fate
of the whole economic system belong to the class of single decisions, or recurring
but non-comparable ones. Therefore, they cannot be described adequately by the
model of preference ordering.*

Certainly the comprehensive reform which came into effect in Hungary on
January 1st, 1968 is an example of this; the general price adjustments carried
out in the socialist countries every three or four years and the periodical acceptance of the five-year plans are also decisions of this type.

Let us consider five-year planning.

1. Obviously, the set $\mathcal{B}(t)$ of implementable alternatives essentially changes
between five year plans. The government had to choose from completely different sets of alternatives in 1949, 1954, ... and in 1969, before accepting a final
version of the five-year plan. The product-mix of the country's production is
always different, technology has advanced, the qualifications of the labour
force have changed, and so on.

2. In addition to the economic decision alternatives in the narrower sense,
over the course of five years, drastic changes have also taken place in the international situation, in the world market situation, in the internal political situation.

3. The relative position in the world of the decision-makers—in our example, Hungary in relation to other countries—also changes.

4. Government decisions are also affected both by domestic and international "public opinion". Here too, we may find "reference groups" whose behaviour is imitated, at least to some extent. These include the countries with whom the state in question is in political alliance. On the other hand, irrespective of political sympathies and antipathies, the economic patterns developed in the economically more advanced countries are generally followed by the more backward ones.

5. Although the fundamental political structure, the major institutions of the socialist countries, remain essentially unchanged for fairly long periods of time, the members of the institutions change. The changes in personnel are both consequences and causes of political changes.

6. If it is true of any field, it is certainly true of politics that every prevailing trend is built upon a compromise among the interests of various groups. In the case of a stable government, the compromise expresses lasting power relations. But there may also be shifts in power relations and in this case the compromise will be modified.

7. There may have been governments over the course of history which neither learned nor forgot. But governments generally do learn; they learn from their historical experience and they modify their policies accordingly. The Hungarian government, for example, has learned a great deal from the mistakes of the first five-year plan which was launched in the late forties.

In the above seven points I have followed the line of reasoning presented earlier in the chapter. Some phenomena have been described only in the briefest fashions, but a more complete exposition would necessitate a separate volume. Perhaps even these short references will suffice to show that in the case of fundamental decisions the behaviour of a government controlling an entire country cannot be described with the aid of a fixed preference ordering. The more dynamic the country and the more frequent are radical political changes, the less possible it is to speak about the constancy of a preference ordering over time.

At the same time, what was said in Section 11.8 about lower-level decisions, holds also for high-level ones, namely, that the mathematical models based on the calculation of conditional extreme values, on "optimization", should be used to increase the "cleverness" of government decisions. They may contribute to the cognition process involved in preparing fundamental decisions. The series of calculations performed with the aid of the models may help the government to obtain a better survey of alternative actions and clarify its own intentions and desires as well as the possibility of their realization. It is in this spirit that we use mathematical programming, "optimizing", models in Hungary in preparing government decisions.[12]

Finally, let me add a concluding remark on the normative application of the preference ordering model.

One of the basic ideas of the GE school is that the economic system should be constructed in such a way that production, the utilization of resources, is adapted to the needs of the consumer. Production should serve man and not the opposite. This is a beautiful, humanistic idea, the importance of which cannot be stressed sufficiently. It would be a grave error if, in the course of criticizing the GE school, this significant idea were forgotten. It is, however, my conviction that *this* idea *is not identical* to the requirement that every consumer should find the consumption program which maximizes his utility function. The utility maximizing condition is neither a necessary, nor sufficient condition for allocating resources in a way that serves human needs.

This idea will be pursued in greater detail in Part III of this book.

12. ASPIRATION LEVEL, INTENSITY

12.1. Notion of the Aspiration Level

In Chapters 10 and 11 we have presented a critique of the preference ordering model. Now, let us return to the theme abandoned at the end of Chapter 9 and continue investigating the decision processes of an organization, applying our own conceptual framework.

We shall deal exclusively with those decision processes in which the "aspiration level"—to be defined later—has a role. Thus, our discussion does not embrace all decisions but only one class (which is, itself, broad enough).

We shall restrict the scope of our discussion to *recurring* decisions (see Definition 10.4) and we shall treat the case of a single elementary decision process solving a single decision problem p, within an organization. For the sake of simplicity we drop subscript p indicating the particular decision problem at hand.

Each decision alternative is described by an indicator vector of K components. These vectors are elements of set $\mathcal{A}$, a subset of the $\mathcal{L}^K$ linear space of K dimensions.

To ease interpretation of the indicator vectors we introduce the following convention; indicators having the character of a result, return, are assigned a positive sign and those having the character of expenditure, outlay, a negative one. Accordingly, an increase in the value of an indicator will be judged as a favourable development.[1]

In *Figure 12.1* we present some relations which will be expounded below. To simplify the presentation, let us assume that $K = 1$, that is, we have only one indicator. Our discussion holds, however, for the case of $K > 1$. The horizontal axis shows time, the ordinal number of the period. The vertical axis shows the value of the indicator.

Consider the following decision problem. A productive firm is about to introduce a new product. One important indicator is the number of units that will be sold in the first year.

---

[1] Choice of the signs, strictly taken, implies a partial preference ordering. The decision-maker is not expected to have a *complete* preference ordering considering *all* indicators characteristic of the alternatives. But if there exist two alternatives which differ only with respect to a single indicator, he must be able to tell which he considers to be more favourable—that is the alternative in which this single indicator takes a bigger value or where it takes a smaller one.

Actually, the firm faces a series of similar decisions recurring at regular intervals. First there is a decision related to the introduction of the first new product; a second decision involves a second new product, and so on.

Let us first look at decision 1. The decision preparation process begins at time t_1 and ends at time $\bar{t}_1$, the time at which the decision is made. The decision initiates a real action—in our example the beginning of sale of the new product.

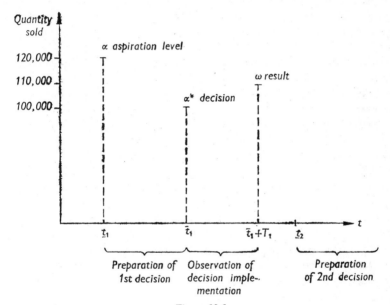

Figure 12.1

Aspiration, decision, result

Implementation of the real action decided upon is an observable process. Let us denote by T_1 the number of periods after the decision is made when information is available about its fulfilment. In our example this might be one year.

Definition 12.1. The information concerning the real action started by decision a^* is the r e s u l t. The result is an indicator vector of K components, available to the decision-maker, T_1 periods after the decision. It will be denoted by: $\omega \in \mathcal{A}$.

In our Figure a^* equal 100,000 units; ω equals 110,000 units, indicating the plan has been overfulfilled.

Preparation of the second decision in the same series begins after the result of the first decision is available, that is $t_2 > \bar{t}_1 + T_1$. Thus, the previous result can be taken into account when preparing the present decision. This case is shown also in the figure.[2]

The elementary decision process begins with a first formulation of intentions or wishes. Decision a^*, appearing at the end of the process (time $\bar{t}_1$) is preceded by the *aspiration level* α.

In our example, the firm begins to plan the introduction of a new product one-half year before the final decision. At that time the technical development department and the sales department submit their first ideas. On the basis of earlier experiences and their knowledge about the absorbing capacity of the market they suggest that they will be able to sell 120,000 units of the new product annually. This first figure is the aspiration level, shown in Figure 12.1 at time t_1.

We shall define the notion of aspiration level in two stages. (The preliminary character of the following definition is indicated by a prime after the serial number.)

Definition 12.2′. The a s p i r a t i o n   l e v e l is an indicator vector; it is an element of the set of possible decision alternatives; $\alpha \in \mathcal{A}$.

It arises at the beginning of the elementary decision process taking place over period $[\underline{t}, \bar{t}]$ and expresses the decision-maker's first ideas about the decision to be taken at the end of the process.

The notion of aspiration level must not be confused with that of an objective function or utility function. For example, let us consider a group of high-jumpers. The common "objective function" of each of them is to jump as high as possible. However, the aspiration level does not express only the direction of the effort ("as high as possible"), but a level to be attained which can be represented by a *real number*. Its value depends on the endowments of the decision-maker. A secondary-school student preparing for the school-championship may aspire to attain 170 cm; the prospective participant in the olympic games might aspire to attain 220 cm.

Let us return to the example of the productive firm. The aspiration level of 120,000 units expresses the wish of the decision-maker. "It would be fine to sell 120,000 units..." His wish is not merely a dream; usually it expresses a real possibility. He reckons approximatively with the firm's *internal* endowments; knowing that it could not produce more than 130,000 units, even if more could be sold. In addition, he makes efforts to account for *external* endow-

---

[2] Our assumption does not impose any special restriction. We have here a series of recurring decisions. If it happened that preparation of a new decision must begin before knowing the result of the preceding decision, an observation on the result of the next most recent decision may be available. In this case: $t_3 > \bar{t}_1 + T_1$. The analysis in the following parts of the chapter can be easily re-drafted for this case.

ments. Perhaps it is optimistic to forecast the sale of 120,000 units. But that is not so absurd as a forecast for the sale of 200,000 units.

Attainment of the aspiration level cannot be *ab ovo* hopeless. A forecast for which the decision-maker, himself, knows that the probability of attainment is zero cannot be considered as an aspiration level. However, a forecast can be considered as aspiration level if, according to the subjective forecast of the decision-maker, the probability of its attainment is slight, but positive. Thus, in forming an aspiration level, the decision-maker might be thinking: "If it is up to me, and circumstances are favourable, I should like to attain this..."

A Hungarian citizen, even if he enjoys considerable income, cannot aspire to buy a private aeroplane. This is an unrealistic desire. But he may aspire to buy a good car at some desired date. True, only every fifth person intending to buy a car can procure it at the date desired. This is known to all in Hungary who wish to buy a car. Thus, the probability of fulfilling the aspiration at a desired date is not too great. But it is not hopeless. If our decision-maker is fortunate, his aspiration may be fulfilled.

Now we are in a position to give a complete definition.

Definition 12.2*. The a s p i r a t i o n   l e v e l is an indicator vector; it is an element of the set of possible decision alternatives[3]: $\alpha \in \mathcal{A}$. It arises at the beginning of the elementary decision process taking place over period $[t, \bar{t}]$. It expresses the first ideas of the decision-maker about the decision to be taken at the end of the process and takes into account his wishes and internal expectations. According to the subjective estimates formulated by the decision-maker at the beginning of the elementary decision process it is not impossible that the event $\alpha \leqq \omega$ should happen; that is, attainment of the aspiration level is not excluded in principle. According to the decision-maker, the internal conditions necessary for attainment of the aspiration level probably can be satisfied; if conditions are favourable, the external conditions also can be met.

The concept of aspiration level is a broad *comprehensive* concept. It requires precise *specification* for practical applications. For every aspiration level we must specify an associated elementary decision problem or set of problems, a time period, anterior information and subjective probability distributions on which it is based, etc. We shall apply the concept of aspiration level to

---

[3] As a matter of fact, a more exact description of the psychology of decision-making could be produced by considering the aspiration level as a subset—with more than a single element—of the set of possible alternatives. Namely, at the beginning of the decision process the aspiration level appears less "sharply". However, for the sake of simplicity, we shall disregard this and define the aspiration level as a single element of set $\mathcal{A}$.

many specific cases: national planning, buying and selling intention on the commodity market, investment decisions.

The notion of "aspiration level" first appeared in the field of mathematical psychology.[4] Later, it was adopted by sociologists and economists. The notion is not interpreted uniformly by all authors. Many call aspiration level what we have called "bounds on acceptance" in Definition 8.7. Others simply equate it with the optimum decision taken with a given preference ordering.[5] In the following we shall use the notion strictly in the sense of our own Definition 12.2.

The aspiration level is an element of the control sphere; it constitutes a class of information types and cannot be observed in the real sphere. Purchasing aspiration is not identical with actual purchasing, nor is a selling aspiration identical with actual selling, nor is production aspiration identical with actual production. The aspiration level expresses intention or wishes—the first outlines of some later decision.

However, this does not mean that the aspiration level is "not palpable"—that it is inaccessible to empirical research. The aspiration levels appearing in the decision processes of a firm, government, or office generally are laid down in documents (e.g. first proposals, drafts, etc.).[6] Additionally, interviews with the decision-makers can reveal their intentions or aspirations before they would crystallize in final decisions.

The precise specification of the concept of aspiration level for the purpose of a concrete application is more or less equivalent to the exact formulation of the questions for the interview. As in the case of all surveys, we must provide a background for the questions; what are the assumptions, the subjective probability considerations, etc. of the interviewed person, revealing his intentions, aspirations?

Finally, one more remark. When formalizing the aspiration level, we disregarded the uncertainty of the decision makers. Deterministic formalism is justified solely for the sake of simplicity; we are only at the beginning of forming a new conceptual apparatus. In the course of its further development we shall have to describe with stochastic variables the aspiration level and the other indices deduced therefrom.

[4] This concept has been introduced by *K. Lewin*. See [151].
[5] See, e.g. *Siegel* [230].
[6] In Section 12.5 a detailed example is presented for the case of high-level plan decisions.

12.2. Notations of the Extensive Indices

In the following a group of indices will be described; comprehensively, they will be called "extensive indices." The exact sense of the term extensive will become clear in the course of our presentation.

In the following definitions we shall introduce symbolic notations. However, at present we shall not use the notation to deduce mathematical theorems in the framework of a formal model. Why, then, do we bother the reader with a whole series of symbols? The purpose of the symbolic description is twofold. Partly it facilitates and renders more compact the definitions of later concepts derived from these clarified earlier. Secondly, the symbolism renders the tasks of observation and *measurement* more unequivocal. This is most important since we discuss such magnitudes as have not yet been observed in standard economic statistics.

In Chapters 12, 18 and 19 we shall have to deal with two types of extensive indices. One type comprises those indices expressed in "*absolute terms*"—vectors whose components are measured by the same unit of measurement as the components of the aspiration level and the decision. For example, if the first component of the aspiration level is expressed in tons and the second in forints, in this first type of extensive indices the first component also will be measured in tons and the second in forints. This follows from the fact that these indices represent the *difference* between two indicator vectors. In forming this difference, the aspiration level and the decision appear as minuend or as subtrahend.

The second type of indices includes those of "*percentage*" character. Here, the indicators are quotients and the aspiration level and the decision appear as dividend or divisor.

Let us introduce the following notation convention:

1. Let $r_i = p_i - q_i$. If r_i is the difference, $\hat{r}_i$ will be the quotient: $\hat{r}_i = \dfrac{p_i}{q_i}$.

2. Let $\hat{r}$ denote the following vector:

$$r = \begin{bmatrix} \hat{r}_1 \\ \hat{r}_2 \\ \vdots \\ \hat{r}_n \end{bmatrix}. \tag{12.1}$$

The i-th component of $\hat{r}$, $\hat{r}_i$ is the quotient of the i-th components of vectors p and q. The vector $\hat{r}$, that is $\hat{r}_i = \dfrac{p_i}{q_i}$ ($i = 1, \ldots, n$) will be denoted

$$\hat{r} = \frac{p}{q}. \tag{12.2}$$

For the sake of completeness three special cases must be mentioned:

$$\hat{r}_i = +\infty, \quad \text{if} \quad p_i > 0 \quad \text{and} \quad q_i = 0;$$
$$\hat{r}_i = -\infty, \quad \text{if} \quad p_i < 0 \quad \text{and} \quad q_i = 0; \qquad (12.3)$$
$$\hat{r}_i = 1, \quad \text{if} \quad p_i = 0 \quad \text{and} \quad q_i = 0.$$

A further convention is related to the treatment of time. In the previous paragraph we specified precise dates associated with the aspiration level, the decision and the result: $\underline{t}$, $\bar{t}$, and $(\bar{t}+T)$. In further treatment we assume only that $a^*(t)$ has been taken sometime in period t. To *this* period t we also shall ascribe the $\alpha(t)$ aspiration level and the $\omega(t)$ result; the former was produced at the beginning of the decision process, prior to making the decision in period t; the latter is available to the decision-maker only after the implementation of the decision, observation of its fulfilment, and the arrival of the report on the observation. These time lags are subsumed in argument t.

Having clarified the necessary notation, we may pass to a discussion of the extensive indicators.

12.3. Tension

Definition 12.3*. Let us call t e n s i o n o f a s p i r a t i o n and denote by $\varepsilon(t)$ the following vector of K components:

$$\varepsilon(t) = \alpha(t) - \omega(t) \qquad (12.4)$$

Let us call degree of t e n s i o n o f a s p i r a t i o n and denote by $\hat{\varepsilon}(t)$ the following vector of K components:

$$\hat{\varepsilon}(t) = \frac{\alpha(t)}{\omega(t)} \qquad (12.5)$$

Similarly, let us call *tension of decision* and denote by $\zeta(t)$ the following vector of K components:

$$\zeta(t) = a^*(t) - \omega(t) \qquad (12.6)$$

The degree of tension $\hat{\zeta}$ — as a "percentage" index — can be calculated in a way analogous to the calculation of the degree of tension of the aspiration.

The term "tension" as defined here is used in a similar manner as it is in Hungary. Hungarian planners call a plan too tense if $\hat{\zeta}$ is greater than unity; the greater it is, the more tense is the plan.

In general, the indices of tension are characteristic of the soundness of the decision preparing processes and of the reliability of forecasts.

12.4. On the Emergence of the Aspiration

Formation of aspirations is a complex process within economic organizations. One of the most important components of the system of response functions in the control of the economic system is the aspiration function. To the present time the process has been observed in but few fields. Study of aspirations should be a common task for economists, economic sociologists, and economic psychologists in the future. One area in which aspirations have been considered carefully is the study of consumers' buying intentions or aspirations there is a valuable literature here, and we shall discuss it in Part III of this book. At present, we shall restrict ourselves to a few remarks concerning the emergence of aspiration.

In formulating his aspiration the decision-maker draws on two major sources of information. One is *his own memory, his own past experience;* a comparison with his earlier aspirations, decisions and their results. In the framework of the decision algorithm, formulation of the aspiration level takes place according to simple rules of thumb, some of which are based on the indices listed above. For example, an industrial firm may formulate the aspiration level of sales according to the growth rate attained with earlier sales. If it has succeeded frequently in attaining an annual rise of 15–20 per cent, the firm might let $\alpha(t_1)/\omega(t_0) = 1,2$. If last year's result was 100,000 units, this year's aspiration level might be 120,000 units.

This formulation of the aspiration levels is a manifestation of the cognitive processes taking place in the organization.

Another source of information available to the decision-maker is the study of the behaviour of other organizations, offering a sample to be studied or perhaps example to be followed. This phenomenon has been dealt with at length in sociology. There it is called *imitation,* although usually the phenomenon is not mechanical imitation but a more flexible pattern of behaviour.

The aspiration levels deeply characterize the psychology of the decision-maker and ultimately affect his actual behaviour. In a dynamic organization ambitions are strong. However if ambitions are strong and aspirations high and if the organization is impotent in implementation, great tensions may arise.

The preceeding theory also may be related to sub-systems composed of several organizations or of entire systems.

12.5. From Aspiration to Decision

Having made some general remarks on the emergence of aspirations, let us proceed from the aspiration to the decision, that is, from the beginning of an elementary decision process to its completion.

Recall that before the end of the decision process that $\mathcal{F}(t)$, the set of *eligible* alternatives emerges: $\mathcal{F}(t) = \mathcal{B}(t) \cap \mathcal{D}(t)$. (See Definition 8.8.) $\mathcal{F}(t)$ is the result of the cognitive process occurring in the course of decision preparation. In one or more steps the decision-maker assesses what can be implemented, taking into account the real possibilities. (This is reflected in the shape of the set $\mathcal{B}(t)$). Besides, the decision-maker gradually formulates expectations and attempts to harmonize the internal and external expectations; in addition he must consider the financial conditions of normal functioning, that is, he accounts for the effect of the C-sphere. (These will influence the shape of the set $\mathcal{D}(t)$).

By the end of the process it may turn out that $\alpha(t) \in \mathcal{F}(t)$, that is, the aspiration level is eligible in setting the final decision. It may, however, also turn out that the aspiration level has proven to be unrealistic, owing to either R- or C-possibilities: $\alpha(t) \notin \mathcal{F}(t)$.

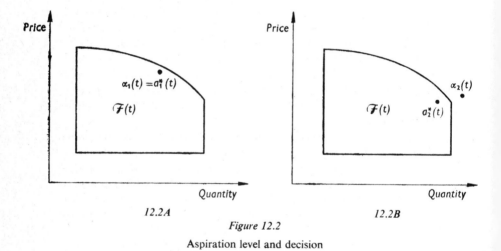

12.2A   12.2B

Figure 12.2

Aspiration level and decision

The two different situations are shown in *Figure 12.2*. Let us return to our previous example,—that of the firm planning to introduce and market a new product. We have two indicators; one is the quantity to be sold in the first year and the other, the average annual price. In our figure the set $\mathcal{F}(t)$ is limited from right and left by the upper and lower bounds on saleable quantity. From below it is limited by the minimum price, the lowest one which the various organizations of the firm are willing to accept. From above we see a forecast of the demand function; the saleable quantity can be increased only if accompanied by price reduction.

In part A of the figure the aspiration level $\alpha_1(t)$ is an internal point of set $\mathcal{F}(t)$.[7] The final decision of the decision-maker coincides with the aspiration level: $a_1^*(t) = \alpha_1(t)$.

In part B of the figure we show the case where the aspiration level has proven to be unrealistic. Therefore, the decision provides for both indicators at a lower level: $a_2^*(t) < \alpha_2(t)$.

In the course of decision-making, the aspiration level plays the role of a norm. When preparing a decision the decision making organization is guided by the following goal: "Let us try to attain the aspiration level, if possible..." In making the decision, the decision-maker makes efforts to attain the aspiration level. On the other hand, this effort is conditioned. If it turns out that it is impossible to attain the aspiration level, the decision will deviate from it.

For example, a Hungarian consumer might decide to buy a car of a certain make. This is his aspiration level. He will indeed buy the desired car if possible. But if he is unable to obtain it, he will decide to buy another make of similar quality. A firm might decide to establish in two years a new plant with imported machinery of a certain make. It starts gathering detailed information. Perhaps its plan can be carried out. But it may also happen that it will be compelled to modify the original ideas. Even in this case efforts will be made to set a date not much longer than the two years originally proposed, etc. At the time of the final decision, it may, of course, also turn out that an alternative can be found that is more favourable than the aspiration level (shorter term of delivery or better technology).

The idea that efforts are made to produce a final decision near to the aspiration level may be formalized in the following manner: Let us recall the decision distribution described in Section 8.4. According to this function the decision is chosen at random from the elements of the set $\mathcal{F}(t)$. The *decision distribution* $\xi(\mathcal{H})$ will give the stochastic rules of this random selection. Characteristically a "densification" of the decision distribution can be found precisely around the $\alpha(t)$ aspiration level.

There is no need to formalize this characteristic feature of the decision distribution. We refer only to the substance of the idea; there is a greater probability for the finally accepted decision $a^*(t)$ to take a value near to the aspiration level $\alpha(t)$ than of its taking one farther from it. This idea is illustrated in *Figure 12.3*, which can be considered as an expansion of Figure 12.2. We present here, however, only one of the indices—the target figure of sales. On the horizontal

---

[7] α purposely has been made an internal point of the set and not one on the boundary. It is not certain that the decision-maker will strictly "optimize". For example, because of the contradictory advice given by various advisers, the general manager may have some doubts that the demand function is realistic. Therefore, "for the sake of safety", at prices conforming to a_1 he may provide for the sale of a somewhat smaller quantity than could be sold according to the demand function.

axis the lower and upper bounds on sales are clearly seen. The sales target now is considered as a random variable. Which value between the lower and upper bounds will be accepted by the decision-maker depends on change. The figure shows the *density function* of this random variable. In part A of the figure we see the case where the aspiration level is an element of the set of eligible alternatives. It can be seen that $\alpha(t)$ is the place of maximum (the mean) of the density function; the alternatives in its neighbourhood have a much greater chance of selection than those farther away.

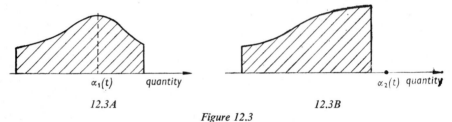

12.3A   12.3B

Figure 12.3

Density function of decision distribution

In part B of the figure we see the case where the aspiration level is not an element of the set of eligible alternatives. Thus, it cannot be accepted. On the other hand, the alternatives in its neighbourhood (that is, the higher targets) have a greater chance of acceptance than those which lie farther.

Definition 12.4. Let us call the c o r r e c t i o n   o f   a s p i r a t i o n and denote by $\varkappa(t)$ the following vector of K components:

$$\varkappa(t) = a^*(t) - \alpha(t). \tag{12.7}$$

Here we can also apply an appropriate relative index ($\hat{\varkappa}$, degree of correction).

With standard decisions and simple decision algorithms, the aspiration level and the decision will more or less coincide since the decision-maker will not devote greater energies to careful examination of the aspiration. With fundamental decisions and complex decision algorithms $\hat{\varkappa}$ may substantially deviate from 1.

Before proceeding any further let us introduce a new collective term already referred to in Section 12.2.

Definition 12.5. Let us call the e x t e n s i v e   i n d i c e s of the decision process the following vectors of K components: aspiration (α), decision (a^*), tension of aspiration (ε), tension of decision (ζ) and correction of aspiration ($\varkappa$).

The extensive indices are complemented by the corresponding percentage indices (also vectors with K components).

12.6. The Example of National Economic Planning

To illustrate what has been said above about the aspiration level and the extensive indices of the decision process, let me present an example—the history of the work preparatory to the third five-year plan of the national economy, which covers the years 1966 to 1970.

The five-year national economic plan embodies the *decision* of the highest political and economic organs. Drawing up the plan may be considered as the preparation of the decision. In our example the decision process began in 1963; by June of that year the first document relating to the subject was completed in the Department of Long-term Planning, National Planning Office. For all practical purposes the decision process ended in May 1966 when the government's proposals were worked out. These were then only slightly modified by Parliament. The decision process took thus more than three years.

The first documents expressed much more than the personal opinions of a few officials at the National Planning Office. From the beginning, experienced leading planners with a profound knowledge of the wishes and intentions of the directive organs took part in the work; they also maintained constant contacts with the highest political and economic authorities. Therefore, it is reasonable to consider the targets worked out at the start of the preparatory planning process as *aspiration* levels. A more detailed knowledge of the possibilities and the bounds on acceptance (i.e. the ascertainment of sets $\mathcal{B}$ and $\mathcal{D}$) finally led to a decision differing considerably from the starting aspiration level. The decision accepted in 1966 was both implementable and acceptable—in the belief of the heads of the National Planning Office and of the government.

We present two tables to summarize the preparatory process.[8] *Table 12.1* follows up the whole decision process. It surveys one by one the most important documents drawn up in the course of the process. As not every document contains all indices, there are several blank spaces in the table. The 1966 government proposals are considered as 100; earlier targets are given as a percentage of these proposals.

Table 12.2 presents extensive indices of decision preparation. At the present time these results were not yet available; instead, we have based the comparison on the so-called expectable results for 1970 or in the case of investments on those for 1966–1970, as estimated by the National Planning Office.

Detailed analysis of the characteristics of Hungarian planning as a decision process lies outside the scope of the present book. This is a task for further research; Tables 12.1 and 12.2 have the sole purpose of illustrating the concepts

---

[8] The documentation was prepared for this book by *Pál Pálinkás* (National Planning Office).

TABLE 12.1

Targets in the various preparatory phases of the third five-year plan*

(last document = 100)

Date of document / Character of data	June 1963	July 1964	September 1964	January 1965	October 1965	January 1966	February 1966	March 1966	May 1966	1970 (expected)
Production (national income) in 1970	115.4		104.7		101.4				100.0	112.3
Consumption in 1970	103.6				98.5				100.0	110.6
Investment in 1966–1970**	113.7	104.1	97.5	99.6	92.9	96.7	96.1	100.0		126.4
Export to socialist markets in 1970	110.0			104.2	99.8		99.8	100.0		107.9
Exports to capitalist markets in 1970	92.5			94.4	99.6		99.9	100.0		103.6

* The effect of price changes have been removed from the data, the time series of the table are thus based on comparable data. The table shows only the date of the documents: it is unnecessary to mention the planning organs which drew up the documents or the purpose of the latter and the higher organs to which they were submitted.

** We present here the expected data for total investment in the period 1966–1970 (and not the 1970 data).

TABLE 12.2

Some extensive indices of the preparation of the third five-year plan

Index symbol	$\hat{\varepsilon}$	$\hat{\zeta}$	$\hat{\varkappa}$
Denomination of the index / Character of data	Degree of tension of aspiration	Degree of tension of decision	Degree of correction of aspiration
Production (national income) in 1970	102.8	89.0	86.6
Consumption in 1970	93.7	90.4	96.5
Investment in 1966–1970	89.9	79.1	87.9
Export to Socialist Markets in 1970	101.8	92.6	90.9
Exports to Capitalist Markets in 1970	89.3	96.5	108.1

introduced in the present chapter. Nevertheless it is worth while to point out some of the characteristics of the process.

In the case of most indicators the aspirations are markedly more ambitious than the decisions. The aspiration usually is corrected "downwards". This reflects an essential difference between current practice and that of the fifties, when in the course of the decision process the plans generally were "tightened". At that period, Hungarian planning was characterized by overstraining and unrealistic ambitions. At present, it is considerably more realistic and more cautious—probably too cautious. In the cases of several indices, the performance promises to turn out more favourably not only than the decision but even than the aspiration.

12.7. Intensive Indices: Introductory Examples

Physics describes several phenomena by using *extensive* and *intensive* quantities together. The extensive quantities are usually physical quantities, the numerical value of which depends on the extension of the given material or physical system.

On the other hand the intensive quantity does not depend on the extension of the same material system. Characteristically extensive quantities in physics are mass, energy, volume. A characteristically intensive quantity is temperature. If we unite two physical systems, the mass of the united system will be the sum of the masses of the partial systems. If, however, the two partial systems were of the same temperature before the unification, their temperature will not change after it. For example, if we pour water into a common vessel from two separate bottles of one litre and of 20°C temperature each, the result will be 2 litres of water of 20°C temperature (not 40°C). Extensive quantities are additive, intensive ones are not.

Let us be content for the time being with a single extensive magnitude: the aspiration level. It expresses *what* the decision-maker would like. But it does not express how strongly he would like it. It indicates the intention, but does not express the "drive", the "serious-mindedness", the intensity of the intention.[9]

Let us return to the example of the high-jumper. Three students of a secondary-school prepare for the school-championship to be held in a month's time. Each of them is able; all three set themselves the objective of jumping at least 170 cm. Thus, the aspiration level is the same for each of them. The first student trains himself for two hours five times a week; he has a strong desire to win. The second one trains for two hours twice a week; his intention is moderately strong. The third one hardly makes any effort to prepare for the competition; he relies on his natural ability. In this example the aspiration level of 170 cm is the extensive quantity. The subjective importance of attaining the level—the seriousness of intention—the "drive" is the intensive quantity related to it. *It is the intensity attached to the aspiration level.*

The example of the high-jumpers demonstrates the fact that *the intensity of aspiration cannot be measured directly, but only indirectly by measuring the efforts made in the interest of attaining the aspiration level.* The aspiration level of the student who promotes with higher efficiency the attainment of that level is more intensive. Thus, the aspiration level of the first high-jumper is the more intensive, that of the second moderately intensive and that of the third the least intensive.

Using the example of the high-jumpers, let us introduce the following notation:

The aspiration levels α_1, α_2, and α_3 all equal 170 cm.

Let us call *promoting activity* the efforts made by the decision-making and implementing unit in order to attain the aspiration level. In our example this is the training. In this case, the promoting activity can be simply measured with a single real number, the hours spent in training. Let it be denoted by $z_i : z_1 = 40$, $z_2 = 16$, $z_3 = 2$.

---

[9] *R. Hoch* discusses the *intensity* of consumers' needs in a similar sense [91], p. 348,

Let us finally denote the intensity of aspiration by w_i ($i = 1, 2, 3$).

The volume z_i of the promoting activity depends on the α_i aspiration level and on the w_i intensity; it is a monotonically increasing function of the latter. Let us denote this function by f_i:

$$z_i = f_i(\alpha_i, w_i) \tag{12.8}$$

Let us assume that function (12.9) takes the following simple form:

$$z_i = \alpha_i w_i \tag{12.9}$$

Accordingly, in our example the dimension of intensity is hours/cm. Numerically: $w_1 = 40/170 = 0.24$; $w_2 = 16/170 = 0.1$ and $w_3 = 2/170 = 0.001$.

12.8. Economic Examples

Let us turn to the economy for further examples. Part III of this book will frequently use the concepts of aspiration level and intensity. There, the examples will be taken from the sphere of purchase and sale. In this chapter let me quote three examples in anticipation of Part III.

First example. In an industrial branch with many firms, two firms aspire to become the leading firm of the branch, from the point of view of technical development. Their aspirations can be described with the aid of the $\alpha_1(t)$ and $\alpha_2(t)$ aspiration vectors. The components of the vector are targets such as the following: when should we appear on the market with a definite new product; when should we introduce a new technological process; to what level should we raise the various technical parameters of our earlier products, and so on. Let us assume that the aspiration is expressed in both firms by indicator vectors of K components. Each applies the same type of indicator, although the numerical value of the indicators differs.

Attainment of technological leadership may be promoted by several activities. The volume of each such activity can be measured by a real number. For example: how much (money) is given to external research institutions for new research; how many man-hours of engineers are spent by the development department of the firm on the attainment of the objectives; how much time do the general manager and the chief engineer devote personally to these activities.

Connections between two categories of the decision process also appear in the example. In the *primary* decision process the decision-maker keeps in view the *final results* of technical development, its main indicators. In the other category, the *partial measures* promoting the final result are provided for.

The partial measures are tools, and thus subordinated processes, to attaining the aspiration level.

The firm for which attainment of the aspiration level is more important will make greater efforts in its interest and accept greater sacrifices. The greater efforts indicate a greater intensity of aspiration.

Second example. In the case of the two firms compared in our previous example we have made a "cross-section" analysis. But also dynamic "time-series" investigations may be instructive. This time let us describe the yearly plan of a Hungarian firm say in 1955 and 1969. Our subject now will not be aspiration but rather decision: the annual plan of the firm. We may neglect now the facts that in 1955 the firm was legally required to make a plan and that it now draws up the plan on its own initiative. We may also disregard the fact that in 1955 many indices had to be planned which no longer occur in the annual estimates of the firm and, conversely, that it now considers indices which it did not consider at the earlier time. Thus we shall consider solely those indices which appeared in the annual plan both in 1955 and in 1969. For example, we shall observe the total value of production, total value of sales, total wage bill of the firm, the profit rate and its volume. These are six types of indicators.

Naturally, the *extensive* indicators of the firm's aspirations have changed between 1955 and 1969. Thus the two plans, a^* (1955) and a^* (1969) may differ considerably. However, the nature of differences between the two situations are not described satisfactorily by extensive indicators alone. One of the most characteristic differences is, namely, *that the intensive indicators* belonging to the individual *types of indicators have substantially shifted.* Earlier, a greater intensity was associated with the indicators of production volume and wage-bill and a relatively slight one with the others; now the situation is reversed, a greater intensity accompanies profits, prime costs and sales.

The shifts in material incentives and in motivation explain the relative shifts of intensity.

In both situations there are definite promoting activities. In 1955, such activities were efforts for continuity of production, good utilization of fixed assets, but at the same time, also "shock-work", neglect of maintenance, overtime, deterioration of quality. In 1969, promoting activities included increased thriftiness, more flexible adaptation to demand but perhaps also open or concealed price rises.

Third example. Let us assume that an engineer of a firm always has been interested in introducing innovations. In the past he could obtain for his minor or major innovations at most an annual increase in income of 2000 forints. Thus he has made certain efforts to innovate, though not too many.

Suppose that his situation changes. E.g. he makes a major invention, and caknowledgement and reward are due to him in his capacity as inventor. If his innovation is really introduced, his reward may reach 200,000 forints. Thus, his material interest has risen a hundredfold. He will certainly make much greater effort to aid the success of the invention; wherever possible he will promote its application; he will convince, mobilize, organize, and try to remove obstacles. His aspiration is, as it was earlier, to introduce innovation. But now he promotes it with more "drive"—with an intensity many times greater than formerly.

12.9. Definitions of Intensity

Now, let us turn to a more general treatment. In the first approach we shall neglect the dynamics of the problem and accordingly, the argument t, indicating time, will not be used in describing variables.

We shall examine a complex decision process of the C-unit of some organization.

First, we consider (out of context) a single elementary decision process related to the solution of the *primary* decision problem p_0. At the beginning of the elementary decision process the primary aspiration level $\alpha^{(0)} \in \mathcal{A}^{(0)}$ emerges. (The set $A^{(0)}$ is a subset of the linear space $\mathcal{L}K^{(0)}$ the space of the possible indicator vectors related to problem p_0.)

From the examples shown in the preceding section, in the first one technical development, in the second one the annual plan, and in the third one the introduction of innovations all qualify as primary decision problems.

Definition 12.6. Attainment of $\alpha^{(0)}$ p r i m a r y a s p i r a t i o n l e v e l, and the implementation of $a^{(0)*}$ primary decision are aided by p r o m o t i n g a c t i v i t i e s (outputs and changes in state). The C-unit determines the levels of the promoting activities by solving the decision problems $p_1, p_2, \ldots, p_N$. Solutions to the decision problems are the decisions $a^{(r)*} \in \mathcal{A}^{(r)}; r = 1, \ldots, N$. (Set $\mathcal{A}^{(r)}$ is a subset of the linear space of $K^{(r)}$ dimension, the space of indicator vectors related to problem p_r.)

In the three examples used previously in this chapter, the promoting activities are the commissions given to external research institutions, the man-hours of engineers used in the factory [first example], the good utilization of fixed assets, "shock-work", thriftiness, price-raising, etc. [second example] and the agitative and organizing activities of the engineer in the interest of the success of the new invention [third example].

Definition 12.7*. The subjective importance for the decision-maker of $\alpha^{(0)}$ primary aspiration level is expressed by an intensity indicator related to the extensive indicator vectors mentioned: $w^{(0)}$ is the vector of the i n t e n s i t y o f a s p i r a t i o n comprising $K^{(0)}$ components.

The decision regulating the promoting activities depends on the primary aspiration level and its intensity:

$$a^{(r)*}_{\text{в мв}} = \varLambda^{(r)}(\alpha^{(0)}, w^{(0)}, \ldots) \quad r = 1, \ldots, N \qquad (12.10)$$

The function (12.10) will be called p r o m o t i o n   f u n c t i o n.

The three dots in formula (12.10) after the two arguments indicate that, beside the aspiration level and intensity, the promoting activities may be influenced by other factors also.

The three examples shown earlier in the chapter illustrate the promotion functions. In the second example the indicators of intensity changed as a consequence of the reform of economic administration, and the promotion functions changed accordingly. In the third example the degree of interest of the innovating engineer increased and so did his activity.

The notions introduced in Definition 12.6–12.7 can be transferred to processes over time, i.e. to the more general case when we have to deal not with a single p_0 primary problem, but with a series of repeated primary problems $p_0(t_1)$, $p_0(t_2)$... taking place in the same indicator space $A^{(0)}$. But we do not deal with the formal description of the dynamic case at this point.

The formal description of intensity presented above requires further improvement. We do not consider Definitions 12.6–12.7 to be entirely satisfactory. For instance, the following question remains open. How should the intensities of different organizations be compared if indicator types are identical but aspiration levels differ? (E.g., can some normalization be performed in such cases?) However, it is not the present form of the definitions, nor the formal language used in our discussion that is most important, but rather the general notion of intensity. The basic idea may be summed up as follows:

Statement 12.1. Certain activities of economic organizations and the efficiency of those activities depend on two major factors: the extensive level of aspirations of the organization (the level to be attained) and the intensity of aspiration (its subjective importance).

Ideas and concepts regarding intensity will be used in this book first in Section 21.9.

We hope to refine the definitions and improve formal description after we gain further experience in working with the concept.

The intensity vectors cannot be directly observed. As in the case of the theory of "revealed preference"—which has been sharply attacked in this book only recently—intensity can be observed only through its revelation. This means that promoting activities must be observed regularly (their extent, practicability, and efficiency must be determined) and from the observation one must deduce, *in an indirect way,* a psychological phenomenon which is the intensity of the aspiration.

Today sufficient empirical material is not yet available. This is only natural; in general a relationship—at least in its broad outlines—first is formulated in a hypothesis and only then is there an attempt to observe the variables and parameters playing a role in the relationship.

primary aspiration level, and its fulfilment, the result ω---. Similarly, the decisions $a^{(r)*}$ relating to the promoting activities and their fulfilment, the results $\omega^{(r)}$ are tangible quantities.

It would not be correct to specify *a priori* the shape of the function (12.10). This must be done on the basis of empirical observations. Obviously, their mathematical nature will not be uniform but will depend on the character of both the primary problem and the promoting activities.

When sufficient empirical material is available, two kinds of analyses will be required. Cross-section analyses are needed to compare similar organizations in identical periods. In addition, dynamic, time-series analyses are needed to examine changes in the behaviour of the same organization.

12.10. Comparison

1. *Complex motivation.* Already in Statement 7.2 we have shown that the behaviour of institutions is characterized by complex motivation. Of course this is reflected also in the behaviour of organizations constituting the institution as a subsystem.

The concepts introduced in Chapters 8, 9 and 12 offer a suitable formalism for describing complex motivation. First: the various motives are expressed in the various bounds on acceptance. Second: they are expressed in the extensive measures of the aspiration level. And third: they are expressed by the intensity indicators attached to the aspiration level; in the final analysis the motives are expressed by the promotion processes exerted in the interest of the various motifs.

Those who try—in the spirit of the GE school—to describe the behaviour of the organization with the apparatus of the utility function, would like to state how the decision-maker "weighs" his various motifs, objectives and interests. In a firm e.g. the 1st objective is to maximize short-term profits, the second, to maximize long-term profits, the 3rd, to raise the share of the firm in the market, the fourth, to attain a leading position with respect to technical development, the fifth, quiet and safe operation, and so on until the Mth objective.

It seems an unnatural formalism, and one foreign to the psychology of decision-making, to make efforts to use a *single* maximand—by calculating a weighted average or any other mathematical form. *Instead, we shall use throughout indicator vectors of many components.* In the example above, the first objective of the firm can be described with the first component of the indicator vector, the second objective, with the second component, and so on.

The decision-makers usually think in terms of absolute targets, not of relative preference weights. "I demand at least as much", "I want at most as much" (bounds on acceptance), "I should like to have as much" (aspiration level) —these are the characteristic schemes of reasoning in the decision process.

In addition, we may characterize the behaviour of the organizations by describing the efforts they are inclined to make in order to achieve the targets, requirements, aspiration levels. (Promoting activities of aspirations and decisions).

2. *Consistency.* In the formalism advocated in this book we *do not* demand that the components of the aspiration level or of the decision should be in strict harmony with each other or with the earlier aspiration levels and decisions. On the contrary, in the real operation of economic systems simultaneous attainment of the aspiration levels attached to the individual types of indicators is frequently impossible. In the course of the decision process the decision-maker moves from a unrealistic aspiration level to an implementable decision; and even the decision may prove to be unrealistic—one that cannot be implemented.

The conceptual framework proposed here also comprehends the possibility that the decision-maker may not be consistent over time. Owing to various influences—either because external circumstances have changed or because he has learned from earlier experiences—the decision-maker may change the aspiration levels and the decisions from period to period, in doing so, he violates the requirement of consistency taken in the sense of Definition 10.3.

3. *Explanation of deviations.* According to the GE school, all capitalist firms behave in an identical manner: they maximize their profit functions. But, without more precise knowledge about the behaviour of the firms, how can we explain the fact that some firms are strong and others weak; that there are some firms which grow rapidly and others that grow less rapidly, some of which may even fail. Is the difference that the one "maximizes more" and the other less?

If we cannot explain the *differences* between the development of the firms, we also cannot explain one of the most important phenomena of modern capitalism: why does *concentration* take place in the economic system? The stereo-typed explanation (that all firms maximize profits) might perhaps be acceptable in a world where there are hundreds of thousands of atomistic homogeneous firms. But it tells us nothing about the real world where big and bigger firms emerge from among the mass of small firms and in many an important industry a few giant oligopolies play a dominating role.

It is not the task of the present chapter (and not even of the book) to analyse the causes of the emergence of oligopolies, concentration or, in general, differences between firms. Our task is much more modest. Rather we attempt only to propose a new formalism capable of replacing the old formal, conceptual framework which today is obscuring understanding of the differences between firms. *We propose a formalism—concepts and schemes of reasoning—that enables description of different forms of behaviour, and that even expressly calls attention to the necessity of explaining these differences.*

In our framework differences in the behaviour of two firms may be explained by observing:

a) the types of indicators that the firms use most in the decision and control process,

b) the regularities or trends that characterize the extensive indices of their decision processes (aspiration, tension of aspiration and of decision). What are the trends of these indices, to what extent do they fluctuate around the trend,

c) the intensity of the efforts of the firm striving after implementation of its aspirations).

With the aid of language we can describe the differences between the dynamic and the stagnating firm, the progressive and the conservative firm, the initiating and the imitating firm, the ambitious and the ambitionless firm, between the one which fights with a strong will for its objectives and the one which is satisfied with small efforts. It is our belief that such differences in the spirit and attitude of the firm explain to a great extent[10] the differences in their development.

[10] Obviously, they do not offer full explanation, development or decline, success or failure depend also on the luck or misfortune of external circumstances.

13*

13. AUTONOMOUS FUNCTIONING OF THE SYSTEM

"It has never been so,
that it would not have
been somehow."
(Common saying in Budapest)

In the last five chapters we have dealt with the decision processess taking place within an organization (in fact, within their control units). To continue the simile used in Chapter 4, we now end after our "low flight" and we begin to fly higher. In Chapters 13–14 the *combined functioning and interactions* of the organizations within the economic system will be examined from various viewpoints.

13.1. Analogy of the Living Organism

In living organisms of higher order (particularly in man) biology distinguishes two classes of functions: *the autonomous and the higher functions*. (The terminology in Latin calls the first "vegetative" functions.) Let us consider the human organism. Autonomous functions take place almost independently of will. Control is exercised partly by the autonomous nervous system, which is to a certain extent separated from the central one. Autonomous functions are controlled also by chemical, hormonal processes. Here belong, among others, the activities of the heart, the vascular and the digestive system and the various processes of metabolism. Autonomous functions generally involve the simple self-maintenance of the organism.

Total separation of the autonomous functions and the processes controlling them, is not warranted; the autonomous and higher functions and their control systems interact. It is well known that transitory disturbances or lasting diseases of the heart or the stomach may be caused by the state of the central nervous system: by tensions, shocks, psychic conflicts. Conversely, disturbances of autonomous functions may have repercussions on higher ones; e.g., chronic bodily diseases may influence mood, capacity to work, etc. However, in a normal state of good health close interactions are less obvious and relative separation may be observed.

A separation analogous to the one discussed above may be observed in the economic system. The "autonomous" and the "higher" functions, as well as the control processes of the two kinds of functions, are separated from each other to a certain extent both in individual organizations and in the economic system as a whole.

176

13.2. Autonomous Function—First Approach

In the remainder of this chapter we shall try to separate the two kinds of function in "pure theory". We should like to stress — and shall return to the question — that a "pure" separation is, of course, an abstraction, and the bounderies are arbitrarily drawn. In reality the two functions are unseparable in many aspects, or are at least in close interaction.

As a first, rough approximation we may state *that the major characteristics of autonomous functions are uniform throughout all modern economic systems. It is in the higher functions that these concrete economic systems differ from each other.*

Let us return to the analogy of human organism. The heart, vascular, stomach and digestive functions of an average citizen, a great scholar and a criminal show identical rules of autonomous function. At the same time, these people differ radically with respect to intellectual capacity and moral behaviour, that is, in their higher functions.

In order to point out the autonomous functions of economic systems we must find these functions that are essentially similar in *every* modern economy based on division of labor.

Consider our own economic system, the Hungarian economy. Informally economists and specialists working in various fields of the economy often express the following idea: "Our economy really must rest on strong foundations if it could weather all the mistakes made in economic policy—if it could function continuously in spite of the distorting effects of the old mechanism, prior to the reform of economic administration."

Most of us were inclined to overestimate the actual effect of central plan instructions on production and management. Those adhering to the system of central plan instructions overestimated the favourable effect of the instructions while its critics overestimated their adverse influence. However, in the final analysis, it only appeared that all economic activities were controlled by central plan instructions; in fact, many activities occurred "by themselves", that is, as autonomous functions.

A factory was given an annual plan each year and a quarterly plan each quarter. It produced, but *not because* it received plans and instructions. It would have produced even if it had received no instructions and on the whole it would have produced the same quantity as was prescribed by the plan. A brick-yard can produce only bricks, and roughly as many as its capacity allows.

The autonomous function is rooted in simple social and socio-psychological facts. Consider a factory which is capable of functioning. Let us forget for the moment that the factory itself had to be produced —this aspect *does not* belong to the category of autonomous functions. In a normally functioning factory a siren indicating the beginning of work is sounded each morning and the workers take their places at that time. They feel that it is natural to work, to prepare the product wanted in the quantity desired. Of course, there may be lazy workers, but the majority would feel odd if the working time passed without work being done. Therefore, they wait for materials and tasks. The material is issued by the depository, where the material purchaser sees that the stocks thus depleted are replaced; the foremen, engineers, and factory administration assign the day's tasks, take stock of the finished products, prepare the tasks of the next days and weeks, and so on. Finally the finished products must be sold; they must be forwarded to the consumer, where they are needed.

At the beginning of the chapter we quoted the common saying in Budapest: "It has never been so, that it would not have been somehow." Although this does not sound very scientific, it is still the substance of autonomous functioning. *Autonomous function is based on the average diligence, conditioning, routine, identification with one's job of the people working within the system.*

Beyond that, it also relies on the fact that, as in every material system, a certain inertia prevails also in economic systems. Until some force disturbs the system, it will repeat itself. One might lament the fact that it conserves the old, or one might rejoice that it promotes stability. But however one feels about it inertia exists and acts, and one of its effects is evident in the autonomous functions of the system.

13.3. Stocks and Reserves

In autonomous functioning stocks and reserves play a central role. Therefore, they will be treated separately in this section. In Part III of the book their role will be discussed farther.

Stocks and reserves have many forms.

Product stocks are deposited at each point of production, turnover or consumption. Product stocks accumulate with the releaser; if the product is sold by a retailer stocks also accumulate there, and finally they accumulate with the user.

"Reserves" are unutilized parts of resources or fixed assets having a function similar to that of the stocks, e.g., less than fully utilized machinery or buildings, unemployed or partially employed labour, fallow land area, or unexploited natural deposits.

Every economic system has a considerable amount of reserves and stocks. Stocks and reserves are not an inessential aspect of a real economy but an indispensable component, having at least two important functions.

A) One function of the stock or the reserve is to promote smooth, undisturbed operation of the system.

There are reserves in every living organism. For example the human organism accumulates reserves of water, sugar, fat, iron, etc.

The economic system cannot operate with an empty, "cleared" market, without stocks and reserves. Stoppages, disturbances in both purchasing and selling, may occur and the presence of stocks diminishes the danger of reverberating effects. In the case of a sudden expansion of production, purchasing may not be able to procure the necessary materials quickly enough; thus such expansion could not take place without stocks. For sudden expansions reserves of resources are needed also.

Operations research literature includes many inventory models. It is unfortunate, however that economic theory has become separated entirely from the branch of operations research dealing with stockbuilding; economic theory almost entirely neglects the role of stocks and reserves. Other branches of operations research, (e.g., its allocation models), have developed in close cooperation with economic theory. This separation is the fault not of operation researchers but of theoretical economists.[1]

Under the preceeding section we have talked about the role of stocks and reserves in promoting smooth functioning in the *real sphere*. Let us turn now to the role of stocks and reserves in the *control sphere*.

B) An additional function of stocks and reserves of resources prevails in the control sphere: they are to operate as a singal system.

Observation of changes in stocks, reserve resources and reserve capacities yields important information to the productive firm:

With respect of output: too great stocks—production can be reduced; too small stocks—production can be increased.

With respect of input: Too great stocks—purchasing can be diminished; too small stocks—purchasing of resources may be increased.

Statement 13.1. Changes in stocks yield outstandingly important information of non-price character. They are signals that are most economical of information and they can be observed within the firm. They are extremely sensitive to momentary changes and analysis of their trend indicates more lasting tendencies as well.

Passing from the producer to the consumer; the household also relies to a considerable extend on stock signals. If the stocks of fat, sugar, soap, etc. sink

---

[1] The truth of this statement is not refuted by the fact that distinguished economists adhering to the theory of equilibrium also dealt, even in their major works, with inventory models. (See, e.g.: *Arrow-Karlin-Scarf* [17].) This is a peculiar "personal union" between two *separate* empires, but it does not imply their integration.

below a critical level in a household, the housewife will care for their being replenished.

A signal system based on the observation of the movement of stocks and reserves is at least of equal rank with the signal systems based on prices. It is not certain that a firm will respond to a change in individual prices, but it will surely respond to changes in its own stocks. This is true both in capitalism and in socialism. Even in the overcentralized socialist economy the signal system of stock changes had a huge role. Even then it was this signal that determined both the momentary actions of the firms and the steps taken by higher control bodies.

It is worthwhile to return to the analogy of the living organism. Several processes producing certain materials necessary for the life of living beings are regulated according to the stocks accumulated in the organism. If the water, salt, sugar, and other stocks of the organism sink below definite lower levels, or rise above higher ones, control processes become activated that will raise or reduce the stocks within normal boundaries.

Control based on the observation of own stocks and reserves belongs to the sphere of simple decision algorithms, in that of standard decisions. It may have several algorithms. Only one characteristic example is given hereunder.

In the organization there develops a norm for stocks conforming to the average necessary for normal functioning. Output and purchasing are adapted to the norm. In the case of an output stock this may be formalized in the following manner:

$$\vec{x}_i(t) = \vec{x}_i(t-1) + \Delta\vec{x}_i(t), \tag{13.1}$$

where $\Delta\vec{x}_i(t)$ is a modification of the product output given a product output of $\vec{x}_i(t-1)$ in an earlier period. The response function takes the following form:

$$\Delta\vec{x}_i(t) = g_i\left(\Gamma_i - \frac{y_i(t-1)}{\sum\limits_{\tau=t-1-T}^{t-1} \dfrac{\vec{x}_i(\tau)}{T}}\right) = g_i(\Gamma_i - G_i(t-1)). \tag{13.2}$$

The decision-maker observes his actual production through period T and calculates the number of periods of average production to which the stock in the last period is equal? The fraction $G_i(t-1)$ in the formula is the stock divided by average production per period. Its measure is, therefore, time: e.g. to how many months of average production does the last actual stock correspond.

The stock normative Γ_i has been given in the same measure. The positive coefficient g_i is a correcting factor which defines the extent of modifying production. In the case of a small g_i the firm will respond with relatively small modifications to deviations of stocks from the norm; in the case of a big g_i the response will be stronger.

The formula insures that in cases exceeding the norm $\Delta \tilde{x}_i$ will be negative; that is, production will decline; in cases lagging behind the norm, $\Delta \tilde{x}_i$ will be positive and production will grow.

The rule relating to the purchasing of product inputs necessary for the production of the firm can be formulated in an analogous manner.

Further research is required before we can describe formally actual algorithms based on stock signals and examine their actual influence on the functioning of the economic system. However, the role that this regulation can play in operating the economic system could be analysed theoretically with mathematical models.

13.4. Delimination of Autonomous and Higher Functions

A more intensive study of the problem probably will lead to distinguishing more than two categories. Modern physiology also has distinguished more than two stages. For the time being though, as a first approximation, let us remain satisfied with the present categorization.

The autonomous and the higher functions will be differentiated by five major criteria.

1. *Real processes.* We include here all those real processes which involve a simple repeating of real processes in earlier periods. In the terminology of Marxian political economy; simple reproduction belongs in this category. In the case of production, output would be set at about the earlier level, with the earlier technology and with a product-mix corresponding to the earlier one; also there would be a provision for maintenance and simple replacement of fixed assets. In the case of turnover, transactions would occur between the usual sellers and purchasers. In the case of consumption, consumption would be established at the accustomed level and in the accustomed pattern.

More radical and deeper changes in real processes are not classified here but belong to the category of higher functions. To the latter must be grouped investments, major technical development, releasing of new products substantially differing from earlier ones, radical changes in the volume and pattern of production, and also essential shifts in consumption.

The two spheres of real process are sometimes more, sometimes less distinct. For example, they may be separated functionally within complex institutions, (within a modern industrial big firm) as has been treated in detail in Chapter 7. They may be distinguished in time; some organization, or a whole system, may stagnate on the whole over some longer period, and on such occasions the real processes characteristic of autonomous functioning dominate its operation. Periods of stagnation may be then followed by periods of slower or quicker development.

In other situations and at other times the two spheres may overlap substantially; in these cases, the autonomous and the higher functions are not separate —at least as regards real processes; they may be distinguished only in an abstract way.[2]

2. *Type of decision.* Let us now turn our attention to the control processes. Autonomous functions are always regulated by standard decisions, by simple algorithms. (This notion has been clarified in Definition 9.2.) Frequently, the decision-makers do not even perceive that they are making a decision. For example, they simply repeat their earlier behaviour. The decision-maker's main source of information is his own memory. If some kind of decision process takes place, it is short. In these cases the decision is made on the basis of a simple response function, a simple decision algorithm and little information input is required.

Higher functions are controlled to a considerable extent by fundamental decisions. These decisions are frequently preceded by a rather complicated response function, decision algorithm, and by longer preparation for the decision. (E.g., in firms introducing an important new product.) The same happens in a household. The usual purchases of food will not cause too much headache (autonomous function), but the buying of a family home certainly will (higher function).

3. *Individual motivation.* In the functional organization of complex institutions (mainly in productive firms) those participating in the control of autonomous functions are motivated mainly by identification with their functions. (See Section 7.3 on this subject.) The leaders of the production department want the plant's production to be rhythmic, continuous, and free from major fluctuations; the material purchasing department wants to secure normal supply of the plant with materials; and so on.

Division of the complex institutions into functional organizations insures that some of those active in the field of production and turnover specialize expressly in the control of autonomous functions. This explains the "psychic affinity" between plant engineers, foremen, etc, functioning in the most diversified economic systems. They have similar problems to cope with; the identity of their work, their tasks, their "existence" creates similar consciousness in them.

People controlling higher functions also are motivated by identification with their tasks. Here, however, other motives play a much greater role. Poli-

---

[2] It is open to argument how to classify—from the point of view of an abstract separation of autonomous and higher functions—the expansion of production which takes place on an unchanged technical level.

tical and moral incentives, desire to increase social power and prestige, accumulation of wealth and the desire to bequeath wealth to their descendants, the thrill or fright of risk-taking—all are motives that influence decision-makers in controlling higher functions.

4. *Characteristics of information flows*. The information flow of autonomous functions has two major components. One of them,—the role of observing the own real stocks-has been described in the previous Section 13.3. The other main component is an immediate oral or written communication between the producer and the user. This aspect will be treated in detail in Part III of the book.

The information flow of autonomous functions may be characterized using the notions introduced in Chapters 5 and 6 as follows:

a) The information either *reflects* real variables directly (e.g. report on stocks), or comprises at most a *single transmission* (e.g. direct information exchange between seller and purchaser).

b) The information most characteristic of autonomous functions is of a *non-price character*. This is sufficiently obvious in the case of own stocks. "To how many months of production does the product stock correspond?" – "To what degree do we utilize our machinery?"

In the immediate communication between deliverer and recipient, information of price-character and non-price character is, of course, intertwined. Since concrete systems differ profoundly in the strength of price effects and in the regularities of price formation, price-type information cannot be classified among the characteristic features *common* to the autonomous functioning of all systems.[3] However, the non-price-type information exchanges between deliverer and recipient to belong to this category. "I need so many pieces..." – "I should like to get as much from the product of this quality...". This type of information flow can be found in every economic system.

Non-price information appears in a war economy based on rationing and direct allocation as well as in the peace time economy relying on sale and purchase. It dominated all information flows between socialist productive firms

[3] One should question whether it is justifiable to exclude price-type information from the information flow of autonomous functioning. The main argument for excluding them is that autonomous functions are identical in every system, but the functioning and effects of the price system may differ considerably from each other in various systems. The rigidity of delimitation may be eased if—as has been already mentioned—the functioning of the economy is broken, into more than two stages. E.g. in the functioning of higher organizations, among them in that of man, we know conditional reflexes. On the one hand, these do not belong to the group of autonomous functions with the organization from the time of its birth, since they are based on conditioning and on habits acquired in the course of life. On the other hand, neither can they be ranked with higher functions controlled by deliberate decisions. Perhaps the effect of prices on the market also belongs to some similar intermediate category.

in the highly centralized, old way of controlling the economy, when the trans-action price of the product was practically immaterial for the firms. But it has a great role also in any capitalist economy in the relations between sellers and buyers.

c) The anterior and the posterior time horizons of information are short; they are *approximately simultaneous* with the real event.

On the basis of what has been said, it can be stated that *the structure of information related to autonomous functions is relatively simple.*

On the other hand, *the structure of information related to higher functions is complex.* In the order of the points above:

a) Higher functions are frequently controlled by multi-phase information flows.

b) Information of price-character plays a major role. As we have pointed out before, the differences in the higher functions between various concrete, econo-mic systems are found to a considerable extent in the diversities of *the price systems.*

c) In the control of the higher functions the time horizon is relatively great. Appearance of anterior information relating to a long time horizon—that is of planning—is one of the most important signs of a higher function. One of the major differences between concrete economic systems is whether they make plans covering a longer period, and embracing the whole of the system; whether the influence of the plan on the real processes of the economy is great, whether its methods are the same, etc.

5. *Levels of control.* The control processes of autonomous functions are single-level; they all take place on the lower level of the system. The flow of information is horizontal. On the other hand, a large part of higher control takes place on higher levels, the flow of information is mostly vertical.

The biological analogy again is a good one. As has been pointed out in Chapter 6, multi-level control prevails in the living human organism. The autonomous functions are controlled on the "lowest level", separated from the higher-level, central nervous system.

Now we may state a summary definition.

Definition 13.1*. The a u t o n o m o u s   f u n c t i o n s of an economic system generally involve the simple repetition of real processes. They are controlled by standard decisions. Their motivation is based mainly on identi-fication with the function. The information structure is relatively simple; it is mainly single-phase and not of a price type, consisting of almost simultaneous information. An essential component of control is based on stock reports and the immediate information relayed between the deliverer and the recipient of the product. Control takes place exclusively on lower level; the information flow is horizontal. All functions of the economic system that cannot be classified within the scope of autonomous functions generally belong to the scope of higher functions.

The specific nature of the autonomous control depends on the degree of development and on the technology of information and control within the economy. For example the organization of stock reports is different in a XIX. century plant and in a modern firm with computerized inventory book-keeping. But the features of the autonomous functions do not depend on the political and ownership relations of the system; these affect only the higher order functions.

13.5. Major Statements: Comparison

The models of the GE school do not distinguish the autonomous and the higher functions; however they deal with some details of both kinds of function, taken out of their context, leading to a curious mixture.

Let us deal at this point only with static (and as well, stationary) Walrasian models. On the one hand, these models attempt to explain the real processes belonging to the scope of autonomous functions. This has been pointed out in previous chapters of this book. The models basically treat *current* production and consumption. They are restricted to this by the stationary character of the models. In the framework of their own assumptions and concepts they are hardly capable of explaining such phenomena as technical transformation, changes in the tastes of consumers, etc.

At the same time, however, when treating the autonomous *real* processes, they do not describe the related "autonomous nervous system" controlling them; they do not describe the primitive control processes (e.g. the decisions relying on the observation of stocks, the non-price-type information flows accompanying the transactions, and so on). Instead, they focus attention on a major, but not exclusive component of the higher control, namely, the price system.

What is needed is a more detailed description of the functioning of economic systems. In the following some further statements are made concerning the distinction between autonomous and higher functions.

Statement 13.2. A considerable part of the real and control processes of all economic systems are autonomous functions.

We believe that the truth of this statement does not need much argument, since it already has been expounded in Sections 13.1–13.4; It is presented here only as a summary.

Statement 13.3. The autonomous functions of organizations change little over time. The inertia acting against changes produces the stationary character of autonomous functions. This is one of the stabilizers of the functioning of economic systems.

In our view, this statement is of great practical importance. Theoretical economists, among them the representatives of the GE school, are constantly looking for stabilizers of the economy. They would like to show that there exists a price system that alone is capable of sustaining the equilibrium of the economy. (Nevertheless, they can prove it only if they make very restrictive assumptions.) In reality, however—in our view—one of the most stabilizing factors is the inertia dominating the functions of the economy, the natural inclination of functions to repeat themselves. It is the autonomous functions that create continuity in the functioning of organizations and entire economic systems.

The reform of the economic control and management system in Hungary provides remarkable, indirect proof. Overnight, January 1, 1968, the system of plan instructions, which allegedly deeply influenced the everyday life of the firms, was abolished. Simultaneously, radical changes were put into effect concerning prices, material incentives, finances, and planning—in one word, involving all forms of higher control. In a few years the effects of these factors will grow, but, obviously, the reforms can not exert their full influence on the behaviour of firms and economic decision-makers immediately.

Although neither the "old" nor the "new" had asserted itself with full vigour, no vacuum ensued; economic life continued to function smoothly. We think the explanation must be based on two facts. The first is that the autonomous functions of the system were continuous before and after January 1, 1968 and after that date, *this* is what made the life of the system continuous. The other fact is that the preparation of the reform was wise and circumspect enough not to interfere with autonomous functions.

These ideas leads to the following statement:

Statement 13.4. Although the higher and the autonomous functions are to some extent separated from each other, the deficiencies of the former may cause disturbances with the latter.

Autonomous functions have a "self-defence"; they do not allow themselves to be too greatly influenced by small fluctuations of the higher control processes. However, should grave disturbances arise in the latter, this causes shocks in the autonomous functions and deflects them from their normal course. This is similar to the situation when a critical psychic state causes a heart attack or colic of the stomach.

Such disturbance occurs in a capitalist economy in time of crises or depressions. The trouble begins in the spheres of investment and savings, financial and credit movements, and purchasing power that is, in the sphere of higher control. From here, however, it also spreads to autonomous functions—to everyday production, turnover and consumption. Neither individuals nor organizations are capable of simply repeating their routine decisions and accustomed behaviour.

Disturbances also occur in socialist economies. For example, in Hungary deficiencies of higher functions—of central government economic policy, of five-year economic planning—have caused disturbances in autonomous functions; for example, lack of materials and energy prevents firms from producing smoothly.

Statement 13.4 is one-sided; it calls attention only to the negative effects of wrong higher functions. It must, therefore, be complemented with the following.

Statement 13.5. Autonomous functions, by themselves, can secure only the stationarity of the system. Development of the system depends on the success of the higher control processess.

In other words; the system could exist without higher functions, but it "would vegetate". Expansion of the real processes, technical development, and ever fuller satisfaction of the material and cultural needs of society depend fundamentally on the motivation of the decision-makers, on the price system, on the quality of planning, on high-level decisions, and on government economic policy; essentially, development depends on the quality of higher control.

14. ADAPTATION AND SELECTION

14.1. Adaptation and Selection in the World of Living Organisms

In biology adaptation[1] is a notion having fundamental importance. Living beings adapt to their environment and the changes therein. Nature is not merciful; if the living being does not adapt, it will perish. The *individual* may perish untimely if less than a minimum of adaptation occurs. And the *species* will become extinct if too few of the individuals adapt sufficiently.

In the functioning of human society and, within it, in the functioning of the economic system, we find many phenomena which are analogous to the biological adaptation processes. Social institutions adapt to their environment. An industrial enterprise adapts to the resources available; if land is scarce it economizes on land; if labour is scarce it economizes on the latter. It also adapts to the wishes of the buyer. Even households adapt to the output of the producers; their needs will develop along with the products released by new technologies and everyday production. The economy of a country adapts to natural resources, to neighbouring countries, to the wishes of trading partners. The functions of economic systems change fundamentally if a war or an economic crisis breaks out. Such shocks also start adaptation processes.

Selection also occurs in economic systems; organizations and even entire systems are born and perish. Different forms of behavior, and functional rules ("mutations") arise; some of them become entrenched ("become inherited"); others prove to be unviable and disappear.

Since *Spencer* many sociologists and economist have become aware of the biological analogies of socio-economic processes.[2]

We do not wish to push the analogy *ad absurdum*. No "economic biology" or "biological economics" is necessary. We consider the biological phenomena only as a point of departure. Reasoning must be continued further along economic lines, analyzing special features which are characteristic of the adaptation taking place in economic systems.

---

[1] On the concept of adaptation, see *Bellman*'s book [30].

[2] *Alchian*'s [4] work on the subject is remarkable. A summary survey of the literature treating the problem can be obtained from the work by *Winter* [276].

14.2. Primary and Secondary Adaptation

Two stages of adaptation can be distinguished. Through primary adaptation the organization or the system of several organizations secures its mere existence, its survival, its preservation. With the *secondary* adaptation it strives to achieve more than mere survival, it attempts to achieve a life in which its aspirations, expectations, norms are fulfilled.

Primary economic adaptation is relatively close to the notion of biological adaptation, the criterion of which is the preservation of the individual and the species. But, secondary adaptation is connected with peculiarly human, social phenomena which have no analogies in the world of non-human beings.

There are many fields of economic life where secondary adaptation occurs. In Part III of this book we shall discuss in detail one of the special control sub-systems, the *market*; the control process of production and consumption, their mutual adaptation is a characteristic example of secondary adaptation. However, it would be a mistake to believe that adaptive processes take place in the economy exclusively through the market. For example, the ensemble of the government, the banking system and the industrial firms also constitute a particular sub-system within which the secondary adaptation of *monetary and credit turnover* takes place. Similarly, a secondary adaptation takes place in the planning sub-system, the planning office, the institutions, organizations engaged in planning mutually adapt their aspirations, expectations and decisions.

In Chapter 13 of this book we presented a pair of concepts: "autonomous and higher functions." (See Definition 13.1) Now we are going to introduce another pair of concepts: "primary and secondary adaptation." These pairs of concepts appear to be *symmetrical*. Closer study reveals that they are not identical; we do not have mere synonyms.

Statement 14.1. The autonomous functions of an economic system are usually sufficient to secure primary adaptation.

Statement 14.2. The higher functions of an economic system are always necessary to secure secondary adaptation.

Statement 14.2 is based on well known empirical facts; there is no economic system, sub-system or organization without higher control functions where secondary adaptation takes place.

Rationalization of Statement 14.1 is more difficult. This cannot be accepted on the basis of direct empirical observation, since the autonomous function in a "pure" form is just an abstraction. In reality, it is always intertwined with higher functions to a smaller or greater extent. For example consider the effects of the price system. Even in such an extraordinary circumstance as a war economy under socialism—where the higher control related to prices is very much pushed to the background—other kinds of higher control, such as central, government directive guidance, are enforced even more strongly than is usually the case. However, the truth of Statement 14.1 can be proved rather in an indirect way, with the aid of intellectual experiments and theoretical models.

At any rate, it is important to see that the autonomous functions of economic systems and, within that, of organizations—fundamentally serve primary adaptation, securing survival.

14.3. Adaptive Properties

In this section without pretending to be complete, we shall discuss some properties and criteria for comparison of the adaptive processes taking place in various organizations and systems.

1. *Reaction to slow or sudden changes in environment.* There are two characteristic types of changes in environment: the relatively slow, gradual, "quiet" changes, and the sudden, dramatic, revolutionary, shock-like changes. To the first group belongs, for example, the increase of population; to the latter group belong war and natural calamities. Obviously, there are many intermediate cases (e.g. in technical development the slow, gradual and the sudden changes are intertwined). Yet it still is interesting to ask how each organization and system adapts to "quiet" and how to "dramatic" changes in environment?

It is an advantageous adaptive property of the system based on central directive control that generally it can adapt to shock-like changes in environment with success. The hierarchical structure of the system, the rapid downward movement of central instructions makes quick mobilization possible. Such mobilization was experienced in the second world war when the peaceful production of the Soviet Union had to be quickly transformed according to the necessities of war.[3] This is evident also in Hungarian experiences (e.g. when quick transformation is necessary because of natural calamities).

But even in systems which rely in normal times mainly on horizontal relations, the role of vertical relations (mainly of central directive control on part of the government) will grow in war or at times of natural calamities.

The adaptive properties of a highly centralized system are less favourable with respect of the slow, the "quiet" continuous changes in environment.

2. *Preparation, planning.* A considerable part of the changes that will occur in environment in the future is, at least to a certain extent, foreseeable. The adaptive properties of the organization or the system depend on the extent to which it is capable of foreseeing future changes and preparing for them. This

---

[3] See the book by *Vozn'esenski* [266].

depends heavily on the higher functions; autonomous functions are, by their nature, "blind", they do not foresee the future.

This adaptive property depends primarily on the level and development of the medium and long-term planning activity of the organization or system. The more reliable the plan and the more efficient its implementation, the more quickly and more smoothly is the organization able to adapt to foreseeable changes in environment.

3. *Sensitivity, thresholds of sensation.* According to experiences, not every change in environment is followed by reaction. The organization, i.e. the system does not respond at all to smaller changes; but if the change exceeds certain limits the organization does respond — perhaps in a shock-like manner.

Let us call *threshold of sensation* the limit which the change must exceed in order to provoke a reaction. The *sensitivity* of the organization is measured by the thresholds of sensation.

For example, the firm does not respond to small price changes, but it does to big ones. The firm does not follow small technological changes, but it does follow radical ones, and so on.

Depending on the character of changes and on what characterizes the sensitivity of the firm, the threshold of sensation may be formalized in several ways. A few examples follow.

—The threshold of sensation may be an absolute measure of the change between the starting and the end points of a given period. (e.g. the firm will react if the unit price is $5 higher at the end of the quarter than at its beginning.)

—The threshold of sensation may be the relative size of the change between the starting and the end points of the period. (e.g. the firm will react if the selling price rises by 2 per cent by the end of the period from the price at the beginning.)

—The threshold of sensation may be a limit given for the definite integral of the variable as a function of time, calculated from a given date. (e.g. the firm will react if the cumulated amount of its additional expenses reaches Ft 10 million above a given level calculated from a given date.) (See *Figure 14.1*. As soon as the striped area—that is, the additional outlays calculated from a given starting date—has reached the critical amount of Ft 10 million, the firm will react.)

There are, of course, also other possible formalizations.

Usually a lower and an upper threshold of sensation exist simultaneously. For example, the production of a firm will react to price changes if the price falls at least by 3 per cent or rises at least by 5; movements within this interval will be disregarded.

The smaller the distance between the lower and the upper thresholds of

14*

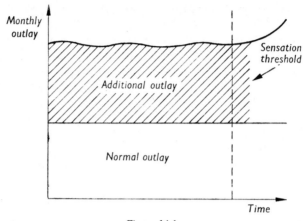

Figure 14.1
Sensation threshold

sensation, the smaller will be the difference between the discrete, "jumping" adaptation and continuous adaptation. One of the most important adaptive properties of the organization or the system is the breadth of intervals between the lower and the upper thresholds of sensations.

There exist processes in a living organism which continuously react to changes in environment (e.g. the functioning of the equilibration organ). However, other processes (as e.g. the pigmentation) are initiated only when definite thresholds of sensation are exceeded. This is the natural self-defence of the living organism against superfluous shifts. The function of the thresholds of sensation is similar in economic systems.

As long as the intensity of the sensation does not exceed the values of the threshold, the organization will repeat its earlier reactions. In this case, the main source of information in the control and decision process is the organization's own memory.

Statement 14.3. One of the stabilizers of economic organizations and systems is the circumstance that certain processes will start, or deviate from their previous volume, only if the change in environment has exceeded certain thresholds of sensation. Too narrow thresholds of sensation lead to hypersensitivity, superfluous fluctuations and shifts; too great threshholds of sensation, however, to a weakening of adaptation and to rigidity.

In this respect it is worthwhile to discuss prices.

An economic system where—as in the GE theory—valid prices react continuously to the market situation and similarly production continuously con forms only to prices would be characterized by hypersensitive adaptation.

However, real, present-day economic systems generally function in another way: price changes only if the change in the market situation has exceeded certain thresholds of sensation; also price changes affect production only when they exceed certain thresholds of sensation.

In addition we find both kinds of unfavourable phenomena. In the capitalist economy, particularly in sectors where atomized markets still predominate, hypersensitivity may prevail. For example superfluous fluctuations in agricultural production follow in the way of fluctuations in agricultural prices. (Consider the frequently mentioned pig-cycle.)

On the other hand, in the socialist economy (particularly at the times of exaggerated directive control) a problem of opposite nature is met not infrequently. The socialist economy did not prove to be hypersensitive but exceedingly insensitive. Only under the effect of sharp and grave signals were production lines changed and adapted to changing needs. In other words, the thresholds of sensation were too great, or the gap between lower and upper thresholds too wide.

4. *Reactivity.* One of the most characteristic of the adaptive properties of organizations and systems is the relation between the change in environment and reaction to that change.

Let us call *reactivity index* those elasticity indices where the numerator gives the percentage change of the reaction (change in state or output) and the denominator that of the change in environment (change in input).

The reactivity index is a generalization of the notion of elasticity indices, well known in traditional economic theory. For example the elasticity of demand with respect to price is a special case of the reactivity index. With the notational system of our book this index of elasticity may be written:

$$\frac{\frac{\bar{x}_i(t)-\bar{x}_i(t-1)}{\bar{x}_i(t)}}{\frac{\bar{u}_i(t)-\bar{u}_i(t-1)}{\bar{u}_i(t)}} \tag{14.1}$$

where $\bar{x}_i$ is the product input of the i-th consumer, (that is, his consumption) and $\bar{u}_i$ his information input (in the present case, therefore, the price of the products consumed).

We can include among the reactivity indices several not treated by traditional economic theory because the information input is not of price character. For example we can use the technical parameters of some product in the course of development as the denominator and the sum of investments preparing the marketing of the product as the numerator. Or, the information input (the denominator) may be the expected bying intention of the buyer and the change in output (the numerator), the modification of the production program.

How "fervent" or "slack" the reactions of an organization or system are to changes in environment is one of its important characteristics. The reactivity indicators testify to the sensitivity of the organization or the system, similarly to the thresholds of sensation. There are "indifferent" organizations and systems which produce hardly any reaction even to strong input changes (that is, their reactivity indices are low) and there are "hysterical" ones which product forceful reactions even to small input changes.

5. *Length of the reaction time.* Adaptation is a chain of four processes:

1st process: change in environment,
2nd process: observation, perception of change,
3rd process: preparation of decision on an adaptive answer,
4th process: implementation of the decision on the adaptive answer; that is, the reaction.

The time elapsing from the change in environment to the completio n of reaction will be called the *reaction time* of adaptation.

Those processes are illustrated in *Figure 14.2.* In Figure 14.2A the four processes strictly follow each other; each one begins only when the preceding one has been completed. In Figure 14.2B each process begins before the preceding one has finished and this shortens the reaction time. This is possible particularly in the case of recurring decisions associated with repeated changes in environment and especially in the case of rhythmically recurring decisions.[4]

In the comparison of organizations and systems, reaction time is a very important criterion. There are "lively" organizations and "lazy" organizations; their reaction times may play a decisive role in determining whether they perish or survive, and if they survive, whether they grow slower or faster. Similarly, the length of reaction times exerts great influence on the development of entire systems. For example, in the Hungarian economy, the fact that we adapt ourselves too slowly to world market requirements has caused and still is causing many difficulties, not to mention the difficulties caused by all too slow responsiveness to domestic consumer demand.

6. *Smoothness, monotonity.* If environment changes, the organization and the system will adapt itself to the change after some reaction period. But how smooth is this adaptation? Let us assume that the result of the adaptation is an increase in production of product "A" and reduction in the output of product "B". During the reaction time, has there been a monotonic growth of product "A" and a monotonic reduction of product "B"? Or has production reached the new level in a fluctuating manner, with alternating growth and reduction?

[4] See Definitions 9.3 and 10.4.

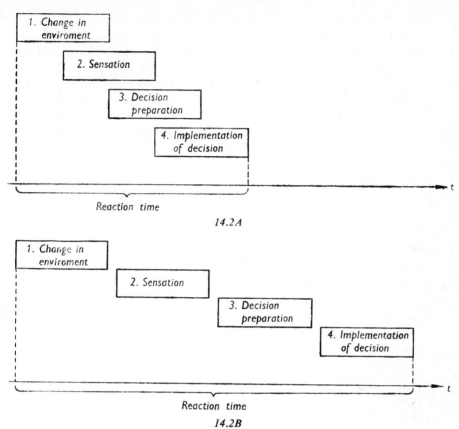

Figure 14.2

Reaction time of adaptation

Planning, the circumspect control of real processes, may have several advantages with respect to this adaptive property. The more adaption is left to trial and error methods the greater the danger of cyclical, fluctuating, "vibrating", non-monotonic, non-smooth adaptation.

7. *The costs of adaptation.* The most common approach to adaptation investigates only the situation before and the one after the adaptation. According to this view, the organization or the system has adapted itself well if adequate changes in state and output have followed the changes in environment. But the problem is not so simple. The shift itself involves costs. It is important that the level of production be in harmony with the demands of the purchasers of the products. But also *changes* in production level involve costs. To raise produc-

tion new labour must be employed and trained, and, in addition, new investment may be necessary. If production must be reduced later, the increased capacities either remain idle or they are scrapped, involving losses. Thus, rigidity may hinder adaptation but it does save the costs.

On the basis of the preceeding, we introduce comprehension.

Definition 14.1*. The major features characteristic of the primary and secondary adaptation of the control processes taking place in an organization or system will be called a d a p t i v e p r o p e r t i e s. These are, primarily: 1) readiness to adapt to slow or rapid changes in environment, 2) readiness to prepare for future changes in environment, 3) the thresholds of sensation of adaptation and the discrete or continuous character of adaptation processes, 4) the reactivity indices of the organization or the system, 5) the length of reaction times, 6) the smoothness, monotonicity of adaptation and 7) the costs of adaptation.

14 4. Selection

The biological analogy leads on from adaptation to selection. If we do not limit our view to a single organization but regard a multitude of organizations, in generations following each other over time, an additional question arises, what kinds of selection take place among organizations.

a) In a functioning economy new organizations are established continuously (e.g. new administrative-information institutions, new productive firms, etc.). Many kinds *could be* established, but only certain ones actually come into existence. What is the social process that regulates this phenomenon? What are the criteria for *selection?* Are the fittest established, or does the selection of new institutions depend on mere chance? What objective or subjective factors are used to decide which of the possible organizations should be formed?

b) The organization already existing may develop variously. It may stagnate or grow, contract, divide, or amalgamate with other units.

Again *selection* takes place. The economic system selects certain organizations for stagnation, others for fading away and again others for growth. One will divide; another will amalgamate. By what criteria does this selection take place? Why is it that unit "A" contracts and unit "B" grows, and why not conversely?

c) In the course of its life the organization develops new features and new properties (similar to biological mutation). This is manifest mainly in technical development—in the development of new products, new procedures, new management and administration methods. But also the changes in the internal structure of the firms, the modification of their decision algorithms and "rules of thumb" may be classified here. Some new features are accepted, "adopted" by the environment and others are discarded. The ones that are adopted will be inherited; those future organizations continuing the life of the present organization will acquire the new properties.

d) There is additionally a final selection: survival or death. Units not only are brought into being but also perish; they are ruined, or their liquidation is ordered in an administrative manner. Why does any particular unit perish? Why not others? By what criteria are they sentenced to death?

Which are the selection criteria characteristic of the system? How does selection take place, according to the above points? These are empirical questions and economics still owes a large part of an answer. We only shall list alternative hypotheses:

The birth, development and death of an organization may be affected by the properties of that organization—by the degree of its viability and capacity to develop. Recall the concepts of aspiration that we discussed in Chapter 12: ambition of the organization, the tensions inherent in aspirations and decisions, intensity of aspirations and decisions, etc. In addition we must consider the adaptive properties: sensitivity, reactivity of the organization, its reaction times, the smoothness of adaptation, etc.

The life of an organization also may be affected by other organizations. For example, in a socialist economy institutions are frequently created or liquidated by administrative action.

Finally, there may be also an unexplained element; the effects of unexplained factors may be considered the working of chance.

Selection (particularly according to points b) and c)) induces the emergence and perhaps the growth of differences (e.g. difference grow between productive firms in respect to capital strength, technical level, etc.). An essential question of economic systems theory asks what is the cause of such differentiation and what are its consequences? And what are its manifestations?[5]

Selective and differentiating processes may involve also *concentration;* the processes of the economic system—both the real activities of production, distribution and utilization, and the control activities of information processing, transferring administration—may become concentrated in the hands of fewer organizations. Concentration does not necessarily take place and not in every field; signs of deconcentration, fragmentation and diversification can be perceived. Another basic question of economic systems theory asks in what fields concentration takes place? Under the effect of what factors? What counter-tendencies emerge? What is the influence of concentration or the lack of it on the control and the real spheres?

[5] Literature is scarce on the problem of differentiation. Some important phenomena and problems are pointed out in the study by *Simon–Bonini* [239].

One of the great merits of *Marx*'s theory is that it gives concentration an important place.[6] This has become a tradition in Marxist thought. *Lenin* and other Marxian economists also treated the phenomenon of concentration in detail. Marxian economists have rightly pointed out that one of the fundamental technical sources of concentration in increasing returns; the advantages of mass production drive us to establish ever greater units.

Marxian political economy was—understandably—interested mainly in the socio-political effects of the concentration phenomena. Thus it has not treated in detail the chain of effects directly producing this process. The latter would be the subject of economic systems theory.

14.5. Comparison

The idea of adaptation is not a foreign one to the GE school. On the contrary, its attention is focused on one adaptive process: the mutual adaptation of production and consumption in given environment, with given primary resources and with a given level of technical development.

Also there are dynamic versions of the GE theory which describe the functioning of the market as an adaptive process. A critique of the market models of the GE school will be undertaken mainly in Part III of this book. Here we only wish to show that *the adaptation described by the GE school is a special case of the general adaptation of economic systems*. This is reviewed in *Table 14.1*.

Table 14.1 shows that *the models of the GE school are narrow and poor if we conceive of them as descriptive-explanatory real science theory. Precisely on this account, they are unacceptable as normative theory.* Let us present a few examples:

—It would not be wise to adapt ulteriorly to all changes in environment; it is worthwhile to prepare for foreseeable changes with planning.

—It would be a mistake to propose a continuous, hypersensitive adaptive process for the conscious formation, transformation and reforms of economic systems. One factor in the stability of systems is the threshold of sensation which cushions superfluous fluctuations; for many processes discrete adaptation is more favourable than continuous adaptation. Only that threshold of sensation that is too big—exaggerated rigidity—must be eliminated.

—Selection, by healthy criteria, is one of the propelling forces of systems. It would be a mistake to insure the eternal life and unchanged dimensions of all existing systems.

[6] See *Lenin* [150] and *Hilferding* [88a].

TABLE 14.1

Characteristics of adaptation

General Case	Special Case Described by GE School
1. Adaptation may take place	Adaptation only in form
A. with given response functions	A. with given response functions
B. by modifying the response function	
2. Reaction occurs to slow or sudden changes in environment	Stationary or slow changes in environment
3. Preparation for foreseeable changes in environment or adaptation without preparation	Adaptation without preparations
4. Discrete or continuous adaptation; in the former case there are thresholds of sensation	There are no thresholds of sensation; continuous adaptation
5. Reaction times of differing length	There is no reaction time: changes in environment and reaction coincide. (In some models and regular tags)
6. Adaptation involves costs	Adaptation does not involve any costs
7. According to various criteria, a selection takes place among the organizations	It does not treat selection: the number of organizations is constant
8. Organizations become differentiated owing to their own properties, administrative interference and random factors	It does not describe or explain the differentation of organizations
9. Concentration	It does not describe or explain the concentration of organizations

We should say a few words separately about the GE school's handling of concentration. True, in recent decades important works have appeared in the scope of the GE school on monopolies, oligopolies, and imperfect competition. These works, however, do not abandon the system of basic assumptions of the equilibrium theory; they only loosen them somewhat.[7] Thus, they cannot provide a satisfactory explanation of the processes regulating monopolization and concentration itself, etc. They only record what happens if imperfect competition, oligopoly or monopoly has already appeared.

14.6. "Quantum-Economics"

At this point we should like to make a small digression in connection with the *continuity* of the variables and connections in the model. Now sufficient material for a short discussion of this question has been assembled.

Statement 14.4. A considerable part of the real and control processes of economic systems is characterized by discrete variables and by breaks or jumps in the functions.

1. *In the scope of real processes*, particularly investment, creation of new, fixed assets is frequently characterized by various *indivisibilities*. In our age, one cannot create a plant of whatever small size. Plant size has a sound lower limit; we cannot organize a smaller one. It is impossible to found an aircraft factory releasing annually 5 planes, an automobile factory with an annual output of 100 cars, and so on.

Also where technology is developing, there occur indivisible, jumplike changes. True, in the case of fabric wool and man-made fibres of different quality can be mixed at will. But, in the case of a part of machinery it must be made of metal or plastic. Or, should the firm shift to conveyor-belt production or not. Or, should the firm produce color television sets, as well as black and white ones.

There also are continuous variables in the life of a household; e.g. the consumption of vegetables and fruits can be combined in any proportions. But even here there may be discontinuous variables. For example, a family living in a rented flat eventually might consider buying a house. They either buy it or not; and if they do decide to buy, they have to commit themselves to shouldering a whole series of related consequences.

Basic assumptions 6 and 8 of the GE theory are the convexity of the production and consumption sets. Indeed, part of the productive and consumptive real processes can be described well by convex sets. But another part cannot.

The phenomenon of increasing returns is also closely related to the phenomenon of indivisibility. Let us use again the examples of the aircraft factory on the automobile factory. Precisely because there are indivisibilities with both the

---

[7] The theory of imperfect and monopolistic competition never really could be integrated into the traditional equilibrium theory. On this problem, see the study by *Bain* [21].

initial investments and day-to-day operation (some inputs are fixed or hardly vary as a function of the volume of production) relative economies are achieved by increasing plant size. This well-known phenomenon—economies of scale, related to mass production, bigger series and bigger plant size—again involves production sets that are non-convex.

2. *In the scope of control processes* we distinguish the standard and the fundamental decisions.[8] The fundamental decisions usually (though not exclusively) are related to the phenomena of indivisibility, treated under 1. In a productive firm, a fundamental decision precedes the establishment of a new plant, the introduction of an entirely new product or technology, the application of a new method in administration, information processing, decision preparation, and so on. Most of the fundamental decisions are of the "yes or no" character.

Also there are non-continuous response functions belonging to the non-continuous variables. "We can implement the first action only if simultaneously we begin the second one . . ." " . . . The third action is impossible since the fourth is already in process and these two exclude each other . . ."

3. *Again in the context of control processes*, we have pointed out the importance of *thresholds of sensation* in adaptation. Here too, there are non-continuous relations between changes in surroundings and reaction, there is no response to small sensation, and there is a jump-like reaction to a sensation larger than the threshold value.

It is a tradition in mathematical economies to approach the description of these discontinuous phenomena with the aid of continuous variables and continuous differentiable functions.

There seems to exist a strong analogy with the development of physics.[9] Classical physics worked with continuous variables and differentiable functions; with their aid physicists were able to describe several important features of physical reality. But later, despite the refinement of this mathematical apparatus, it simply became impossible to describe the world of elementary particles, which are characterized precisely by indivisibility, with non-continuous quanta. A new road had to be opened, and the mathematical apparatus of quantum-physics was created.

The mathematical apparatus of classical mechanics is applicable to macro-physics, micro-physics must, however, be approached from the quantum aspect. The situation is similar in economics: the great, aggregate processes of macro-economics can be well described with continuous variables, while many phenomena of microeconomics have a "quantum" pattern. In spite of this, the application of continuous variables and of continuous, differentiable functions is still predominant in economic analysis.

[8] See Definition 9.2.
[9] *Tamás Lipták* called my attention to the analogy.

This problem has been "in the air" for some time. In answer to this problem, methods of *discrete programming* have been developed. In operations research, however, these methods were not useful because of their cumbersome computational requirements. Yet these models are much better suited to reflect economic reality than are the continuous models.

Few results can be expected from the discrete programming procedures from the point of view of research into economic systems theory. These procedures break with the usual assumptions of the GE school only at a single point—though at a very essential one. They depart from tradition in the question of continuity (and with it, in that of returns to scale). But invariably optimization is still involved, and the computational difficulties are due to this fact. They are not satisfied with *describing* the regularities of processes and phenomena. (Quantum-physics is more modest, it is satisfied with as much.) The GE economist wishes to optimize as well—and he fails.

In order that we be able to work out an economic systems theory, the optimization approach should be discarded both on the grounds presented here and on those emphasized elsewhere in this book. Should we do so, a way will be opened to describe and explain economic processes which are characterized, among other things, by indivisibilities, fundamental decisions, and adaptation with discrete thresholds of sensation.

15. CLASSIFICATION AND AGGREGATION

At the beginning of Part II, we used the simile of a flight to outline the further course of the book. First, we were to view our subject from a great altitude; then we were to descend to a height where even house-blocks and houses would be discernible; eventually we would re-ascend for a second look at the grand, comprehensive view.

Our flight reached its lowest level in Chapters 8 to 12 where we discussed the decision and control processes taking place in the organizations and within them, in the control units.

Then, we started to ascend. In Chapters 13 and 14 we dealt with the *joint* operations of organizations, with autonomous and higher-order operations, with adaptation and selection. Now, we shall present an even more comprehensive picture. In this chapter, we shall discuss how to classify and to aggregate the multitude of organizations, institutions and processes.

15.1. Individual Description and Aggregation

The operation of the elements of the system—the institutions, organizations and units—can be observed individually through case studies. For example, one such study might investigate the operations of the investment department of a firm, describing its characteristic decision algorithms.

However, from the scientific point of view, by itself a single study allows but few conclusions to be drawn. Theoretical conclusions can be reached only through generalization, on the basis of the simultaneous observation and joint analysis of a multitude of institutions, organizations, units and processes identical in character. Accordingly, when we build a theory, we make general statements referring not to individuals but to their ensembles, groups, categories. In other words, we analyse *aggregates*.

To emphasize the necessity of aggregation is commonplace. Nevertheless, we are prompted to a few remarks by our feeling that the concepts of aggregation accepted by our discipline involve general methodological errors. Essentially the error is in *uniformization*. To illustrate the problem, let us present an example.

15.2. Example: Investment Functions

For example, consider the description of investment decisions of large firms operating in capitalist countries. The problem may be outlined as follows, using the conceptual framework of the book.

In each firm there is an investment organization and within it there is a control unit which regulates the real investment activities. The operations of the C-unit in question is characterized by a response function. The output of the response function is the investment decision, the instructions controlling the real investment processes. What are the inputs, the ingoing information, to this response function?

Let us write down the general form ϱ_i of the investment function of the i-th firm:[1]

$$\vec{u}_i(t) = \varrho_i(\vec{u}_i^{(1)}(t),\ \vec{u}_i^{(1)}(t-1),\ \ldots,\ \vec{u}_i^{(2)}(t),\ \vec{u}_i^{(2)}(t-1),\ \ldots). \qquad (15.1)$$

where vector $\vec{u}_i(t)$ is an information output: the instructions controlling the real investment processes. The vectors $\vec{u}_i^{(k)}(t)$ are information inputs: the k-th information group influencing the investment processes in the t-th period.

The form of function (15.1) has been a subject of extensive controversies in economic literature concerning $\vec{u}_i(t)$, the output of the response function; in other words, what are the most suitable indicators describing the investment processes. This is a problem of statistical-technical character and does not belong to the subject-matter of this Chapter.

Two questions remain open; a) what information inputs should be included as arguments of the response function, and b) what should be the mathematical form of the relationship (including the formalization of time lags).

Problem b) may be disregarded here. To simplify the statistical estimation of the parameters most authors employ simple linear forms of the relation with the usual "distributed lags".

The crucial problem is determining the information input actually used in the investment decision; what is the economic content of $\vec{u}_i^{(1)}, \vec{u}_i^{(2)}, \ldots$?

From the enormous literature on the subject we discuss only one work: the article published by *Jorgenson* and *Siebert* in 1969.[2] This recent work is of a

---

[1] When formulating function (15.1), the output of period t was made dependent directly on the inputs not only of the period t, but also of the period $t-1$, the period $t-2$, etc., and the memory content was not made to figure among the arguments of the function. The effects of information arriving in periods before t could be reformulated according to (4.2); this means that these effects would influence the output of period t as components of the memory content of the period $t-1$.

[2] See the already mentioned article [104] by *Jorgenson–Siebert*.

high standard and is based on thorough-going investigations; yet it provides a typical example of the methodological problem which we desire to discuss.

The authors describe four investment theorems as follows:

A) The "accelerator theory", according to which desired capital is proportional to the volume of production.

B) The "liquidity theory", according to which desired capital is proportional to the firm's liquid funds available for investment purposes.

C) The theory of "expected profit", according to which desired capital is proportional to the "market value" of the firm (i.e. to the discounted present value of profits to be expected in the future).

D) The "neo-classical theory", according to which desired capital is proportional to a fraction, the numerator of which contains the value of production and the denominator, the price of the "capital service". Practically, this means that investment depends on the price index of investment goods, the rate of replacement, the rate of interest, the rate of profit tax, and the rates regulating tax exemption for depreciation.

In the case A), B) and C) the form of function ϱ_i is linear, whereas in case D) the function is non-linear.

Based on mathematical statistical methods the authors examine a sample of 15 large U.S. corporations which they consider representative. On the basis of this data and the usual econometric, mathematical statistical criteria, they desire to determine which of the hypotheses and theories listed above can be considered more, and which less, correct.

The authors regard the four theories, the four hypotheses, as *alternatives mutually excluding each other*. Their intention is to establish unequivocally which of the four is the most acceptable. Having done so, they would then regard the accepted one as valid for all 15 firms, or, rather, for the whole set represented by the sample, i.e. for all U.S. (or, perhaps, for all capitalist) firms.

The authors main conclusions are as follows:

1. The most "correct" explanation is furnished by the neo-classical theory; the corresponding investment function fits the empirical data best.

2. The "second-best" theories are the "expected profit" and the "accelerator" theories. In each case, the empirical fit is approximately identical. Both are less close than in the case of the "neo-classical theory" but considerably closer than in that of the "liquidity theory".

The conclusion outlined in 1 above, is considered a substantiation of the neo-classical theory of the firm.

We are unable to rule on the validity of the conclusions of these authors; it is impossible to form a judgement from Budapest as to the true behavioural characteristics of large American corporations. However, we would like to state our misgivings concerning their methodology.

First of all, why do the alternative response functions mutually exclude each other? It is only in the case of comparatively simple information structures of low complexity that the existence of mutually exclusive alternatives can be assumed. (Recall Chapter 5, where the complexity of information structures is described.)

However, according to all indications the American economy is character-ized by a highly complex information structure. The large US corporation will not decide on major investment projects solely on the basis of a single informa-tion type or even a small group of information types received over a single channel. Multiplicity of information is much more likely. In other words, the response function ϱ_i will include in its arguments the volume of production in physical units (wherever this is measurable at all) as well as the value of pro-duction based on prices, the available fund of money, the expectations concern-ing future profits, the price index of investment goods, and the rate of replace-ment, the rate of interest (and probably also other factors).

We have emphasized the fact that in the modern economy, information of price and non-price character occur side by side, complementing each other. Decisions are influenced by ulterior information (reaction to past production volume) and anterior information (reaction to profit expectations).

It is likely that if all information inputs influencing investment were taken into account, the investment function obtained would contain too many para-meters. The econometrician working with a small sample cannot cope with this problem; he would be unable to separate from the joint effects of all factors the effects due separately to the individual information types. However we must not confound the problem of the true nature of the investment function ϱ_i, with the difficulties of mathematical statistical estimation. We may use for practical applications simple single-variable functions (or functions with few variables), if their predictions are acceptable approximations of reality. How-ever, we must not draw too far reaching *theoretical* conclusions from these simple models. For example, we must not state that the firms actually rely on one or another signal, exclusively.

Econometrics has as yet not produced a methodology of the correct observation and parameter estimation of complex information structures and of multiple information flows. Still, the complexity of information structures remains a fact; incomplete econometric description must not be taken as a proof against the existence of complexity.

This becomes clear—if only in indirect form—from the paper of Jorgenson and Siebert. Hypothesis D) is not much less correct than hypotheses A) and C); there is but a shade of difference. In the case of several important firms—with such giants as General Electric, Reynolds, Dupont, Anaconda, US Steel, IBM among them the fit of hypothesis A) B) or C) appears equally good and in some cases even better than that of the neo-classical hypothesis D).

This suggests the following idea:

It is not wise to give one *uniform* aggregate function ϱ for the American economy as a whole. A few function types—say, $\varrho^{(I)}$, $\varrho^{(II)}$, $\varrho^{(III)}$—would appear to be expedient. In the first type, investment would depend more on production and less on expected profits or on the rate of interest; in the second one, more on the rate of interest; and less on production, and so forth. The large US firms could then by *classified* according to their behaviour in investment decisions. There may be firms of the "accelerator type" which, although forming their investment decisions on the basis of several factors, are *mainly* guided by earlier production; firms of the "expectable-profit type" where profit expectations constitute the main information input, and so forth. The task of classification in the present case does not fall within the province of a Hungarian economist; however, the article quoted clearly indicates its possibilities and advantages.

One additional remark is in order before we proceed to comments of a more general character. In our opinion, and contrary to the statements put forward by Jorgenson and Siebert, the paper does not substantiate the neo-classical theory of the firm (as represented by the GEs chool). What it does prove is nothing but —to use the terminology of this book—the *existence of investment response functions. There is a stochastic relationship between the investment decision as an information output on the one hand and the various parallel information inputs on the other.* The latter include information both of price and non-price character; posterior observations as well as anterior expectations. The econometric description does not prove (nor disapprove) that the firm carries out some kind of optimization or maximization; it simply indicates that the firm is reacting to definite information. The facts presented in the Jorgenson-Siebert paper are wholly compatible with the behavioural model outlined in Chapters 4 to 12 of this book.

15.3. Typology, Classification

Let us now turn to more general conclusions.

There is hardly a branch of real science which would not devote a considerable part of its intellectual efforts to the classification of the phenomena observed. In several branches of science — suffice it to mention only botany and zoology — it was exactly this that gave the first impetus to progress.

In the beginning, classification was rather primitive. Gradually the classifica-

15*

tion became more sophisticated. The scientists of antiquity believed in the existence of only four elements: fire, water, earth and sky. The periodic system of elements emerged many centuries later and new elements are still being discovered. Microphysics first believed the atom was indivisible. Later on, two or three elementary particles were distinguished within the atom. At present, nuclear physicists know already 70 different elementary particles, but the world inside the atom has yet to be fully explored.

Classification is evident in economics in descriptions of the real sphere. Tables of real inputs and outputs are broken down into sectors, in the knowledge of the fact that the real response functions are not identical in metallurgy and in the textile industry. *However, as soon as the mathematical economist, trained in the spirit of the GE school, passes into the control sphere, he is apt to forget the reasonable practice of classification and aggregation by classes and types, employed by all sciences.*[3]

For the economist of the GE school, 1 seems to be the magic number. In the model of the firm, he always assumes the existence of but one type of firm. He assumes that all firms act on the impulse of a single motive. All firms set prices on the basis of the same single rule. The decision algorithm is identical and uniform for every organization. When formalizing the information type influencing the organization, only a single type of information is considered.

The economist of the general equilibrium school might refute this criticism with the following argument: "Abstraction is the right of science. To work out an individual model for each individual firm and for each individual decision process would be senseless; what would be the use of a thousand model types."

However, there are not only the two extreme possibilities of a single type or thousands of types. To take into account 3, 4 or 5 types of firms instead of a single one would not be contrary to the principle of abstraction. The behaviour of the set of firms can be described also by indicating their distribution by types. Instead of taking into account a single motive only, we might consider five or eight main motives and characterize the complex motivation of the organizations by the relative weight of the component motives. (See Chapters 8 and 12 dealing with motivation.) Instead of a single decision algorithm or a single behavioral rule, there may be five or twenty types and classes; the set of decision processes could also be described by distribution patterns.

It is unnecessary to give any further examples. The general methodological lesson is the following.

We ought to guard against the aggregation effacing essential differences, against the uniform description of the behaviour of organizations, against the

---

[3] *Orcutt* has called attention to this problem, especially from the viewpoint of generating micro-analytical simulation models. See [197] and [199].

oversimplification of structures. *Rather, let us carry out the classification of the various behavioural regularities, response functions, decision algorithms. Let us characterize with separate response functions* $\varphi^{(1)}$, $\varphi^{(2)}$ ..., *with separate algorithms* $F^{(1)}$, $F^{(2)}$, ..., *the groups of C-units, organizations differing essentially in behaviour from one another.* Let us describe the system after classification and aggregation of the groups characterized by different behavioural regularities, by *indicating the distribution of the groups over types of behaviour.*

16. THE COMPREHENSIVE CHARACTERISTICS OF THE OPERATION OF ECONOMIC SYSTEMS

16.1. The Desiderata

We have finished the task of compiling the basic "vocabulary" which we require to describe economic systems within our conceptual framework. There remains one further subject: how to evaluate the operation of economic systems.

Definition 16.1. Let us call those comprehensive operational characteristics which essentially affect the lives of those living in a system the d e s i r a t a[1] of an economic system.

Listing the desiderata which might arise in connection with any system is nearly as great a task as is editing an encyclopedia. Therefore, let us only outline the main groups of desiderata.

1st group of desiderata: the real growth of the economic system. Growth of economic systems has an extensive character. Production, national income, consumption and the volume of fixed capital should increase at the quickest possible rate.

The basic nature of this group of desiderata is obvious. However, the problem has been given so much prominence in economic science that instead of emphasizing the importance of these desiderata, rather we should warn against overestimation of their importance. A system must not be judged exclusively on the basis of the growth rate of per capita national income or some similar index of real growth; the other groups of desiderata, discussed below, also should be taken into account.

2nd group of desiderata: technical progress. The system should be creative; it should be a pioneer in introducing the greatest possible number of significant inventions employing them before they are used in other countries. (By inventions, we mean both novel products and novel procedures and technologies.) It should accept at the earliest possibility the inventions of other systems. New inventions (both those originating from the system itself and those taken from other systems) should be brought into general use speedily. Further development of new inventions should proceed with the greatest possible speed.

---

[1] The term "desideratum" was taken from a study by *Koopmans–Montias* [125]. Many of their ideas have been used in Chapter 16, although our analysis differs from theirs in the grouping of the desiderata and in some other concepts.

This desideratum is of great importance. It will be one of the main subjects treated in Part III of this book; therefore, we shall not go into the details here.

3rd group of desiderata: the adaptive properties of the system. These have been surveyed in Section 14.3; here we summarize the previous results:

—The system should be able to adapt both to slow and sudden changes in the environment.

—It should be able to prepare for future environmental changes.

—It should be neither too sensitive nor too insensitive, i.e. its thresholds of sensation should be of suitable dimension.

—Its reactions should be neither "too vehement" nor "half-hearted".

—Adaptation should be quick.

—Adaptation should be smooth and free of fluctuations.

—Adaptation should be cheap.

The adaptive processes of the system include the adjustment to each other of production and consumption and the coordination of the control processes of production, trade and consumption. These aspects will also be dealt with in Part III.

4th group of desiderata: the selective properties of the system. Although closely related to the adaptive properties, these desiderata deserve special attention.

One of the fundamental questions concerns the criteria for selection of the *organizations* coming into existence, going into operation, terminating their operation and ceasing to exist.

The fundamental question concerns the ways and means of selecting the *persons* occupying the leading positions in the organizations.

Here, the desideratum is that the selection should encourage the best personal abilities and of the most suitable common properties of the collectives. These properties are initiative, ingenuity, the ability to act quickly, organizing ability, firmness, discipline, and the like. In the case of erroneous principles of selection, inefficient men without executive ability, but servile and flattering towards the superordinate organizations or persons, will advance and unviable organizations with low efficiency will subsist, whereas those better qualified will be driven into the background and succumb.

5th group of desiderata: income distribution and employment. This group includes a great number of desiderata which can be justified on the basis of various political, social, cultural and moral viewpoints. Let me point out only a few:

a) Income distribution should provide incentives prompting better performance. However, incentive income distribution usually produces income inequality.

b) Income distribution, to a certain degree, should be egalitarian in character; inequalities should not be too great. This contradicts desideratum a).

c) There should be no unearned incomes.

d) Income distribution should assume definite social and moral requirements. It should compensate those burdened with bringing up children; it should help the disabled and the old. This is to a certain degree contradictory to desideratum c).

e) Income distribution should stimulate thriftiness, the saving of part of personal incomes. This, too, contradicts to a certain degree desideratum c). As a matter of fact the inheritability of personal property may also stimulate thriftiness, but this in turn results in unearned income for the heirs.

These are the desiderata of income distribution; now let us list some desiderata relating to employment:

f) Everybody desiring work should be given the opportunity to work.

g) Everybody should be able to perform that type and amount of work that he or she chooses.

h) Everybody should perform the type and amount of work that the interest of society requires him or her to perform. This may be in contradiction with desideratum g).

i) The time spent working should diminish over time, leisure should increase.

6th group of desiderata. Cultural and social development. This group includes several concrete desiderata: schooling should be extended and educational standards raised; development of science and art should be fostered; public health services should be improved, and so forth.

All this will affect the work performance of those actively participating in the economic system. Nevertheless, it cannot be regarded as part of the 1st and 2nd group of desiderata. Man is not simply a "factor of production" or a resource. Culture and health are in themselves valuable and not merely a means to develop the economic real processes.

7th group of desiderata: decision, property, power. A great variety of desiderata may come under this heading, many of which may be contradictory.

One group of the desiderata involves the centralization or decentralization of the spheres of decision. The desideratum is partly subordinated to the requirements of adaptation; to what extent does centralization or decentralization facilitate or impede the adaptability of the system. However, to many persons, a high degree of centralization or its opposite, extensive decentralization—an economy controlled vertically or horizontally—may be desirable *in itself.*

Another group of desiderata is connected with the social and the class character of the spheres of decision. What are the individual's rights to make decisions and (in accordance with he specific subject of this book) to control the economic processes? Should the owners of the means of production be entitled to carry out this function? Or should those rights be given to those directly and actively participating in the processes concerned—i.e. should we call for social "self-administration"? Or should society as a whole be given these rights? And, if so, how should the will of society as a whole be determined?

The development of power and ownership relations and of the spheres of decision will affect the achievement of all desiderata listed in the foregoing. Nevertheless, it cannot be assumed that the latter simply should be subordinated to the former desiderata, or even to the first two desiderata concerning the real development of the economy. The desiderata which can be included in the 7th group are not only means but also aims in themselves; each of them represents a conflict between requirements and possibilities, a social endeavour and political struggle. The different social strata will insist on a more or less complete fulfilment of this or that desideratum. For example, many people would be prepared to relinquish some material advantages (i.e. some of the claims connected with the 1st, 2nd and 3rd group of desiderata) if the system would adhere to the points of view of the 4th, 5th, 6th and 7th group of desiderata (i.e. the political, social, cultural and moral points of view).

16.2. The Performance of the System

Let us suppose that we wanted to compare the performance of several systems —say, those of Hungary, Austria, Yugoslavia and Romania.

Let these systems be denoted $E_1, E_2, \ldots, E_S$.

The basis of the comparison is the achievement of the desiderata outlined in Section 16.1. Let us suppose that we wanted to take into account a total of N desiderata.

The first difficult problem in making the comparison is that of mesurement. With some of the desiderata, there is an index number that is obviously appropriate. This is the case e.g. with the 1st group of desiderata, where achievement can be measured by the growth rates of national income, consumption, etc.

In other cases it may be more difficult to find indicators which are both observable and genuinely characterize the achievement of the desiderata; sometimes, it may be necessary to resort to arbitrary solutions. In practice, simple indicator numbers would have to be used to characterize, for example, the public health situation (number of doctors per thousand of population, number of hospital beds, number of those vaccinated, etc.) or of the cultural situation of the system. However, it is clear that it should be possible to measure (however imperfectly) a number of the phenomena which are at present not yet comparable. It should be possible to work out units of measurement, let us say, for the characteristic properties of the processes of selection, for the adaptability of the system, for the degree of centralization or decentralization, and so forth. This sets a difficult task for statisticians, econometricians, sociologists and psychologists.

However, most probably there will remain desiderata where no scale expressible in numerical terms can be designed to measure achievement. In such cases, the following compromise is possible.

Characteristic degrees (typical stages) in the achievement of a desideratum can be worked out. Let the desideratum be, e.g. the participation of workers and employees in company management. Some typical degrees are: 1. Workers and employees have no influence whatever on the firm's economic activities. 2. They have some influence but this is exercised indirectly only, through the intermediary of their organs (e.g. party organs, trade unions). 3. They have no right to appoint the managing staff but they can veto their appointment; otherwise they cannot interfere in management affairs. 4. They have the right to appoint the managing staff, but then the latter have the right to make decisions until their mandate expires; the collective does not participate in management. 5. The major questions must be submitted to the collective for decision (as in the cooperatives where the members meet to decide on economic questions). In such cases, the degree of the achievement of the desideratum will be expressed by ordinal numbers.

Achievement of a desideratum should be observed over some longer term T consisting of several periods.

Definition 16.2*. To every system and every desideratum there can be ordered a real number $d_{ij}(T)$ which measures the degree $(i = 1, \ldots S: j = 1, \ldots, N)$ of achievement of the desideratum for the term T. Vector $d_i(T)$ composed of N components $d_{ij}(T)$ will be called the p e r f o r m a n c e of system E_i.

Vector $d_i(T)$ summarily characterizes how some system performs, all the desiderata being taken into account.

In analysis of the performance vector it must be remembered that the satisfaction of some wishes are not independent of each other. Among some of them we find a positive correlation; for example, the material welfare and the cultural level of a country usually rise together. There are, however, also "competing" wishes; e.g., increasing the volume of production may push qualitative improvement into the background. This will be dealt with later in greater detail.

16.3. Comparison of Systems

If carried out with suitable objectivity, evaluation of the performance vector $d_i(T)$ *constitutes a real-science problem free of value judgement.* The researcher must work out the list of desiderata as well as the index numbers and rankings suited to measure the fulfilment of the desiderata in an unprejudiced and unbiased way, and as completely as possible.

There is little work comparing different economic systems by means of suitably defined performance vectors including all or part of the desiderata listed above and based on careful observation.[2] The most popular procedure consists of arbitrarily selecting some groups of desiderata and leaving the others entirely out of account.

Let us, however, suppose for a moment that we have the vectors $d_1(T)$, $d_2(T)$, ..., $d_S(T)$ describing objectively the performance of systems E_1, E_2, ..., E_S. Further steps to work out a *comparative evaluation* of the systems cannot be free of *preconceived value judgements*.

Evaluation of whether system E_1 is better than E_2 depends on the relative importance attributed to the various desiderata as seen by the person performing the comparison.[3]

We do not suggest that researchers can analyze the economic systems without inner political and moral conviction, giving the impression of some false objectivity. However, we want to call attention to the fact that we should *separate the objective* basis of the comparison (the definition of N desiderata, and the measurement of the fulfilment of the desiderata, or in other words, the establishment of performance vectors $d_i(T)$) from the *value judgements* on which the final comparison is based, (i.e. from the evaluation of the relative weights, of the relative importance of the different desiderata).

---

[2] Two valuable exceptions are *Denison* [51] and *Adelman–Morris* [2].

[3] Formally, this could be expressed as follows: System E_i is "better" than system E_2, if

$$\sum_{j=1}^{N} \pi_j d_{1j}(T) > \sum_{j=1}^{N} \pi_j d_{2j}(T) \tag{16.1}$$

where π_j expresses the *weight, the relative importance* of requirement j according to the conviction, political standpoint of the research worker. For the sake of simplicity in this illustrative example we use a linear function in evaluating the two systems.

16.4. The Questions Posed by the GE Theory

Now let us return to the phenomena with which the GE school is concerned. As expounded in Section 3.4, these phenomena are the following:

—The equilibrium between production and consumption, and the stability of equilibrium,

—The optimization of the position of producers and consumers as determined according to their respective preference preordering.

Both phenomena belong to the sphere of the 3rd group of desiderata, i.e. to that judging the adaptive properties of the system. *But they do not even exhaust the 3rd group.* This fact must be evident already from the discussion of the information structures and of the adaptive processes above. Further critical comments will be contained in Part III of this book.

What we want to point out here is the fact that the GE school have not raised any significant questions pertaining to the other groups of desiderata.

Let us now consider for a moment the theories of the GE school as a *normative* theory. Let us forget the fact that it actually offers advice which cannot be followed, and let us assume that there is a country where all hypotheses of the GE school are realized, meaning that the economy is—in the sense of the theory—in a state of equilibrium and Pareto optimum. The system may nevertheless be functioning badly; its real processes may fail to expand or even stagnate technically; adaptation may be too expensive, oversensitive and highly fluctuating; the selection of organizations and individuals may be imperfect; income distribution may be inequitable, or it may be reduce incentives; there may be less than full employment; political and power relations may be anti-democratic and ownership relations exploitative. Also, the reverse may be true. The system may be developing adequately from several points of view; extensive growth may be rapid, technical progress satisfactory, adaptation quick, elastic and cheap, the political, social and cultural desiderata of groups 3 to 7 favourably fulfilled—while it is not strictly meeting the requirements of equilibrium and optimality.

In summary: the questions posed by the GE school are narrow in scope and one-sided.

When we have put forward this criticism in discussions with mathematical economists, over and over again they have raised the following argument: "If it is *not* the questions of equilibrium and optimality that are stressed, then no question can be formulated in a manner sufficiently precise that theory can answer..." It is my belief that there are a great number of questions which lend themselves to exact formulation and that are yet awaiting an answer.

In connection with system E as a whole, the following fundamental question may be raised:

The performance of system E over an historical period $T = [t_0, t_1]$ may be measured by performance $d(T)$. This depends on the characteristics of the system: on the organizations of which it is composed (O), the products it turns out $(\mathcal{G})$, the information types it employs $(\mathcal{S})$, and its response function systems (Φ, Ψ). Furthermore, it will depend on the initial state of the system: the stock of products $(y(t_0))$ and the memory contents $(v(t_0))$.

$$E[O, \mathcal{G}, \mathcal{S}, \Phi, \Psi, y(t_0), v(t_0)] \rightarrow d(T) \qquad (16.2)$$

The fundamental question of the economic systems theory is to determine the above relationship for the various systems $E_1, \ldots, E_S$.

To provide a complete answer to the fundamental question is an enormous task which will require the combined work of several generations of economists. Meanwhile, partial answers may be attempted. Individual components may be selected from both the dependent and the independent variable of the relationship described in (16.2), and the relationships between these components may be described.

In Part III of this book, we have endeavoured to indicate that every component of relationship (16.2) can be described formally. The relationship (16.2) can, therefore, be analyzed by means of deductive models and intellectual experiments. In addition, existing economic systems $E_1, E_2, \ldots$ can be investigated empirically; both their characteristics and their performance can be systematically observed and described.

In Part III, which now follows, we deal with one "segment" of this wide sphere of questions—with the market and the processes of buying and selling.

PART III
PRESSURE AND SUCTION IN THE MARKET

17. THE MARKET

17.1. Delineation of the Subject

In Part II of this book we introduced a "language", a conceptual appara tus Actually, it must be left to the reader's judgment and intuition whether this proposed language makes it possible for him to describe more easily, more completely and more accurately the functioning of economic systems as he knows them than is possible with the traditional conceptual framework. And beyond the reader's immediate immagination further detailed research and a multitude of monographs and case studies analysing a number of economic systems will be required before the contents of Part II really become convincing.

Here, the proposed conceptual framework, or at least some of its elements, will be tested in the investigation of a single problem—the description and expla-nation of the *market* (the relationship between sellers and buyers). We limit ourselves to a single example not because the importance of every other problem is dwarfed by this one, but because it would be an impossible task to treat the market and other problems of equal rank in the same book. From among several groups of problems we had to choose a *single* one. We have chosen the market as the theme of Part III because of its special importance in Hungary. The title of this book, "Anti-Equilibrium", indicates that we are attacking general equilibrium theory. That theory concentrates its attention on the market (probably exaggeratedly so). Therefore, it is only proper that we "play the match" on GE theory's "own grounds". The schools of thought wishing to replace GE theory must not shy away from the territory where the latter is really at home.

"The market" is one of the terms most frequently employed by economists. However, on close scrutiny it turns out that everyone interprets the market in a different way and that there are many rather vague associations connected with it.

According to a frequently employed interpretation, the market is a black box. Its inputs are demands and supplies, and prices; its outputs are the agree-ments between sellers and buyers, and the fulfilment of these agreements.

However, this interpretation does not explain what is taking place in the market, inside the black box.

An anonymous process of price formation and sale-purchase organization is taking place there; only the final result is expressed in the price and in the volumes of purchase and sale. The "black box" interpretation does not deal with questions asking *how* the process works, what types of selection take place to eliminate the "superfluous" sellers or buyers, and so forth.

The terminology used in present day Hungarian practice is not any clearer than that used in the sphere of theory. Terms such as "market forces", "market effects", "value judgement of the market" are often heard, without there being any clear understanding of what these terms exactly mean to those using them.

17.2. The Elementary Contracting Process

Let us try to penetrate the market's black box. In the following, we shall speak of the "seller" and the "buyer". Each party constitutes an *organization*. Thus, when a firm (i.e. an institution composed of several organizations) wants to sell—one of its functional organizations, the sales department, acts as the seller. The case is similar when the firm is a buyer; then its purchasing department acts on its behalf. In other markets the households may appear as buyers.

In the present and in the next chapter, we shall choose our examples from the sphere of inter-firm purchase-and-sale relations. Suitably employed and modified, all results also hold for the relationship between the selling firm and purchasing household. However, it is simpler to demonstrate and to discuss the problem if we restrict attention to inter-firm relationships.

Definition 17.1. The agreement between a buyer and a seller concerning a transaction of purchase and sale will be called the c o n t r a c t.

The question whether the contract is actually carried into effect will be disregarded in the following discussion.

The concluding of the contract is preceded by an *elementary decision process*.[1] During the decision process, the organization releases and receives information: offers, bargaining, modifications of offers, etc. Information is always connected with sub-decisions; to whom should offers be submitted; what should the offers contain; what should be accepted from the other party's offer, and so on.

Let us consider one of the parties—either the seller or the buyer—from the moment it drafts the first offer to be sent out or receives the first information to the moment in which it concludes the contract.

Definition 17.2. For each party to the transaction, the special elementary decision process which leads to the concluding of the contract will be called the e l e m e n t a r y c o n t r a c t i n g p r o c e s s. The contract constitutes, accordingly, the common final point of two elementary contracting processes, the unanimous decision of two organizations.

[1] See Definition 8.2.

The two elementary contracting processes leading up to the mutual contract need not begin at the same time. For example, the seller may begin drafting the first offer; the buyer may begin after receiving this first offer; but he might as well submit an offer in the second wave of offering only, and so forth.

Let us take an example.[2] Let us observe the behaviour of the seller; that of the buyer is symmetrical. In *Figure 17.1* the perpendicular axis represents, downward from above, the passing of time. The arrows starting from the axis represent the information released by the seller; the arrows directed towards the axis represent the information received by the seller. The decision events taking place on the part of the seller are represented by rectangles; the ellipses represent the addresses of the information released and the addressers of the information received by the seller.

On the basis of Figure 17.1 it is easy to understand the elementary contracting process. It begins with the firm's decision—on basis of its memory, i.e. earlier experience— of what to offer and to whom to offer. Then the offers are sent out. Next the answers are received. Now the original list of addresses is revised; a new offer probably is not sent out to all former addresses, and new addresses are added to the list. At the same time, the contents of the new offer is reviewed and probably modified. Then the new offer is sent out, bringing new answers, new revisions, and so on. Thereafter, the final decision is made concerning the answer to be accepted and the contract to be concluded.

Let us now leave the concrete example and list the characteristics which are necessary to describe an elementary contracting process.

1. What is the starting event? In our example, the first offer is sent out by the seller. But the reverse case is also possible: the buyer's offers may arrive first. Or, both may take place simultaneously, in parallel.

2. If the process begins with an offer, what are the criteria for selecting the first addresses. Already, selection is taking place; who should be contacted and who not.

3. What are the contents of the first offer. Usually there is not just a single volume along with the price belonging to it, but several other characteristics of the product offered for sale such as terms of delivery, credit terms, etc. Possibly, alternatives are presented. For example, earlier delivery might depend on the payment of a surcharge; a discount might accompany ordering well in advance, and so forth. These alternatives implicitly define functional relationships between the various indicators (the discrete points of the functions). For example, the time of delivery might be a function of price.

[2] In examining the problem, we have derived great help from the simulation model of *Balderston–Hoggatt* [23] which described the operation of an American lumber market. However, we have tried to present a more general formulation than the specific model described there.

16*

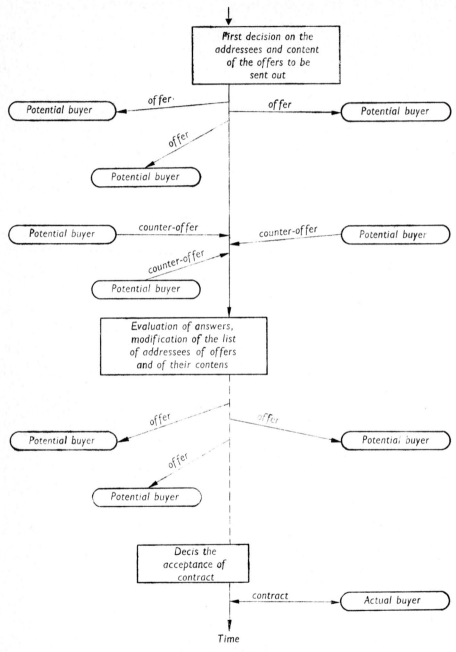

Figure 17.1

Elemantary constracting process

The offers of potential buyers are determined in the buyers' elementary contracting processes. Here, we will assume that these have been received by the seller.

4. What principles will **guide** the seller in sorting out the offers he receives? To what extent will selection depend on the concrete contents of the offers received; on earlier connections with the potential buyer; on outward circumstances, etc.?

5. What principles will guide the seller in modifying his original offers?

6. What are the criteria for agreement on a contract? Does running out of time for information gathering make selling inevitable? Or is there unconditional acceptance of the most advantageous offer? Or must some norm be fulfilled: (e.g. that only the offer ensuring a definite profit would be accepted)? Will the offer of a well-proven business partner be accepted in any case?

7. What are the characteristic time spans between the beginning and the end of the contracting process, between concluding the contract and effecting actual delivery? For example, does the firm produce exclusively to fill previously contracted orders; or does it produce partly on its own initiative and partly to fill previously contracted orders; or does it sell exclusively from stock, etc.?

The characteristics listed under 1–7 can be defined independently. Usually, however, they are interrelated. Together, characteristics 1–7 describe the seller's decision algorithm[3] and the information flow connected thereto in the elementary contracting process.

The buyer's decision algorithm can be expressed in an analogous manner.

The characteristics of the elementary contracting process depend on a multitude of factors. Let us mention but a few.

a) On the technical character of the firm. The practice of offering and order-taking is different in a shipyard (ships would not be produced for stock) than it is in a textile factory (fabrics can be produced for stock).

b) On the business habits and traditions of the firm. (E.g. whether the firm is a rather "conservative", preferring steady partners or an aggressive undertaking, determined boldly to conquer new markets etc.)

c) On the strengths of supply and demand. The seller will behave differently in the elementary contracting process if he is the "master" because of shortages in the market, and differently if the buyer is the master and the seller is at his mercy. For example, in the former case he may work with a shorter list of addressees, in the latter case with a longer one; in the former case he may be more rigid in setting prices, in the latter one more flexible, etc. The phenomenon will be dealt with in greater detail in later chapters; we mention it here only for the sake of completeness.

[3] See Definition 9.1.

d) On previous business experiences. Which clients have stood previous tests; which of the alternative strategies of selling and purchasing have proven useful.

e) On the external information received during the elementary contracting process. What is the attitude of the other sellers; is there any change in the business situation, etc.

Although the definitions given for the contracting process are general ones, the examples given in this chapter deal with a special class of this process only: the relatively longer and more complicated information processes. In practice we very often find also simpler cases. When a firm posts a letter with a stamp, it has, in fact, made a contract with the Post Office that the latter will deliver its letter for the price of the stamp. Obviously, this transaction is not preceded by multiple exchange of information since postal tariffs are well known. In this book, we treat the more complicated lengthy processes, since—having surveyed these—an understanding of the simpler cases follows without difficulty.

17.3. The Definitions of the Market

As was mentioned previously, the term "market," without an adjective, is used in this book to denote exclusively the *commodity* market.

Our definition of the market is as close as possible to the familiar associative concept.

Definition 17.3*. The m a r k e t is a sub-system of the control sphere.[4] The organizations participating in the sub-system are linked by a definite class of information types. The market of a product is the set of all elementary contracting processes relating to that product. The market of the national economy is the set of the markets for all products. This market is also a network; the markets of the individual products are not independent of each other.

Let us point out some important characteristics of the market.

The operation of the market constitutes a *process over time*, an ensemble of chains of events overlapping in time. Drafting and mailing first offers, receiving answers to earlier offers, modifying offers, and signing contracts all may take place *simultaneously* in the firm's selling department.

The market constitutes a special process of information, information processing and decision preparation, it is one of the system's *complex decision processes*.

---

[4] Let us remind the reader that in Section 5.5 in this book, we spoke of the special sub-system engaged in purchase and sale. There we already "anticipated" the market concept. We shall revert in this same chapter to the special information of the market.

The commodity market as it is interpreted in Definition 17.3, is a broad, comprehensive concept. The market regulated by free contract prices is a *special* kind of market. This point must be emphasized since many economists do not make the distinction but rather equate this special market with the *general* concept.

17.4. Information Structure of the Market

The market is only one of the control sub-systems; in addition, there are other functioning sub-systems (e.g. the financial and credit system, national economic planning, the distribution of manpower and personnel, etc.). However, let us confine our attention to the market and determine whether it is characterized by a simple or a complex information structure.[5]

1. In market information,—in the offers and counter-offers, and finally in the contracts,—information of price and non-price character appears side by side; the two types of information are closely intertwined. The seller informs the prospective buyer about the technical properties of the product, the terms of delivery and—in addition to these—its price. In the informative relations in the market, non-price information has at least an equal rank with prices. Moreover, in the systems where increasing profits and decreasing costs are given comparatively minor importance, the effect of information of non-price character will dominate the contracting process.

2. In the actual market (as against the "black box" market of the GE models), the information flow is very often not anonymous; the information relations are establised between specified organizations, known addressers known addressees. There is a "multi-channel" flow of information, with every buyer entering into contact with several sellers and every seller with several buyers before a contract is concluded.

3. Although not in the case of all products, frequently a given real action (e.g. the transfer of the product from seller to buyer) is preceded by a variety of information. From the first offer to the final contract, multi-phase preliminary reflection occurs.

4. Although not in the case of all products, frequently in the course of the contracting process there appears information of various degrees of fineness.

On the basis of all of the above, we may make the following statement:

Statement 17.1. The market, the contracting process preceding purchase and sale, is characterized by a complex information structure.

[5] See Definition 5.8.

17.5. Comparison

The above analysis makes the errors in some of the basic assumptions of the GE school seem even more obvious.

The 10th basic assumption of the GE school postulates information of price character and, in general, simplicity of the information structure. In Chapter 5 we already pointed out the complexity of the information structure, and in Chapter 13, the importance of stock signals. Now we have dealt in detail with the exchange of information between seller and buyer (taking place by means of offers and counter-offers). It is not only through rising prices that the seller becomes aware that it is worthwhile to produce more, but also—and not less important—through the offers and orders of his clients.

The 11th basic assumption of the GE school postulates anonymity of market relations. Instead, we find parties linked directly with one another by information networks. Neither the sellers nor the buyers are indifferent as to the organizations whom they enter into contact; sorting out takes place on both sides in the form of repeated selection.

It is not only the theoretical literature of the GE school that must be pronounced guilty of an insufficient analysis of the market. As far as we are able to judge the empirical literature, too, is rather incomplete in this respect. For example, there have been no real explanations of the selection criteria actually employed by capitalist firms in the choice of their business partners.[6] Yet, this process can be observed, partly by recording actual selections made, and partly by directly interviewing those making the selection.

More empirical experience is needed to describe the elementary contracting processes and the market. Observation of the selection criteria is particularly important. Investigations of this type are relevant and feasible for the case of the post-reform Hungarian system. A great amount of similar observations also should be available with respect to the capitalist economy.

---

[6] Thus, in the above-mentioned *Balderston–Hoggatt* model, two alternative selection criteria are assumed:

1. The seller ranks his potential business partners, assuming business relations with them as nearly as possible in accordance with this stable ranking. The motives of the ranking are not discussed here; the ranking is assumed to be given.

2. The partner is selected at random.

18. DEMAND AND SUPPLY, INTENTION TO BUY AND TO SELL

18.1. The Commonplaces

The GE school suggests the following ideas about demand and supply:

1. Equilibrium exists if supply equals demand or, at least, if demand does not exceed supply.

2. Equilibrium is the desirable state of the economy.

3. When the economy is in general equilibrium, all resources with a positive marginal productivity are utilized.

4. If a resource is not fully utilized, its rent is zero.

5. The price of every product is generally positive. If there is a surplus of a product, the price of that product is zero; however, in this case forces aimed at eliminating the surplus will appear.

To sum up the above five ideas:

6. The equilibrium price will clear the market; i.e., it will relieve it of surpluses.

These ideas are so generally accepted now that they appear as truisms. They had been considered truisms even before the emergence of modern mathematical equilibrium theory; the role of theory has been to work out these ideas in a formal and exact model.

In Andersen's famous tale everybody but the king knew that the king was naked; however, a child's naive frankness was needed for the truth to be spoken out aloud.

We believe that the time has come for plain and open speech in economic theory, too. As a matter of fact, everybody knows that the six assertions above *simply are not true*.

The economy is never in equilibrium. There are always "surpluses". Fortunately, the market is never "cleared". A properly functioning market *always* is full of goods, in the morning as well as in the evening; both before and after the season. There are always stocks of products and reserves of resources (such as manpower, capacities). And although supply never equals demand, except for a few exceptional cases there exist neither products with zero price nor resources with zero rent.

Thus, there is a striking contrast between the "theoretical" statement, which has become truism, and actual reality.

229

This problem will be discussed in Chapters 18 to 23. In the present chapter we deal primarily with definitions.

18.2. Distinction of the Control and Real Processes

The trouble with GE theory begins with unclear concepts. Every work on economics uses the concept of "supply"; but it either omits its definition or gives an unacceptable definition.

Usually the GE school simply identifies supply with production.[1]

Let us recall the 5th basic assumption of the GE school, stating that there is no time-lag nor quantitative difference between intended buying and selling, actual buying and selling, production and consumption.

Here, concepts which should be sharply and unequivocally distinguished by economic systems theory are combined and confused.

First of all, let us distinguish—in the spirit of the dualistic description method—the events of the control sphere and the real sphere.

The selling intention of the selling organization and the buying intention of the buying organization appear in the control sphere, within the framework of elementary contracting processes. These are *information* variables. Whenever the seller informs potential buyers of his selling intention, there is an information *flow*, the addresser being the selling organization and the addressee the potential buyer. The buying intention is symmetrical.

The actual transaction, the product's transfer from the selling organization to the buying organization, takes place in the real sphere. Here, there is a *product flow*. The addresser is the actual (now not only potential) seller, and the addressee the actual buyer.

The transaction is accompanied by a transfer of money. This again takes place in the control sphere. It is an *information flow;* however, its direction is opposite to that of the product flow. The addresser of the money is the buyer and its addressee the seller.

There are two related *internal processes*[2] that are separate from the above *flows*; both of them are *real* processes taking place in the real sphere. One is the production taking place in the productive organizations (the transformation of products and resources into products); the other is consumption (the utilization of products).

In the above, we have separated the events of the control sphere from those in the real sphere, and the flows from the internal processes. Let us now consider the time relations of the problem.

[1] See, e.g. *Debreu* [50], p. 38.
[2] See Definition 4.7.

First, we discuss two "pure" cases: A) that of the plant producing exclusively to meet previously received orders and B) that of the plant producing for stock without previously received orders. Examples of the first category are construction firms or plants producing large-size custom made machines (such as individually produced ships). An example of the second category is a factory of canned goods processing seasonal produce. In both cases, our investigations will be confined to the selling firm; the course of events for the buyer is symmetrical.

A. *Plant producing against previously received orders.* The course of events is visualized in *Figure 18.1*. In the figure (as well as in Figure 18.2, below), time is measured on the horizontal axis.

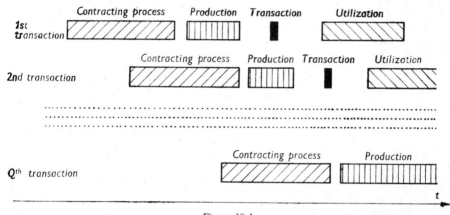

Figure 18.1

Chart of events in the case of a plant producing to order

On the seller's side, several (in the figure a total of Q) elementary transactions (and contracting processes) take place. Let us consider the first one of these. The chain of events starts with the elementary contracting process, which was described in the preceeding hapter. As previously has been pointed out, this takes time. Its final point is the decision taken by the seller (as well as the buyer); in this case it is the production of the product described in the contract. Production is followed, perhaps with another time-lag, by the transaction—the transfer of the product from the selling organization to the buying organization. Following that, probably again with some time-lag, consumption begins.

There is a similar sequence for the chain of events connected with the second, third, ..., Q^{th} transaction. The course of these transactions over time may take place in parallel, simultaneously or with a relative shift over time. In the figure, we have presented the latter case.

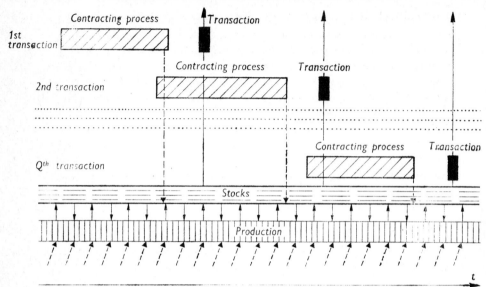

Figure 18.2
Chart of events in the case of a plant producing for stock

B. *Plant producing for stock.* The course of events is shown in *Figure 18.2.* This is more complicated than the chain of events presented in the former figure.

One of the functional organizations of the firm (which is a complex institution consisting of several organizations) is the productive plant. It produces continuously, a fact symbolized in the figure by the lowermost, vertically striped band.

The product output is stocked in the warehouse, also continuously. The product stored in the warehouse (a real state variable) is symbolized by the band above the production band and striped with broken lines.

The scheduling of the elementary decision processes preceding the contracts is basically independent of the course of real processes. Another functional organization of the firm, the sales department, starts elementary contracting processes one after the other and carries them to the point of the contract, as shown in the upper part of the figure. After the contract is made, the goods are "called in" from stock and handed over to the buyer.

The dotted arrows in the figure show the product flows (or, rather, the parts of these flows essential from our present point of view). There are product flows from current production into the warehouse stock, and from the stock to the buyer.

Information flows are symbolized by dotted arrows. Actually, production is not independent of selling; from the contracting processes information flows into the organization controlling production. That organization also received information on the position of stocks.[3] These information flows exert a considerable (although not exclusive) effect on the course of production. The fact that other information flows also exert an effect is indicated by the dotted arrows reaching also from below the lower band of production.

Plants producing exclusively against previously received orders and those producing exclusively for stock constitute two pure cases. They occur in their pure form in practice; however, there are many mixed cases, too. For example, a textile factory may product partly against orders received from the buyers and partly, at its own risk, for stock.

In any case, what has been said above should be sufficient to prove that there is no automatic overlap of buying and selling intentions, transaction, production and consumption. The assumption of overlapping, although appearing to lend to the theory a "simple" and "clear-cut" character, leads to complete confusion. It merges variables existing in different areas, in different dimensions. There is a mix-up of *information flows* (buying and selling intentions), *product flows* (transfer of product), *internal real processes* (production, consumption) and *real states* (product stocks).[4]

18.3. The Maturing of Selling and Buying Intentions

Now let us turn our attention to the contracting processes.

Again, we shall be concerned with the decision process of only one organization, with only one seller. The buyer's behaviour can be described in a symmetrical manner.

We shall use the concepts and notations introduced in Chapters 8 and 12 in the general description of decision processes.

Let us assume that in the period under investigation, Q selling problems have arisen; to these belong Q contracting processes.

All selling alternatives which may arise in connection with the Q problems can be described with the aid of K indicator types. Of these, the first H indicator types indicate the offered quantities of the products intended for sale; the remainder (i.e. the $(H+1)^{th}$, $(H+2)^{th}$, ..., and, finally the Kth indicator) relate to the qualitative and technical characteristics, terms of delivery, prices, credit terms of the products intended for sale.

[3] This has been discussed in detail in Section 13.3.
[4] See the following definitions: information flow (4.16), product flow (4.13), internal real flow (4.7), real state, product stock (4.14).

Accordingly, the possible alternatives are described by indicator vectors composed of K components;[5] these are the elements of set A, the set of possible selling alternatives.

Definition 18.1*. The seller's e l e m e n t a r y s e l l i n g i n t e n t i o n in the ith contracting process $(i = 1, \ldots, Q)$ in period t is described by indicator vector $s_i(t) \in \mathcal{A}$ composed of K components. This vector may change in the course of the contracting process taking place in period $[\underline{t}_i, \bar{t}_i]$. Its *degree of maturity* in period $t(\underline{t}_i \leqq t \leqq \bar{t}_i)$ is given by the difference $\bar{t}_i - t$.

(The concept of e l e m e n t a r y b u y i n g i n t e n t i o n is symmetrical.)

It is important that selling intention be considered as a *process taking place over time*, an element of a decision process. The intention matures gradually, its degree of maturity being characterized by its distance from the end of the decision algorithm.

The aspiration level preceeds gradual maturation of the elementary selling intention.[6] $s_i(\underline{t}_i) = \alpha_i(\underline{t}_i)$. If we should wish to measure the aspiration level, we must ask the seller the following question. "Assume that the market is willing to buy your product in unlimited quantity at the prevailing usual price. Your fixed capital, your capacity, is given; you cannot increase it. What quantity would you intend to sell under their circumstances?" By asking this question to an appropriately selected sample we may estimate the aspiration levels, the original selling intentions, of the whole selling population.

A similar question is required for measurement of buyers aspirations. "Assume that you can find on the market all products in unlimited quantity at the prevailing, usual prices. Your disposable income is given; you must not exceed your budget constraint. What quantity would you intend to buy under these circumstances?"

Later, the aspirations will be modified gradually. Ultimately the conclusion is the decision, which is here, in the process of selling and buying, identical with the contract. Accordingly, $s_i(\bar{t}_i) = a_i^*(\bar{t}_i)$, where a_i^* is the contract.[7]

---

[5] In many cases it is unnecessary to characterize the ith problem by an indicator vector composed of K components. The different elementary contracting processes may offer different products for sale. E.g. in process 1 there may be only products 1, 2 and 3, whereas in process 2 only products 4, 5, and 6. However, it facilitates the formalization if we operate with uniform indicator vectors composed of K components. In our example: in the indicator vector $s_1(t)$ belonging to process 1 the indicator types 4, 5, and 6 will be marked with 0; conversely, in vector $s_2(t)$ the indicator types connected with products 1, 2, and 3 will figure as 0.

[6] See Definition 12.2.

[7] As was already mentioned in Chapter 12, following the definition of the aspiration level, a more exact description of intentions and aspirations really needs stochastic formalism but, for the sake of simplicity, we disregard this point and use a deterministic description.

Between the starting and final point appear elements which have already been discussed in their general form in Chapters 8 and 9. On the one hand, the decision-maker, the selling organization, acquires ever more and more accurate information concerning actual selling possibilities. This information emerges from the buyers' counteroffers and from direct inquiries. Symbolically, $\tilde{\mathcal{B}}_i(t)$, the set of explored selling alternatives takes shape gradually, and increasingly it coincides with $\mathcal{B}_i$, the set of actually implementable alternatives, the actual selling possibilities.

On the other hand, the constraints of selling also take shape. These are partly given beforehand, determined by the selling organization's own conventions, habitual rules of thumb, expectations concerning profits, sales volume, delivery terms. The other functional organizations of the firm—e.g. the financial department or production department, etc.—may have different expectations. Symbolically: $\mathcal{D}_i(t)$, the set of acceptable alternatives takes shape gradually. Ultimately the contract will be one of the elements of $\mathcal{F}_i(t_i)$, the set of eligible selling alternatives:

$$a_i^*(t) \varepsilon \mathcal{F}_i(t_i) = \tilde{\mathcal{B}}_i(t_i) \cap \mathcal{D}_i(t_i) \tag{18.1}$$

The contracting process involves the gradual exploration of the possibilities (including the buyer's buying intentions and propensities) and of the institution's own interests. Because this occurs gradually, the intention to sell may be subject to repeated changes.[8]

Consider the firm's selling activity as a *whole*, at a given moment (say in period t_0). There may be simultaneous selling intentions with highly varying degrees of maturity. This is shown in *Figure 18.3*.

In the figure, there are five elementary contracting processes. Although their starting and final points are different, there is one period when all five are simultaneously in progress. This overlapping, common period is represented by the perpendicular blue band. Let this be period t_0.

It is clear that all five processes have a different degree of maturity. The figures matched with the braces indicate the degrees of maturity: 1, 3, 9, 4, 10, in that order. Thus, the selling intention of process 1 already has reached an advanced degree of maturity, in the next period, contracting may take place. Process 5, on the other hand is still at the aspiration level.

Let us now examine the possibility of *additivity* of the selling intentions. Let us assume that our firm is turning out three types of product: machines of the types "A", "B" and "C". In each case the quantity offered for sale simply is

---

[8] For the sake of simplicity we may assume that with the passing of time selling intention becomes more and more mature. This is true in general, though there may be exceptions. It may be that in the course of the contracting process the intention becomes distorted, becoming "less serious" instead of "more serious".

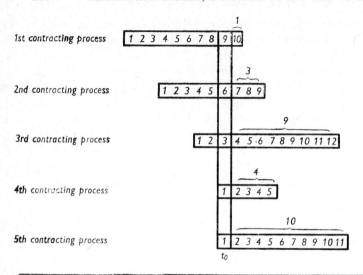

1
1st contracting process | 1 2 3 4 5 6 7 8 | 9 10
3
2nd contracting process | 1 2 3 4 5 | 6 | 7 8 9
9
3rd contracting process | 1 2 | 3 | 4 5 6 7 8 9 10 11 12
4
4th contracting process | 1 | 2 3 4 5
10
5th contracting process | 1 | 2 3 4 5 6 7 8 9 10 11
t_0

Figure 18.3

Maturing of selling intention

measured in terms of units. Thus, in our example, the number of the indicator types identifying the quantities intended for sale (H) is three.

The intentions of selling in period t_0 are described by the K-component vectors $s_i(t_0)$ $(i = 1, \ldots, 5)$. Let us take now the first three components of these vectors and include them in *Table 18.1*.

Because only the estimated numbers of machines within each type "A" can be added only the three row sums might appear in the table: 45 units type "A", 29 units type "B" and 25 units type "C".

TABLE 18.1

Selling intentions

Elementary Contracting Process	1	2	3	4	5	
Degree of Maturity	1	3	9	4	10	
Quantity Destined for Sale						Sum
Product "A"	9	14	3	7	12	45
Product "B"	8	11	7	2	1	29
Product "C"	0	11	2	8	4	25

However, it is not justified to form such row sums. The selling intentions differ greatly in "seriousness" from column to column. Column 1 must be taken quite seriously because here the firm is near to concluding the contract. The contents of column 5, on the other hand, are still rather provisional, constituting a wish rather than a definite plan.

Now we can write down the following definition.

Definition 18.2. The seller's c o m p l e x   s e l l i n g   i n t e n t i o n   in period t is expressed by matrix $S(t)$. The matrix is composed of K rows and $Q(t)$ columns, where K is the number of the indicator types describing the selling alternatives, and $Q(t)$ the number of simultaneous contracting processes in progress in period t. Element $s_{ij}(t)$ of the matrix gives the value of the ith indicator in the jth elementary contracting process. (The buyer's c o m p l e x   b u y i n g   i n t e n t i o n can be defined in a symmetrical manner.)

An example of the structure of matrix $S(t)$ is given in Table 18.1.

Let us sum up the above.

Statement 18.1. The seller's intentions to sell and the buyer's intentions to buy mature gradually from the aspiration level to the final contract. It is not possible to summarize at a given moment all selling (buying) intentions of an organization, because intentions with various degrees of maturity exist simultaneously, side by side.

18.4. Interrelations of Selling Intention, Selling, Production and Stock

Let us go back to the 5th basic assumption of the GE school, and within that, to the coincidence of supply and production. The conceptual differences between supply and production have now been elucidated from various aspects; on the basis of what has been said above, it now is possible to speak also of the quantitative relationships.

1. There is a trivial relationship: the actual transaction, the transfer of the product volume in period t from the selling organization to the buying organizations, cannot exceed the stock of selling organization in period t. Obviously, no more goods can leave the factory than have been accumulated in the factory's stock.

2. There is a less trivial relationship which asserts itself only stochastically, as a tendency, as the average of a period of some length.

The seller's product stock will fluctuate around a normal level. (This has been discussed in detail in Section 13.3.) Should sales increase and, as a consequence stocks be reduced, production usually will be increased sooner or later, and conversely. The differences between actual sales and production will be equalized in the course of time by means of increasing or reducing stocks. On average over a lengthy period, production can neither exceed nor fall short of sales.

3. The more mature the selling intention, the nearer is the firm to actual selling. Accordingly, relationship 2 applies also—less strictly, but at least to a certain degree—to the comparatively mature selling intentions near the contracting phase.

In spite of the three relationships listed above, each of the four variables of the economic system treated here—selling intention, selling, production and stocks—has an "independent life". Each enjoys considerable freedom of movement and a relative independence from the other three. Selling intention (and especially immature intention, the initial aspiration level) may differ essentially from actual selling. The gap between selling intention, selling and production may now widen, now narrow. The relationships and interactions which emerge beyond the three necessary relationships listed above between selling intention, selling, production and stocks, constitute one of the most important characteristics of the specific control processes of concrete economic systems.

18.5. Comparison

Before turning to the relationship between selling (buying) intention and its realization, which is the subject of the next chapter, let us again make a comparison with the GE school. This can be done briefly, since throughout the chapter we have attacked the demand-supply concept of the traditional theory of the market.

In our opinion, the term "supply" is only a loose comprehensive concept. With the terminology introduced in this book, it may comprise any of the following:

1. The seller's selling intention on any degree of maturity; i.e. vectors $s_i(t)$ characterizing the elementary contracting processes or matrices $S(t)$ expressing the total, complex selling intentions of the institution.

2. The seller's actual selling possibilities; in symbolical terms, set $\mathcal{B}$ of all implementable selling alternatives. This set is bounded only by the real constraints of selling: from the seller's side, the actual stocks of products (see relationship I mentioned in Section 18.4), and the buying intentions of the buyers.

3. Set $\tilde{\mathcal{B}}(t)$ of the explored selling alternatives deemed implementable by the seller. This is a subjective reflection of the former set B in the seller's thoughts.

4. The set of eligible selling alternatives which the seller would consider not only implementable but also acceptable from the point of view of his interests: $\mathcal{F}(t) = \tilde{\mathcal{B}}(t) \cap \mathcal{D}(t)$.

In the literature it is not clear and unequivocal which of the above-listed possible interpretations is the correct one. It would be arbitrary on our part to reserve the "supply" concept to denote either $s_i(t)$, or $S(t)$, or $\mathcal{B}$ or $\tilde{\mathcal{B}}(t)$, or $(\mathcal{F}(t)$. Therefore, in the following discussion we shall disregard the term "supply." We believe that if economists intend to continue the use of this term which has become so dear to our hearts, if only for the sake of tradition, we should first come to an agreement on its definition. As long as this agreement is not achieved, it is better to omit the word from our vocabulary in order to avoid any confusion of ideas.

What has been said above also is valid for the term "demand".

Statement 18.2. Because of the lack of conceptual clarification, "supply" and "demand" are not measurable. Therefore, their quantitative relationship (their equality, "equilibrium", difference) does not lend itself to interpretation.

Of course, we are well aware of the fact that the loose concepts of "supply" and "demand" are linked with essential and real problems. Even if it is not possible to interpret the term "equilibrium of demand and supply" in a strict sense, we realize that equilibrium is the "codename" of a whole range of real economic problems. Therefore, we do not want to end the discussion with Statement 18.2 (this would be too easy and too destructive). We do not wish to evade the discussion of the problems of equilibrium.

19. PRESSURE AND SUCTION

19.1. The "Shortage Goods"

Having equipped ourselves with the concepts necessary for a description of the market, now we may proceed to describe and explain some concrete market phenomena.

Let us begin with some experiences gained in Hungary. The concept of "shortage goods" is well known to every buyer there, including not only the housewives but also the purchasing agents of the firms. The concept covers the products which the buyer would need and would like to purchase with money available, but nevertheless, which cannot be obtained.

Shortage goods are not products which never have been obtainable or which no one ever could have purchased. Some people have obtained them. However, there are fewer of them that are obtainable than are needed.

Let us present some examples. There are considerably more people in Hungary who would like to buy a car than there are cars in supply. The buyer must pay in advance a considerable part of the price (i.e. the buyer grants credit to the seller); then he must wait for years—"lining up" behind the customers who had subscribed before him-before he can finalize the purchase. This situation is described in detail in *Table 19.1*.

At present Hungary is on the threshold of the "automobile era"; the number of cars is increasing at a rapid rate. Yet the development of auto repair services is falling considerably behind. To quote an article from the leading Hungarian daily paper *Népszabadság:* "...according to wellfounded calculations, the automobiles awaiting repair would need 11 million work-hours per year, whereas the available capacity is barely 8 million."[1]

Consumers also queue for installation of new telephones and gas heating. The relevant data are presented in Table 19.1.[2]

Shortages in clothing articles occur frequently. *Tables 19.2* and *19.3* present data which characterize this situation.

There are particularly marked disproportions between the capacity of the building trade on the one hand and the demands made on it (including both

[1] See *Fekete* [60].
[2] The meaning of the indicator in the last column of the table is established by Definition 19.3.

TABLE 19.1

*Buyer's fulfilment ratio: motor-car, gas heating, telephone**

Product or Service (1)	Year (2)	Total Valid Orders (3)	Fulfilment in the Cour in the Course of the Year (4)	Fulfilment Ratio (4:3) in Percentages (5)
Motor Car:				
Trabant				
Limousine	1967	14,720	9,179	62.3
	1968	17,406	3,959	·22.6
Wartburg				
Limousine	1967	6,277	3,550	56.5
	1968	7,257	2,463	33.9
Wartburg				
de Luxe	1967	4,432	2,199	49.6
	1968	6,763	1,973	29.6
Moskvich	1967	8,180	2,576	31,5
	1968	6,487	5,460	84.3
Gas heating:	1967	12,800	8,532	66.6
	1968	23,700	14,439	60.9
Telephone				
Station:	1967	94,785	28,893	30.5
	1968	104,758	34,081	32.5

* The data in this table were compiled with the help of *Éva Varga, Pál Csetényi* and *Béla Szabó*. The table was arranged by *Andrea Deák*.

privately financed residential constructions, and government and cooperative investments) on the other. Let us quote an article by Deputy Prime Minister *Mátyás Timár* in the daily paper, "Népszabadság"; "Although some progress could be registered, it is not yet possible to bring about an equilibrium in the field of investments. Both demand and the funds available to the enterprises exceed domestic building capacity and possible machinery imports."[3]

19.2. Compulsory Correction of the Intention to Buy

Let us return to our consumer who wishes to buy a car. For the first time, he develops the intention to buy a car of brand A. He actually has the money or, if it is not actually in his pocket, he is able to secure the necessary loan.

[3] See *Timár* [255].

Thus, an elementary contracting process began with the *aspiration level* as its initial stage. In the given case, the aspiration level is the car of brand A.

According to Definition 12.2, this use of the concept of aspiration is entirely justified. The aspiration level expresses the decision-maker's wishes, internal expectations, about the decision to be taken at the end of the process. According to the decision-maker, the internal conditions of actually attaining the aspiration level (those conditions depending on him) can be expected to be satisfied; in a favorable case the external conditions (those independent of him) can be met also.

TABLE 19.2

Buyer's fulfilment ratio: clothing*

Article	Expected Fulfilment Ratio 4th Quarter 1969 (in percentages)
Flannel piece goods	73
Woollen and wool-type synthetic cloth	93
Man's topcoat	90
Man's greatcoat	92
Woman's topcoat	84
Woman's greatcoat	85
Child's topcoat	85
Child's greatcoat	83
Man's sweater, waistcoat	80
Woman's sweater, jacket, cardigan	80
Boy's and girl's waistcoat, jacket, cardigan	80
Man's shirt, flannelette	70
Boy's shirt, flannelette	75
Woman's panties, cotton	90
Woman's tights, synthetic	86
Girl's panties, cotton	90
Child's tights, synthetic	80
Man's fur-lined shoes and boots	93
Woman's fur-lined boots (incl. leatherette)	94
Child's boots	94
Child's fur-lined shoes and boots	92

* The table is based on a forecast published in the reports of the Ministry of Home Trade on trade in the 3rd quarter of 1969.

In the favourable case our consumer will be able to obtain the desired car of brand "A". Unfortunately, however, affairs may take a turn unfavorable for him. He might be told that no car of this brand would be available in the next few years, or that for brand "A" he would have to wait three years whereas a car of brand "B" could be purchased immediately or within six months. Being mpatient, our friend might make his decision in favour of brand "B".

TABLE 19.3

Lack of assortment in the clothing trade*

	Available	Not available in sufficient extent because shop's stocks are:				Total
		ill-assorted	short in sizes	illassorted and short in sizes	lacking	
Man's clothing articles						
Autumn 1968	56	16	3	13	12	100
Spring 1969	53	19	2	16	10	100
Woman's clothing articles						
Autumn 1968	50	11	2	18	19	100
Spring 1969	46	8	3	25	18	100
Child's clothing articles						
Autumn 1968	54	7	13	14	12	100
Spring 1969	48	9	8	21	14	100

* Source: Central Statistical Office, Report [139] and [140]. Investigations were carried out in 1968 at 213 retail shops and concerning 64 articles in particularly great demand; in 1969, in 229 retail shops and concerning 40 articles. Column 2 in the Table shows the number of shops where there was a sufficient stock of all articles investigated, in every necessary size.

In this case, the intention to buy undergoes a change not voluntarily but under the pressure of circumstances.

The contracting and decision process can be described by an indicator vector whose components contain the characteristics of both car brands "A" and "B" that are essential from the point of view of the purchase . For the ith consumer, at the beginning of the process there develops an aspiration level $\alpha_i(t)$ and by the end of the process, a decision $a_i^*(t)$. As described in Definition 12.17, the difference between the two vectors will be called the *correction* of the aspiration:

$$\varkappa_i(t) = a_i^*(t) - \alpha_i(t).$$ (19.1)

In the present case we have a *compulsory* correction; a *shortage* has compelled the consumer to change his intentions.[4]

[4] The difference between intention and realization may be the consequence of other factors, too: the buyer changed his mind; he got new information, etc. Here we do not deal with these differences.

One of the index numbers suited to characterize the shortage situation is $\varkappa$ or, rather, its percentage form $\hat{\varkappa}$. Observation of the degree of correction and the computation of suitable statistical indices (average etc.) will enable us to characterize the "forced substitutions", the buyer's deviations from his original intentions, desires and aspirations.

19.3. The Tension of the Buyer's Aspiration

Thus far we have been dealing with a single contracting process of a single buyer. Now let us observe the entire market of some product, the positions of all buyers of some product.

We need indices showing how the buyer's aspirations are fulfilled.

As in the preceding section, we take the purchases of motor-car, brand "A", as an illustrative example. For a general denotation, we speak of the jth product.

We examine the aspirations arising in a definite period t_0. For example this period might be the first quarter of 1966.

We have in the system the organizations $o_1, o_2, \ldots, o_m$. Let us denote the set of serial numbers of the organizations aspiring the jth product in this period[5] $\mathcal{J}_j^{(B)}$:

$$\mathcal{J}_j^{(B)} \subset \{1, 2, \ldots, m\} \tag{19.2}$$

Assume that the purchases of the jth product can be measured by a single type of indicator. For example, the quantity purchases of car brand "A" can be measured unequivocally by unit numbers. The quantitative indicator may be a continuous or an integer-value, non-negative variable. In any case for the jth product the aspiration level and the result is measured by a single real number (and not by some multi-component vector).[6]

Let α_{ij} denote the initial buying intention for the jth product in the ith organization in the said period (i.e. the *aspiration level*) and let $\omega_{ij}(t)$ denote the quantity actually purchased in the tth period following period t_0, the *result* ($t = 0, 1, 2, \ldots$).

---

[5] In many further symbols, the superscript (B) will distinguish the buyer. The symbol (S) will play a symmetrical role in denoting the seller.

[6] In order to facilitate the explanation we disregard the problems of measuring quality. It will be assumed that the jth product is unequivocally defined with all its use and qualitative properties. We do not investigate the problem of how the fulfilment of the aspirations changes in the case when the buyer satisfies himself with a pair of less fine shoes than he initially desired. This problem of quality will be treated in detail later on.

There are aspirations which can be fulfilled almost at the moment that they arise. If the aspiration is to drink a cup of coffee, this can be done immediately, even at night, in any coffee-bar. The purchase of a house, on the other hand, even in the case of an ample supply—involves obtaining extensive information, signing a sales contract, securing the help of a lawyer, probably obtaining a bank loan, transcribing the exchange in the land register, etc. All this may take weeks. Thus there exists a period of waiting, a delay, due to the administrative and technical conditions of sale.

Definition 19.1. Let us call the n e c e s s a r y w a i t i n g p e r i o d o f b u y i n g and denote by ϑ_j the minimum number of periods that must elapse between the initial aspiration and the first fulfilment, the contract.

Let us call the l i m i t a t i o n p e r i o d o f b u y i n g a s p i r a t i o n s and denote by Θ_j the number of periods which must elapse before all elements belonging to set $\mathcal{J}_j^{(B)}$ cease their original aspirations, either because the buying intention has been fulfilled or because the original aspiration has been corrected.

Obviously

$$\vartheta_j \leqq \Theta_j \tag{19.3}$$

Definition 19.2*. The t e n s i o n o f b u y i n g a s p i r a t i o n s in the market of the jth product as calculated from the initial period t_0 over an arbitrary period T ($\vartheta_j \leqq T \leqq \Theta_j$) can be characterized by the following index:

$$\varepsilon_j(T) = \sum_{i \in \mathcal{J}_j^{(B)}} (\alpha_{ij} - \sum_{\tau = \vartheta_j}^{T} \omega_{ij}(\tau)), \quad \vartheta_j \leqq T \leqq \Theta_j. \tag{19.4}$$

The degree of tension $\hat{\varepsilon}_j(T)$, the corresponding "percentual" index:

$$\hat{\varepsilon}_j(T) = \frac{\sum\limits_{i \in \mathcal{J}_j^{(B)}} \alpha_{ij}}{\sum\limits_{i \in \mathcal{J}_j^{(B)}} \sum\limits_{\tau = \vartheta_j}^{T} \omega_{ij}(\tau)}, \quad \vartheta_j \leqq T \leqq \Theta_j. \tag{19.5}$$

The tension formula in (19.4) and (19.5) is in harmony with the general definition of the tension of aspiration given in 12.3, of which it is a special case.

What happens to the value of the index over time is presented in *Figure 19.1*. The single first, high column is the aspiration level: 250 units in the first quarter of 1966.

The necessary waiting time, $\vartheta_j = 2$ periods, means that the aspirants presenting themselves in the first quarter of 1966 will be able to begin buying in the third quarter of 1966 at the earliest. This fact accounts for the gap of one period between the first and the second column.

The diagonally striped columns represent the purchases of the aspirants. Let us assume that we wanted to calculate the value of tension index (19.4) for the value $T = 4$. By that time, a total of 110 units have been purchased. By the end of the first year the tension is 140 units and the degree of tension 236 per cent.

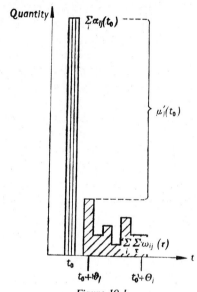

Figure 19.1
Tension of aspiration of the buyer

The limitation period Θ_j in our figure is 8 periods, i.e. two years. After the elapse of this time, no one maintains his original aspirations. This in itself characterizes fulfilment; two years must elapse before everybody either becomes satisfied or loses patience. However, fulfilment also can be characterized by other indices.

Let us call the buyer's *global fulfilment ratio* and denote μ_j the following index:[7]

$$\mu_j = \frac{\displaystyle\sum_{i \in \mathcal{J}_j^{(B)}} \sum_{\tau = \vartheta_j}^{\Theta_j} \omega_{ij}(\tau)}{\displaystyle\sum_{i \in \mathcal{J}_j^{(B)}} \alpha_{ij}} \tag{19.6}$$

[7] The present concept and the degree of tension (percentage indicator) defined in Definition 12.3 are related as follows:

$$\mu_j = \frac{1}{\hat{\varepsilon}(\Theta_j)}$$

Correspondingly, let us call the *unfulfilment ratio:*

$$\bar{\mu}_j = 1 - \mu_j. \tag{19.7}$$

Let us call the *buyer's immediate fulfilment ratio* and denote μ_j the following index:

$$\mu_j' = \frac{\displaystyle\sum_{i \in \mathcal{J}_j^{(B)}} \omega_{ij}(\vartheta_j)}{\displaystyle\sum_{i \in \mathcal{J}_j^{(B)}} \alpha_{ij}} \tag{19.8}$$

The aspiration is *immediately and completely* fulfilled if $\Theta_j = \vartheta_j$ and $\mu_j = \mu_j' = 1$. In all other cases, there is only *partial* fulfilment.

Examples of the fulfilment ratio μ_j can be found in the last column in Table 19.1.

In Figure 19.1 $\sum\sum\omega_{ij} = 200$. Accordingly, the global fulfilment ratio is 80 per cent. The immediate fulfilment ratio, on the other hand, is only 25 per cent. The global and immediate fulfilment ratios are important characteristics of the fulfilment of aspirations, although they do not in themselves yield a picture of the fulfilment's course over time. To obtain that, the value of the tension index must be given for every time point T of the $[\vartheta_j, \Theta_j]$ interval.

To characterize the market situation, the economist engaged in the analysis of the problem may set an arbitrarily (defined *a priori*) period of time which may be termed the *"normal" period of queuing* for the jth product and denoted by T_j^0. Then, he may calculate the value of tension index (19.4) for the case when $T = T_j^0$. In a similar sense, when calculating the fulfilment ratio (19.6) he may carry out the second summation in the numerator not to the limit Θ_j but only to the limit T_j^0. This facilitates the observation necessary for the quantitification of the indices. For example, in the case of car of brand "A", T_j^0 could be one year. The observation may be carried out by interviewing, after the elapse of one year, those who had sent in preliminary orders, whether or not they actually have purchased the car.[8]

The time periods and tension indices discussed up to this point all depend on the choice of the initial period t_0; accordingly, the values of $\mathcal{J}_j^{(B)}$, $\varepsilon_j(T)$, ... will be modified if we change over from t_0 to some other period, say t_1. The

---

[8] *Katona* [113] describes the public opinion reserach carried out by the Michigan Survey Center; among others these include studies of the buying intentions of buyers. A sample of persons was interviewed concerning their buying intentions (including the intention to buy a car) then, after the elapse of some time they were interviewed again about the fulfilment of their intention.

This method is well suited for defining the indices discussed in this section. We shall revert to the question below.

sets $\mathcal{J}_j^{(B)}(t_0)$, $\mathcal{J}_j^{(B)}(t_1)$, ... necessarily form a family of disjoint sets, without any common parts, since, for example, $i \in \mathcal{J}_j^{(B)}(t_0)$ means that in period t_0 the organization o_i has *started* to think of buying the product j. Accordingly, the indices can be separated from each other easily, and it is impossible to count the data of one and the same organization several times.

Thus far, we have examined the purchase starting from a single period of time. Naturally, more information may be obtained if instead of examining a single period we examine the *average* of some longer period $[t', t'']$. The averaging may be carried out by means of some suitable statistical routine method (e.g. calculation or moving averages or the trend, etc.).

In the further course of the discussion the subject will be treated in a general theoretical manner and we shall not deal with the analysis of concrete measurements; thus, it is not necessary to consider the temporal aspects of measuring, nor to indicate the concrete name of the buying organization. Accordingly, *simplified notations* will be employed based on the following assumptions:

1. We deal not with indices connected with the aspiration level of a single period but with the suitable statistical average of a longer period $[t', t'']$;

2. we take into consideration not the whole limitation period Θ_j but a suitable selected "normal" lining-up period T_j^0;

3. we characterize the "average" buying organization.

Taking into account the simplifications mentioned above, the position of buyers in the market for the jth product may be characterized by the following indices:

$\alpha_j^{(B)}$ = the buyer's aspiration

$\omega_j^{(B)}$ = the buyer's actual purchase, the contract (if figuring in the same formula as the aspiration, then with due regard to the suitable time lag)

$\varepsilon_j^{(B)}$ = the tension of the buyer's aspiration

$\hat{\varepsilon}_j^{(B)}$ = the degree of the buyer's aspiration tension (in percentages)

$\mu_j^{(B)}$ = the buyer's global fulfilment ratio

$\mu_j^{'(B)}$ = the buyer's immediate fulfilment ratio.

Now let us return to the actual market and to the phenomenon described in Section 19.1, the existence of shortage goods, the "shortage situation". This situation can be characterized by saying that the market situation is tense; the value of the absolute index $\varepsilon_j^{(B)}$ is high for a great number of products; the degree of tension, $\hat{\varepsilon}_j^{(B)}$, is considerably higher than 100 per cent. The ratio of global fulfilment, $\mu_j^{(B)}$, is much smaller than 1, and the ratio of immediate fulfilment, $\mu_j^{'(B)}$, is particularly small.

19.4. The Seller in the Shortage Situation

When there is a shortage in the market of some product, i.e. $\varepsilon_j^{(B)} > 0$, then the aspirations of the seller are easily fulfilled. This situation is frequently described in literature as the "sellers' market".[9]

To continue the former Hungarian examples, such a situation exists in the automobile trade, the car repair network, and the building trade, where the sellers have no difficulty finding buyers for their output and services.

We have developed an entire series of concepts to characterize the situation of the buyer; it is not necessary to give detailed definitions of the concepts describing the seller's position. Instead, we shall be content with the following definition.

The following concepts should be interpreted as being symmetrical with those defined in Definitions 19.1–19.2, relating to purchase and the buyer: the *necessary waiting period of selling*, the *limitation period of selling aspirations*, the *tension of selling aspirations*, as well as *the seller's global* and *immediate fulfilment ratios*.

In the case of shortage, the seller's position has the following characteristics:

1. The seller need not line up for the buyer; after the necessary and inevitable waiting period, the transaction can be concluded immediately.

$$\Theta_j^{(S)} = \vartheta_j \tag{19.9}$$

2. Between the seller's selling aspiration and actual selling there is no tension:

$$\varepsilon_j^{(S)} = 0, \quad \hat{\varepsilon}_j^{(S)} = 1 \tag{19.10}$$

In exceptional conditions the tension may be negative; the producing firm is compelled to produce more and, consequently, to sell more than it actually desires to produce:

$$\varepsilon_j^{(S)} < 0, \quad \hat{\varepsilon}_j^{(S)} < 1. \tag{19.11}$$

3. The seller's aspiration is completely and (as pointed out in paragraph 1) immediately fulfilled:

$$\mu_j^{(S)} = 1 \tag{19.12}$$

[9] In the Hungarian economic literature, it was *Gy. Péter*'s articles [201] the forerunners of the 1968 reform of economic administration, in which the problems of the "sellers'" and the "buyers' market" in Hungary were first discussed. It is from these articles that we found first inspiration in forming our own ideas on the subject; these ideas were first put forward in an book, *Overcentralization in economic administration*, published in 1957. (See [126], Chapter IV.)

Chapters 19 to 23 of the present work is an attempt at formulating more precisely and completely the ideas presented in our earlier book.

19.5. The Seller Queues

Having discussed in Sections 19.1–19.4 the phenomenon of "shortage products", now let us treat the reverse case, the buyers market. So far, we have taken the Hungarian motor-car market as an example of a shortage situation. The American motor-car market is an example of the opposite situation.

The *buyer* can buy the desired car immediately after the aspiration arises, with the elapse of the technically necessary waiting period ϑ_j. The transaction may occur within a few hours. If he has the necessary cash or a bank account, he can pay for the car immediately. If he wishes to do so, he can buy on credit; then, the seller grants credit to the buyer. Any brand of a wide range is available for choice.

The *seller*, on the other hand, has serious marketing problems. The buyer usually buys his car not from the factory, but from a distributor. Here, however, we disregard the special problems of the decision processes of commerce and focus our attention on the producing firm.

For the sake of illustration in *Table 19.4* we present one series of data: the McGraw-Hill indices of capacity utilization in U.S. manufacturing. The index is based on a questionnaire, sent regularly to a permanent sample of representative American corporations. A characteristic feature of the questionnaire is the fact that it does not give a specific definition to the term "capacity", a concept which usually is defined in very different ways by different economists. The corporation responding to the questionnaire can use its own definitions to determine what constitutes "100%". But even more important, is a second characteristic; the corporations must give two indices, one for the "preferred" rate and another one for the "actual" rate of capacity utilization. In the conceptual framework of the present book the "preferred rate" is nothing else but the aspiration level of capacity utilization. If, for example, the answer is that the preferred rate 90% and the actual rate, 85%—this answer is sufficient for our purpose. With this information we can determine the degree of tension of aspiration:

$$\hat{\varepsilon} = \frac{\alpha}{\omega} = \frac{85}{90} \tag{19.13}$$

For the present discussion we do not need the ratio 85/100 which would reflect an engineering economic interpretation of 100% capacity.

Now we must introduce a concept.

In our opinion, the aspiration level of the producing firm usually seems to be higher than the actual level of production and sales. There are possibilities of production expansion.

TABLE 19.4

Index of capacity utilization in U.S. manufacturing*

Year	Preferred Rate	Actual Rate of Operation**
1954		84
1955		92
1956		86
1957		76
1958		80
1959		85
1960		77
1961		83
1962	92	83
1963		85
1964		88
1965		90
1966	93	88
1967		86
1968		85
1969		83

* Source of data: direct information of the McGraw-Hill Publications Company, Department of Economics.
** Operation in December.

Definition 19.3*. Let us call the p o t e n t i a l   p r o d u c t i o n   i n c r e-
m e n t and denote g_i the maximum additional production volume which the
i^{th} organization (producing firm) could attain within a comparatively short
running in period by means of a more intensive utilization of the available
resources and inventories. (The components of indicator vector g_i belong to the
same indicator type as the components of aspiration level α_i and result ω_i.)

The potential production increment may have various sources. A greater
stock of semi-finished and finished products may have accumulated in the
firm than is necessary. Machinery and space usually are not fully utilized. The
firm may switch over to longer operation hours, e.g. by introducing several
shifts in some workshops etc. Bottle-necks may be widened by means of minor
and quickly realizable investments.

The increases in production which can be attained by means of major investments over comparatively long periods of time can *not* be included in potential
production increment g_i.

From what has been said above it is clear that the g_i/ω_i ratio can hardly
reach 50 or 100 per cent. However, in most plants an increment of 5, 10 or 20
per cent could be achieved, if a buyer could be secured for the additional
products.

Accordingly, the vector g_i expresses a gap between potential and actual production. In the terminology employed in socialist countries, this represents the factor's "concealed reserves". In Western terms, this is the "slack" concealed in the firm.

In connection with this problem we have the following hypothesis:

Statement 19.1. The selling aspirations of the producing firm will crystallize in the neighbourhood of the former level of production increased by the potential production increment.

Therefore, according to Statement 19.1

$$\alpha^{(S)}(t_1) \approx \omega^{(S)}(t_0) + g^{(S)} \tag{19.14}$$

Statement 19.1 is but a hypothesis which ought to be substantiated empirically (e.g. by interviewing decision-makers in the firm). However, its truth does follow partly from the definitions; in so much as it does, it is not an assertion but a definitional relationship only.

Aspiration level α expresses the wish, the optimistic expectation of the decision maker: "If it depended only on myself and if conditions were at least comparatively favourable, then I would like to produce and to sell so much" — this is the meaning of the aspiration level in the producing firm selling its products. It is but natural that this wish, this optimistic expectation, takes into account the potential product increment. If the matter depended exclusively on the producing firm, the latter would be able to turn out a volume $\omega + g$ instead of ω.

Let us now sum up the characteristics of the market when the seller's aspirations are higher than the actual volume of production.

1. The sellers queue for the buyers and not conversely.

$$\Theta_j^{(S)} > \vartheta_j, \quad \Theta_j^{(B)} = \vartheta_j. \tag{19.15}$$

2. There appears a (positive) aspiration tension with the seller but not with the buyer.

$$\varepsilon^{(S)} > 0, \quad \varepsilon^{(B)} = 0. \tag{19.16}$$

3. The seller is not fully satisfied but the buyer is:

$$\mu^{(S)} < 1, \quad \mu^{(B)} = 1. \tag{19.17}$$

19.6. General Definitions and Statements

We have surveyed separately the two types of situation—the shift in favor of the seller on the one hand and that in favor of the buyer on the other. Now let us proceed to the definition of general concepts.

Definition 19.4*. There is p r e s s u r e in the market of the jth product when the sellers queue for the buyers, when a positive aspiration tension appears for the sellers whose aspirations are not completely fulfilled. There is s u c t i o n in the market of the jth product when the buyers queue for the sellers, when a positive aspiration tension appears for the buyers whose aspirations are not completely fulfiled. Common to both pressure and suction is d i s e q u i l i b - r i u m on the market. There is e q u i l i b r i u m in the market when the aspiration levels of the seller and the buyer are equal.

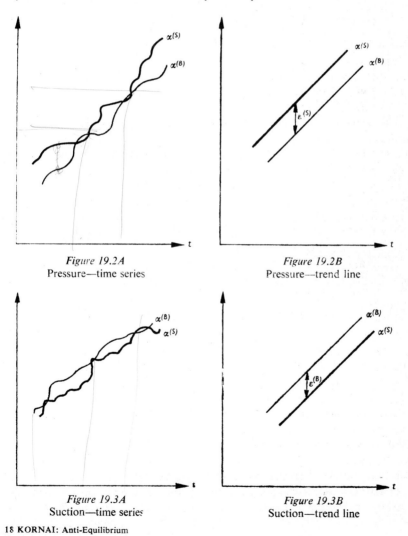

Figure 19.2A
Pressure—time series

Figure 19.2B
Pressure—trend line

Figure 19.3A
Suction—time series

Figure 19.3B
Suction—trend line

The characteristics of the situations of pressure, suction and equilibrium are presented in *Figures 19.2* and *19.3*, and summed up in *Table 19.5*.

Parts A of Figures 19.3 and 19.4 present the actual course of the market situation over time. We shall revert to the B parts of the figures below. The continuous line shows the aspiration level of the seller and the dotted line that of the buyer. In case of pressure the continuous line is above the dotted one; in the case of suction the position of the lines is reversed.

TABLE 19.5

The characteristics of pressure, suction and equilibrium

	Pressure	Suction	Equilibrium
Relationship between limitation and waiting period	$\Theta_j^{(S)} > \vartheta_j = \Theta_j^{(B)}$	$\Theta_j^{(S)} = \vartheta_j > \Theta_j^{(B)}$	$\Theta_j^{(S)} = \vartheta_j = \Theta_j^{(B)}$
Tension of selling aspiration	$\varepsilon_j^{(S)} > 0$	$\varepsilon_j^{(S)} = 0$	$\varepsilon_j^{(S)} = 0$
Tension of buying aspiration	$\varepsilon_j^{(B)} = 0$	$\varepsilon_j^{(B)} > 0$	$\varepsilon_j^{(B)} = 0$
Fulfilment of seller's aspiration	$\mu_j^{(S)} < 1$	$\mu_j^{(S)} = 1$	$\mu_j^{(S)} = 1$
Fulfilment of buyer's aspiration	$\mu_j^{(B)} = 1$	$\mu_j^{(B)} < 1$	$\mu_j^{(B)} = 1$
Relative strength of market forces	$\Omega_j > 1$	$\Omega_j < 1$	$\Omega_j = 1$

Concepts similar to pressure and suction are sometimes employed by the plant engineer in vertical plants. One case is when the foundry is "pressing" the castings to the machine shop and the latter, the parts to the assembling shop. In this case it is the supplier who is impatient, flooding the user with the product. The other case is when the machine shop is sucking the parts from the machine shop. Here, it is the user who is impatient and who is urging the supplier to produce more.

"Pressure" corresponds to the state sometimes called a "buyers' market"; and "suction", to the one called "sellers' market".

Definition 19.5. In the market of the jth product, the ratio of the aspiration levels of the seller and the buyer is the i n d e x   o f   t h e   r e l a t i v e s t r e n g t h of market forces which will be denoted Ω_j:

$$\Omega_j = \frac{\alpha_j^{(S)}}{\alpha_j^{(B)}} \qquad (19.18)$$

The Ω_j ratio determines which one of the trading partners in the market is stronger, which one dominates the market. When $\Omega_j < 1$ we have a "sellers' market", with the sellers dominating; when, on the other hand, $\Omega_j > 1$, we have a "buyers' market" where the buyer is the "master."

Definitions 19.4 and 19.5 treat the market of a single product only. To describe a whole economic system, we must go farther and describe the general situation in the market. This can be achieved by calculating suitable averages of the indices α, ε, μ, and Ω. The calculation of two types of averages is necessary.

Instead of a single period, the average of some *longer period* must be taken into account (e.g. moving averages or trends obtained by regression calculation). This has already been mentioned in the concluding part of Section 19.3; it is generally valid.

Moreover, instead of a single product, the average of all products must be calculated. For weighting, the prices of the products can be used.

For the sake of illustration let us describe such an index.

Let us denote by $\varepsilon^{+(B)}$ a vector; the number of components of that vector is equal to the number of products in the economy. The jth component is determined in the following way:

$$\varepsilon_j^{+(B)} = \begin{cases} \alpha_j^{(B)} - \omega_j^{(B)}, & \text{if } \alpha_j^{(B)} - \omega_j^{(B)} > 0 \\ 0, & \text{if } \alpha_j^{(B)} - \omega_j^{(B)} \leq 0 \end{cases} \qquad (19.19)$$

In this formula, $\alpha_j^{(B)}$ and $\omega_j^{(B)}$ denote respectively the sum of the aspiration levels and the sum of the actual purchases of all buyers in the economy. Accordingly $\varepsilon^{+(B)}$ represents the tension of the *unfulfilled* buying intention. In the vector we find zeros where the aspiration is fulfilled, or where—the following forced substitution—the buyers purchased more than their original intentions.

The total of all buyers unfulfilled intentions can be represented as follows:

$$Z^{(B)} = \varepsilon^{+(B)}p, \qquad (19.20)$$

where p is the price vector. More characteristic might be the relative index $\tilde{Z}^{(B)} = Z^{(B)}/\omega p$, where ω is the vector of all actual purchases.

Symmetrically $\varepsilon^{+(S)}$, $Z^{(S)}$ and $\tilde{Z}^{(S)}$ can be computed for the sellers.

In a given economy in a given time period both $Z^{(B)}$ and $Z^{(S)}$ can be positive. However, it is very characteristic of the state of the economy for one of the two numbers to be significantly larger than the other. If $Z^{(B)}$ is much larger, than $Z^{(S)}$, there is general suction; in the reverse case there is general pressure.

We must emphasize that above we have shown only one of the possible measures of pressure-suction. On the present general level of clarifying concepts, it is unnecessary to take a stand on the concrete mathematical-statistical details of averaging; these will depend, among other things, on the numerical data used and on the aims of the analysis. Therefore, we may content ourselves with a looser definition:

Definition 19.6. The market situation in an economic system will be characterized by g e n e r a l   p r e s s u r e   (or   g e n e r a l   s u c t i o n) when for the products accounting for the major part of social production there prevails pressure (or suction) over the average of a longer period of time.

The relationships described in Table 19.1 also characterize the situation of *general* pressure, suction or equilibrium (the suffix *j* indicating the serial number of the product must be eliminated from the symbols).

With these new concepts we can make the following statement:

Statement 19.2. In most of the actual economic systems there usually prevails either general pressure or general suction. Should the relations of market forces be different in the markets of the various products, with neither general pressure nor general suction prevailing, the markets of the individual products will still be characterized either by pressure or by suction.

And here we have reached one of the most important subjects of our book.

Occasionally, in the market of some individual product, the economic systems may reach the equilibrium situation $\alpha^{(B)} = \alpha^{(S)}$. This equilibrium situation is represented in Figures 19.2A and 19.3A, by the intersection points of the continuous and dotted lines. However, these are, exceptional moments. The characteristic situation—certainly in the market of the individual product but also in the economic system as a whole—is *either* that of pressure *or* that of suction.

In the controversies on equilibrium, many economists willingly admit the following; of course, the market is never precisely in the state of equilibrium, but it oscillates around the equilibrium. In our language, this would mean that although the values of $\alpha^{(B)}(t)$ and $\alpha^{(S)}(t)$ may momentarily differ from each other in period *t*, their *trend* over a longer period $[t', t'']$ would be identical. However, in Figures 19.2B and 19.3B we present what in our view corresponds to the actual reality, namely that the two *trends* are not identical. It is either the trend of buying aspirations that is above the trend of selling aspirations or conversely. Accordingly, *lasting disequilibrium* prevails.

Statement 19.3. In the case of pressure it is the seller and in that of suction, the buyer, who endeavors to reduce the tension, wishing that his aspiration be fulfilled. These forces acting towards equilibrium will assert themselves. However, the tension is continually reproduced.

19.7. The Problems of Observation and Measurement

In our work we have endeavored to separate the variables of the control sphere from those of the real sphere. Now, in speaking of the problems of observation and measurement, we must bear this distinction in mind.

Firts let us consider the buying intentions. When there is pressure in the market, buying aspirations are usually identical with actual purchases: $\alpha^{(B)} = \omega^{(B)}$. Thus from the observation of the real sphere we can draw conclusions concerning the variables of the control sphere. In simpler words, it is sufficient to observe the actual purchases to draw conclusions concerning the buying intentions.

Accordingly, that which is called a "demand function" in econometrics actually reflects not only the purchases but also the buying intentions—but *only in the case of pressure.*

In the case of suction, the situation is different. As a matter of fact, in that case intention and realization may differ essentially from one another: $\alpha^{(B)} > \omega^{(B)}$. Many buyers have to correct their original aspirations. Turnover statistics and household statistics reflect not the buyer's aspirations and intentions (his "demand") but the distortion of those intentions produced by circumstances external to the buyer.

In such cases, we must not draw conclusions about the *control* variables (buying intention) from the *real* variables (purchase), as mentioned in Section 18.3. The buyers' aspirations can be observed only through suitable interviews. The subject of the interview must tell what he would buy if his actual purchase depended only on himself (his purse and tastes). Market research institutions all over the world are engaged in such public opinion polls; their techniques are highly developed and they have withstood the test of practical application.[10] Thus it is entirely possible to organize such observations.

Now, let us turn to the intentions of the *seller*. In the case of suction usually there is an identity of selling intentions and actual sales: $\alpha^{(S)} = \omega^{(S)}$. Thus, in this case, the real variables (production and sales) reflect the seller's intentions, i.e. the control variables.

As has already been pointed out, there is a special case of suction when the seller's aspirations are not identical with actual sales and a negative tension appears: $\alpha^{(S)} < \omega^{(S)}$. This case appeared in Hungary in the early fifties: the central instructions, the too high plan targets, and the urgings of dissatisfied buyers forces the firms to perform above their capacity. In that stituation, it would have been possible to learn something about the genuine aspirations of

[10] In addition to *Katona* [113] cited above, see also in Hungarian, e.g. *Szabó* [246].

the firms from interviews with their managers; probably these would have yielded frank answers such as: "What we would like is a quieter, slower pace of production."

Obviously, in the case of pressure, the sellers' aspirations can be discovered only by means of interviews and opinion research: $\alpha^{(S)} > \omega^{(S)}$. (See Section 18.3.)

However, in accordance with the hypothesis described in Statement 19.1, it is possible to draw conclusions concerning the selling intentions in an indirect manner, namely from estimates of the potential level of production. By adding to the previous actual volume of production the potential production increment (i.e. the additional production volume which can be achieved under the given conditions of the *real* sphere) we can draw conclusions concerning the variable of the *control* sphere which interest us, the selling aspirations.

In this book, the problems of measuring pressure and suction, and the indices relating to them only have been raised. Measurement and observation are usually preceded by the theoretical clarification of the interrelationships; theory, in turn, can be further improved and made more precise on the basis of experiences gained in measurement and observation.

19.8. On Demand and Supply Functions

The problem of demand and supply functions is closely related to observation and measurement; this already has been touched upon in several places.

It appears that there is no necessary contradiction whatsoever between the results attained by econometricians in the formalization and quantification of demand and supply functions on the one hand and the conceptual framework and statements contained in this book. The econometric statements on demand and supply functions are to be considered a contribution to a better knowledge of the system of response functions. The *empirical* statements relating to demand and supply stand alone, by themselves they do not require support from either the utility functions or other elements from the GE school. On the contrary, if econometric observation breaks the threads connecting it with the GE school, it will be able to expand and the empirical observation of buying and selling intentions will become deeper and better grounded. We should like to call attention to a few tasks of such expansion.[11]

1. The explanation of consumer's demand is restricted too much to the

[11] One of such "expansion tasks" already has been mentioned in the preceding section: direct interviews should be used to identify the purchasing intention of the buyer in the case of suction and the selling intention of the seller in the case of pressure. Statistics on turnover do not yield a true picture of these.

examination of price and income effects. This corresponds to the view of the GE school—but in general, reality is more complicated. The effect of prices and income is great indeed, but there are also other, important explanatory variables; informations of non-price character have their influence, too. Some examples are:

—Imitation, fashion, following the lead of reference-groups.

—Continuous re-stratification of the pattern of consumption in favour of *new* products, at the expense of the old ones.

—Consumer behaviour, as a function of social position, e.g. the effect of urbanization or the development of suburbs.

2. There are very few empirical observations available regarding the demand functions of firms. The explanation of changes in the input pattern of production which relies exclusively on prices is a poor one. Changes in factor combinations are explained, in the final analysis, by changes in the volume of resources available, and in this relation by processes of technical progress. To a certain extent, technical progress has a life of its own. Informations of price character reflect more or less these real changes, and also the lags in adaptation to those real changes (the temporary disproportions). Necessity for technical changes is partly conveyed by informations of non-price character.

3. There are few reliable empirical works on supply functions of firms. Real observation is mostly replaced by *a priori* assumption: the supply function should coincide with the marginal cost function. If it is true that price is set independently of the firm, that the firm maximizes profits and the cost function is convex, the volume of production will be always at the point where the price equals marginal cost, provided average variable costs are covered.

In fact—as has been pointed out several times—all these assumptions are rather weak. The response function of the firm is a multi-variate one; both the volume of production, and the intention to sell develop under the influence of many kinds of impulses. (As we have seen, the two are not the same.) We already have discussed these impulses: stock reports, direct information from the buyers, expectations regarding the future, instructions or recommendations obtained from central organs, etc.

Each of the "expansion tasks" outlined in points 1–3 is related to the fact that empirical observation must get rid of the yoke of the GE school.

19.9. Comparison

In Sections 19.7 and 19.8 we began our critique of the GE school. In the following we shall compare mainly the conceptual frameworks.

Between the conceptual frameworks there are essential differences. In Chapter

19 we generally have not employed the concepts of "demand" and "supply" because we found them too imprecisely defined.

There is only one point in common between two conceptual frameworks, and this is the *concept of equilibrium*. This concept is not the feudal property of the GE school but the common property of the major part of the natural and social sciences.

The definitions of equilibrium necessarily deviate because they have been conceived in different "languages", with the aid of different conceptual frameworks. Their meaning is, however, identical: the equality of buying and selling intentions.

Accordingly, it is not in the interpretation of the equilibrium concept that the difference lies but rather in answering the question, *what is the role of equilibrium* in the functioning of the economic system.

According to the GE school—if we consider it as a descriptive-explanatory science—equilibrium is the normal, general and characteristic state of the economic systems. It is the trend around which buying and selling intentions oscillate. The main message of Chapter 19, on the other hand, is that the lasting tendency, the trend of actual economic systems over time, is one of disequilibrium, in the form of either pressure or suction.

In the debates provoked by the earlier drafts of this book the following argument has been voiced frequently.

The present work does nothing but give a *new name* to equilibrium. What we call pressure is the same phenomenon that is called equilibrium by the GE school, since the equilibrium state as described in GE theory implies the existence of some normal reserve capacities.

We do not desire to become entangled in an empty argument about terminology. It is well known how rigidly the deeply convinced adherents of orthodox intellectual schools insist on use of certain *words*. There are words to which are coupled not simply intellectual deliberations but feelings and which have become almost holy. Sometimes the revision of statements, theorems, principles is more easily accepted than that of concepts or words. Whosoever desires this comfort may call equilibrium what we call pressure. It is not the conceptual difference which is important but the difference in contents. Equilibrium expresses in every discipline *equality* between two opposed forces, effects. In contrast, according to the definitions of this book, *pressure* expresses *inequality*. We call pressure the state where there is overweight, preponderance on the side of one of the partners (the buyers, that is, the buyers' market). We believe that most advocates of the GE school consider this state of affairs to be desirable from the point of view of economic policy—but this conclusion of theirs becomes rather blurred in their theoretical works advocating equality of forces.

But here we have come to the normative evaluation of the problem. Up to this point, we have dealt merely with the descriptive-explanatory interpretation of the theories of the GE school. Is it or is it not beneficial from the point of view of the performance of the economic system for disequilibrium to exist for pressure or suction to prevail? The answer to this question, the normative interpretation of the theory, will be left to later chapters.

There is a second comment about our terminology that often arises in discussions of the present book. According to a second view, our concept of "pressure" is identical with the traditional concept of "excess supply". We shall discuss this point in Section 22.9.

19.10. Economic-Historical Starting Points

Further treatment of the subject builds upon Statement 19.2: the functioning of economic systems is characterized by the fact that usually there prevails a lasting state of either *pressure* or *suction*. From the point of view of economic history, the statement can be completed as follows:

Statement 19.4. In most socialist countries, including Hungary, suction has prevailed over long periods of time, more or less generally (i.e. in the system as a whole).

Statement 19.5. In most advanced capitalist countries pressure has prevailed over long periods of time (especially with the exception of war years) more or less generally (i.e. in the system as a whole).

Both statements contain restrictions ("in most countries", "more or less generally...") indicating that we acknowledge the presence of opposite phenomena. In socialist countries—in many branches of industry and frequently—surpluses emerge and unsaleable stocks accumulate. And conversely, in the developed capitalist countries the palpable lack of certain products is not unknown even in times of peace. Still, for a longer period and for the majority of industries, for most countries, the statements hold.

Both statements sum up facts generally known from experience. It is hardly to be expected that the simple statement of facts would be challenged by anyone. (To illustrate Statement 19.5, we presented some US date in Table 19.4.) It is rather the following questions that may give rise to debate:

—What are the consequences of the phenomena described in Statements 19.4 and 19.5.

—What are the causes of the phenomena? Is suction a necessary concomitant of socialism and pressure, one of capitalism?

To the second question, we should like to anticipate our final answer, which is in the *negative*. Suction or pressure do not follow from the fact of socialism or capitalism; they arise from the comprehensive effects of a whole range of historical conditions. Our general view on this question will be outlined later on in greater detail.

The course of the remaining discussion will be as follows.

In Chapter 20 we shall shelve for some time the problem of the relationship between market forces, to which we shall revert in Chapter 21. In order to describe the effects of pressure and suction in the real sphere, first we must examine some aspects of economic growth. We shall deal primarily with technical progress and with the questions of *volume* and *quantity*. (The question has been anticipated somewhat in Chapters 11 and 16, but now we shall deal with it in greater detail.)

The analysis of pressure and suction will be taken up again in Chapter 21. There, we shall deal with the *consequences* of the relations of market forces. Somewhat artificially, we have separated from this the explanation of the *causes*, the conditions and circumstances which give rise to pressure and suction in the market. These will not be treated until Chapter 22.

20. VOLUME AND QUALITY

20.1. The Case of the Motor-Car and Textiles

Throughout the world, two closely interrelated processes take place; *the volume of products turned out increases and their quality improves.*[1]

Let us consider two characteristic products, motor-cars and textiles. The case histories are presented in *Tables 20.1* and *20.2*. In both tables, the first column shows the increase in the volume of world-wide production. Volume is measured in the case of the motor-car in numbers of cars, in that of textiles in tons of the raw material. It can be seen that progress is rather rapid.

The volume time-series in themselves yield little information about the consumer's supply of motor-cars and textiles. Today's motor-car is not identical with that of the twenties, and present-day textile production (or, at least, a considerable part of it) greatly differs in quality from that of fifty years ago. The rest of the columns in the two tables are intended to give an idea of these differences.

In column 3 the major events of technical development are listed for each of the two product groups. Here we present the major inventions and innovations which have contributed to improving the quality of motor-cars and textiles. In columns 4 and 5 we indicated the time when these inventions and innovations were first applied in industry and the country where they were first used. The remaining columns show the trends over time of certain qualitative characteristics of the product in question.

It goes without saying that this data cannot present a complete picture of the dynamics of qualitative development. However, it will give us some idea of the qualitative development, and it does show how one-sided it would be to measure the changes in the economy exclusively by the growth in the volume of production. Equally as important as the volume of production are the entry on the scene of new products, the steady improvement in quality, the internal

[1]Here, as well as in the following, we always speak of "products" in accordance with the conceptual framework introduced in Chapter 4. However, the term also is meant to cover services. This particularly must be emphasized when speaking of technical progress, since one of the most important phenomena of technical progress is precisely rapid growth of the volume of services and the enormous improvement in their quality.

TABLE 20.1

Growth of the volume of world production of passenger motor-cars, and improvement in their quality*

Year (1)	Volume of world production, thousands (2)	Major Inventions and Innovations			Share of cars having automatic speed control in U.S. cars sold (6)	Acceleration in European cars (seconds needed to attain 100 km/h) (7)
		Invention or innovation (3)	Country where it was first introduced (4)	Date of first introduction (5)		
		1. Passing from the "coach-form" to "car" form	Germany	1900		
		2. High-tension magnetic ignition	Germany	1903		
		3. Electric lighting in cars	Germany	1910		
		4. Electric self-starter	U.S.A.	1911		
		5. Engine with short stroke travel	France	1913		
		6. Hydraulic brakes	U.S.A.	1921		
		7. Four-wheel brakes	France	1923–24		
		8. Stop-lights	U.S.A.	around 1925		
1926	4,355	9. Electric windshield wiper	Germany	around 1926	0%	
		10. Built-in radio in cars	U.S.A.	1927		
1930	3,390	11. Automatic speed control	U.S.A.	1930-ies		
		12. Front-wheel drive	France	1933		
		13. Streamlined car	Czechoslovakia	1934		
		14. Body pressed from steel plate	U.S.A.	1934		
1938	3,050	15. Passenger car with Diesel-engine	Germany	1936		
1950	8,170					40 seconds

TABLE 20.1 (continued)

Year (1)	Volume of world production, thousands (2)	Major Inventions and Innovations			Share of cars having automatic speed control in U.S. cars sold (6)	Acceleration in European cars (seconds needed to attain 100 km/h) (7)
		Invention or innovation (3)	Country where it was first introduced (4)	Date of first introduction (5)		
1955	10,950	16. Plastic body	German Democratic Republic	1955		
		17. Power-steering	German Federal Republic	1956		
		18. Foam rubber coated fitting plates	U.S.A.	1956		
1960	12,670	19. Wankel-engine	German Federal Republic	1963	74%	
1965	19,090					14 seconds
1967	18,260					

Sources of the volume data [117] and [262]. Sources of the data concerning the inventions: [95], [196], [202], [270] and [274]. Beside the published sources—we have received information informally from *Ferenc Konrád Urvölgyi* and *György Liner*. The table was complied by *Attila Soós*.

TABLE 20.2

Growth of the volume of world production of textiles, and improvement in the quality of textiles

Year (1)	Index of world production 1910 = 100 (2)	Major inventions and innovations				Share of man-made staple fibres as percentage of world / U.S.A. production (7)
		Invention or innovation (3)	Short description (if not generally known) (4)	Country where it was first introduced (5)	Date of first introduction (6)	
1910	100	1. Viscose yarn		Great Britain	1904	
1920	89	2. Acetate rayon		Great Britain	1921	
1930	119	3. Sanforizing of cotton fabrics		U.S.A.	1928	
		4. Crease-resisting finish		Great Britain	around 1930	
		5. Water-repellant finish		U.S.A.	1930	
		6. Flame-proof finish		G.B., U.S.A.	1938–1940	
		7. Nylon		U.S.A.	1939	
1940	159	8. Polyester fibre		U.S.A.	1941	
		9. Polyvinyl-alcohol fibre	used for underwear and bed linen	Japan	after 1945	
		10. Non woven fabrics: glued	prepared by glueing and pressing together non-woven carded webs	U.S.A.	1948	
		11. Malimo	made from yarn, sewn looped fabric	German Democratic Republic	1949	
1950	162	12. Non woven: tufted	mechanized carpet tufting	U.S.A.	1952	0.4
		13. Stretchable silk-type manmade yarns		Switzerland	1952	0.8 (1948)

TABLE 20.2 (*continued*)

Year (1)	Index of world production 1910 = 100 (2)	Major inventions and innovations		Country where it was first introduced (5)	Date of first introduction (6)	Share of man-made staple fibres as percentage of	
		Invention or innovation (3)	Short description (if not generally known) (4)			world	U.S.A. production (7)
1960	284	14. Acrylic fibres	used for fabrics mostly in mixtures	Japan, Italy	1953–54	4.8	8.1
		15. Polypropylene	used for upholstery and knitwear	Italy	1960		
		16. Polyuretane (spandex)	used for elastic textiles	U.S.A.	1960–1961		
		17. Permanent press	fabrics are soaked with chemical, then subjected to high pressure	U.S.A.	1961		
		18. Polinose-viscose fibre	the wet breaking strength of earlier used fibres was low, the new product's molecules are modified for greater strength	Western Europe, Japan	1963		
1965	334					11.9	17.4
1966	331					14.5	25.7
1967	329					16.7	32.4

rejuvenation, the process which Schumpeter has so aptly termed as "creative destruction".[2]

20.2. The Increase in Volume

After the introductory examples above, we can proceed to a more general clarification of the concepts. In the course of clarification, in Sections 20.2–20.3, we employ symbolical notations. These are not used in the framework of mathematical reasoning or a formal model. The purpose of the symbolical description is mainly to render the task of observation and *measurement* more unequivocal. This is especially important, since many economists are inclined to relegate the scope of problems discussed here into the realm of "unquanti-fiable" phenomena.

Let us begin conceptual clarification with the notion of volume.

As has already been indicated, the term "motor-car" covers not a single type of product but a whole range of automobiles. With "textiles" the case is similar.

Let us call the set of products $g_{j_1}, g_{j_2}, \ldots, g_{j_{n_j}}$ serving similar purposes and measurable in the same physical unit of measurement the jth *product group* and denote it $\mathcal{G}_j : \mathcal{G}_j = \{g_{j_1}, g_{j_2}, \ldots, g_{j_{n_j}}\} \subset \mathcal{G}$. Let us call the volume of the jth product group and denote $V_j(t)$ the total output of the products in the product

---

[2] See *Schumpeter* [224], and [225].

(Fortnote to Table 20.2)

Sources of the volume data: [98], [242], [243], [260], [261], [262]. We had no immediate data regarding the volume, therefore we had to resort to conciliatory solutions in a number of cases to reconcile data that were not strictly comparable and to perform complementary estimates.

The volume data do not contain the porous imitation leather, foam rubber, and other textiles not used for clothing (e.g. hemp, etc.). The summations have been made upon basis of the raw material production, in tons; with the alteration that—in accordance with the general amount of loss—cotton is taken with a weight reduced by 5%, wool with a weight reduced by 65%, and raw silk with a weight reduced by 50%.

In the cases of some years the number shown on the table has been calculated by interpolation, making use of the data of the closest years.

Since here we want to demonstrate only the increasing trend of the volume, with an aim solely illustrative, this inexactness is not to cause any difficulty. Sources of the data regarding the inventions: [44], [52], [119].

Beside the published sources we have received information from *Sándor Fülöp, Walter Endrei, Mrs. I. Rusznyák* and *Ferenc Izmay*. The table was compiled by *Attila Soós*.

group produced in the tth period, as measured in the physical unit of measurement characteristic of the jth product group:

$$V_j(t) = \sum_{i \in \mathcal{G}_j} \vec{x}_i(t) \tag{20.1}$$

Volume indicators $V_j(t)$ will be called *V-indicators*.

When measuring volume $V_j(t)$ we deliberately disregard the qualitative properties of the various products belonging to the product group. We do not care whether the car is a four-cylinder or a twelve-cylinder one, whether its maximum speed is of 30 or 150 km, whether it seats four or six passengers. On an abstract level we separate two aspects of real economic development, describing the growth in volume independently of qualitative changes of the products.

Fortunately, the ideas of this book do not require that we *add up* the volumes of the various product groups. It will be clear from the literature on the subject that such summation involves great methodological difficulties.[3] Here, we need not get entangled in these difficulties. When dealing with volume, we always speak in the *plural*; instead of a single, economy-level volume indicator we speak of the volumes $V_1(t)$, $V_2(t)$ ... of the various product groups and of the factors furthering or hindering the growth of these volumes.

20.3. The Quality Concept

Let us now turn to the other aspect of real economic development, the improvement of quality.[4] First, the concept of quality should be clarified. Before going into detail, we would like to point out that the *philosophical* problems of "quantity-quality" are avoided here. The concept of the "product's quality" will be used in the everyday sense of the word. It is used in the sense of the engineer speaking of lathes of different quality, or the housewife speaking of different detergents.

---

[3] When calculating most of the aggregate volume indices, the individual volume indices are weighted by the prices. In such cases it is problematical whether to use constant or current prices, those of the first or of the last year, domestic or foreign prices, etc.

Other methods of computation would avoid the prices and weight the volume indices of product groups by other indicators (e.g. manpower or work-hour input).

To measure economic development, *F. Jánossy* ([99], [100]) has calculated direct relationships between aggregate indicators (e.g. national income in dollar terms), on the one hand, and the volume of the most characteristic product groups expressed in physical units, on the other.

[4] In this chapter, we have drawn on the works of *Kuenne* [142] and *Griliches* [76].

In the sense of the engineer or the housewife, the product's quality is not some undefinable or unmeasurable property. It is not our intention to become entangled in the problem of measuring the aesthetic properties of commodities; rather we are prepared to accept the principle of "de gustibus non est disputandum". However, the qualitative properties taken separately are quite "tangible" and lend themselves quite well to description and thus also to measurement.

Some qualitative properties obviously can be measured by real numbers. For example the motor-car's maximum speed, normal fuel consumption, trunk capacity, etc. can be described by means of continuous variables. The number of doors or passengers is an integer variable. In the case of textiles, it is possible to describe in terms of continuous variables the fabric's breaking strength, the count of the yarn, the density of weft etc.

Other qualitative properties can be expressed by the presence or absence of some characteristic. For example is the car fitted with automatic transmission, yes or no? In such cases, a variable of the value 0 or 1 may be ordered to the qualitative property in question.

Definition 20.1. The q u a l i t y of the ith product is composed of M_i qualitative properties. Let these be denoted $u_{i_1}, u_{i_2}, \ldots, u_{i_{M_i}}$. Let us denote $\mathcal{U}_i$ the set of all qualitative properties describing the quality of the ith product: $\mathcal{U} = \{u_{i_1}, u_{i_2}, \ldots, i_{i_{M_i}}\}$. To every qualitative property there can be ordered a measure of the degree to which the qualitative property asserts itself. The degree of assertion of the jth qualitative property in the ith product is given by quality parameter q_{ij}. The vector q_i composed of a total of M_i quality parameters q_{ij} is called a quality vector.

There are products, the quality of which can be described by 2 or 3 or 5 quality properties (gasoline is a case in point). In the case of other more complex industrial products, 50 to 100 quality properties may be needed to give a full description of the quality, although even in such cases it will usually be possible to emphasize 10 to 20 as most important.

All quality or technical standard specifications constitute an example of quality vector q_i. The standard specification lists a finite number of properties, giving for each of them a standard quality parameter value q_{ij}^*. For example, what should be the octane number of the gasoline or the breaking strength of a fabric?

From what has been said above it should be clear that the *quality of a product is often a "quantitative"* category, measurable by a vector the components of *which are real numbers, quantities.* (It is for this reason that we avoid, as a counterpart of the term "product quality", the term "product quantity", speaking instead of the *volume* of production.)

We do not wish to produce the illusion that we have succeeded in defining the quality concept in a way that makes measurement simple. Grave problems still remain open. Let us suppose that we compare the quality of two kinds of products with similar destination and comparable. Some of the quality parameters of product No. 1 are better than those of product No. 2, its other parameters are worse. What should the summarizing judgment be, which is the better product? Is it possible to measure the "average quality", valid for the big product groups or even for the full production of the national economy, to measure the dynamics of the average quality? Is it possible to compare, using summarizing indices, the dynamics of the average quality of the production of two different countries? A few interesting experiments have been made in order to calculate price indices that would reflect the quality development of certain product groups. (For example, *Griliches* worked out the "hedonistic price index" of the American car industry, which is to express the changes in quality.)[5] However, I am sceptical about the question as to whether this can be done for an entire national economy, especially for *long, historical* periods. My conviction is that *it is not possible to express adequately the qualitative development of the total product output of an economic system over a lengthy historical period by means of a single qualitative price index.*

As in the case of measuring volume, here, too, it may be stated that from the point of view of the present book's argumentation it fortunately is not necessary to have a summary index over the whole system describing the comprehensive improvement in quality of all products.

Again, we are content with a whole range of indices $Q^{(1)}$, $Q^{(2)}$, ... which *collectively* throw a light on the changes in quality, on the slower or quicker rate and the characteristics of the changes.

Let us call *quality indices* or briefly *Q-indices* the group of statistical indices which serve to measure some characteristic aspect or component factor of some characteristic set of the qualitative changes in all products of some product group or of a whole sub-system or system.

The Q-indices usually are of a *partial* character; they do not represent comprehensively qualitative development as a whole but only one or another of its aspects or elements. Only the definition will be presented here; various Q-indices will be presented in the course of further treatment of the subject.

20.4. The Revolutionary New Products

There are some products—such as television and nylon, the computer and penicillin—which distinguish the economy, production, or the whole way of living of the present era from that of fifty years ago.

---

[5] See *Griliches* [76].

Let us regard *Table 20.3*. Here we have listed a total of 79 revolutionary new products which have been introduced in industrial production in the course of the past half-century. We have endeavored to confine ourselves exclusively to "civilian" goods; accordingly the table does not contain either atomic energy or rocket technique. True, it is rather difficult nowadays to draw the boundary between industrial research serving military and civilian purposes. There are nuclear power stations serving peaceful ends and rockets used in scientific space research. And conversely, transistors are built not only into the radio sets which young people take with them to the beach but into military equipment. Therefore, any delimitation is necessarily arbitrary. To the 79 new products of Table 20.3 let us add some of the inventions shown on Tables 20.1 and 20.2 regarding car and textile production. Thus, a total of nearly 100 new products are treated here.

The tables do not claim completeness. We certainly could find 10 or even 30 additional products which could be termed as revolutionarily new as justly as some inventions in Table 20.3. It is also open to debate whether all products listed in Table 20.3 merit the title of a "revolutionary new product". (Undoubtedly, only some of the inventions indicated in Tables 20.1 and 20.2 can be regarded as revolutionary new ones.) However, we believe that the tables give some idea of the inventions and new products that have contributed most markedly to forming the way of life in our present era.

Why do the products figuring in the table merit the term "revolutionarily new"?

The most important criterion is that they *tend to transform thoroughly the behaviour of the user organizations and individuals*. The computer has revolutionized scientific research, administration and management; the television has revolutionized entertainment and the way of living.

The revolutionary new product usually generates *new needs*. Of course, people had always longed to see things happening far away; this wish had found expression in many a tale and legend. However, it was only after we had television that the wish took the form of a need which realistically could be satisfied. Or take another example. Small-size, portable—so-called detector—radio sets actually existed decades ago. However, the need faded because of the primitive character of the sets. It came to new life with the appearance of transistor sets which ensured excellent reception.

The production of revolutionary new products usually (although not always) gives rise to *new industries*. This is what happened with the production of motor-cars, tractors, air-craft, etc. Within a few decades after innovation, these became leading industrial branches in several countries. The production of man-made fibres and synthetic materials, and electronics as well have grown into independent industries within even shorter periods of time.

TABLE 20.3

*Revolutionary new products in the past half a century**

Industrial sector (1)	Name of the new product (2)	Short description of new product if not generally known (3)	Country first introducing it (4)	Date of first introduction (5)
Machine industry and electronics**	1. Radio tube		U.S.A., Germany	1913–1918
	2. Fluorescent lighting		France, Germany	around 1934
	3. Radar		France, G.B., Germany, Netherlands, U.S.A.	
	4. Television		G.B.	1935–1938
	5. Electron microscope		Germany, U.S.A.	1936
	6. Tape recorder		Germany, U.S.A.	1939
	7. Vidikon	TV pick-up tube	U.S.A.	after 1940
	8. Long-playing record		U.S.A.	1945
	9. Mesh collecting grill	Used in high performance transmitting tubes	German Federal Republic	1948
	10. Transistor		U.S.A.	1948
	11. Fine adjusting device for the working of gear wheels		Switzerland	after 1950
	12. Electronic computer			1952
	13. Halogenized lamps	Injection of halogen elements into the internal space of lamps to attain better colour spectrum, more comfortable to the eye	U.S.A.	after 1950
	14. Silicon transistor	Small-size transistor whose basic material is silicon; used in integrated circuits	U.S.A.	1959
			U.S.A.	1962

Table 20.3 (continued)

Industrial sector (1)	Name of the new product (2)	Short description of new product if not generally known (3)	Country first introducing it (4)	Date of first introduction (5)
Vehicle production (except cars)*	15. Machine tools with digital control		U.S.A.	after 1960
	16. Diesel-electric engine		Sweden	1913
	17. Tractor		U.S.A.	1920
	18. Pneumatic brake		Germany	1938
	19. Helicopters		U.S.A.	1941
	20. Jet-propelled engines		G.B., Germany	1943
Optical and photo-chemical industry**	21. Drop shutter		Germany	around 1925
	22. Colour photography		Germany, U.S.A.	around 1935
	23. "Zoom" lens		France	after 1945
	24. Polaroid land camera	Takes and develops picture within camera	U.S.A.	around 1959
	25. Automatic electric shutter linked with photometer		Japan	1960
	26. Maser	Quantum-optical simplifier	U.S.A.	1953
	27. Laser	Quantum-optical simplifier	U.S.S.R., U.S.A.	after 1960
Printing	28. Xerography	Printing procedure serving as a basis for offset printing	U.S.A.	1952

*This part of the Table does not include the inventions involving cars, which have been shown on Table 20.1. Sources: [181], [196]. Personal information: *Ferenc Konrád Urvölgyi, Zoltán Köröskenyi.*

** Source: [41]. Personal information: *Mrs. M. Lugosi, József Sauer* and *János Pap.*

Table 20.3 (continued)

Industrial sector (1)	Name of the new product (2)	Short description of new product if not generally known (3)	Country first introducing it (4)	Date of first introduction (5)
Other machinery*	29. Cotton-picker		U.S.A.	1942
	30. Sulzer-loom	Weaving machine without shuttle working twice as fast as the traditional one	Switzerland	1950
	31. Mechanization of protecting devices in mining (self-moving protection)	Enables complex mechanization of production practically eliminating danger of fatal accidents	G.B.	1950
	32. Kaplan water turbine		Czechoslovakia	1921
Precision engineering, metal mass products**	33. Automatic wrist watch (self-winding)		U.S.A.	1928
	34. Zip fastener		U.S.A.	1918
	35. Ball-point pen		U.S.A.	1945
	36. Beryllium-bronze alloy	High-strength and elastic material used in telecommunication for small-size parts	U.S.A.	1934
Metallurgy***	37. Cold-rolled textural transformer steel	Basic material for transformer plates with low hysteresis loss	U.S.A.	1942
	38. Industrial production of titanium	Used for aluminium alloys	U.S.A.	1944

* Personal information: *György Hajós* and *János Pap*.
** [102], is the only source.
*** Personal information: *Péter Juhász*.

Table 20.3 (continued)

Industrial sector (1)	Name of the new product (2)	Short description of new product if not generally known (3)	Country first introducing it (4)	Date of first introduction (5)
	39. OFHC-copper	Free of oxygen, good electric conductor, basic material for electronics	U.S.A.	1948
	40. Tungsten-carbide	Better known as vidia: very hard metal used for metal cutting	Germany	1926
Construction and building material, industry*	41. Panel building		U.S.S.R.	before 1940
	42. Multi-story, steel framed building system		U.S.A.	1880–1890
	43. Body-like building system	Heat-insulated walls, built of aluminium panels with drop windows	France, U.S.A.	1951–1952
	44. Shell structure	Used for overbridging big space without propping	France	1910–1915
	45. District heating		U.S.S.R.	1924
	46. Industrial air conditioning		Switzerland	around 1890
	47. Air-conditioning of homes		U.S.A.	around 1917
Pharmaceutical industry**	48. Insuline		Canada, U.S.A.	1930
	49. Penicillin		U.S.A.	1945
	50. Streptomicin		U.S.A.	1948

* **Personal** information: *László Lux.*
** Personal information: *Mrs. I. Kocka* and *Tibor Vidos.*

Table 20.3 (continued)

Industrial sector (1)	Name of the new product (2)	Short description of new product if not generally known (3)	Country first introducing it (4)	Date of first introduction (5)
	51. Cortisons	Used for curing rheumatic and other arthritic deseases	U.S.A.	1948
	52. B_{12} vitamin		G.B.	1948
	53. Chlorocid		U.S.A.	1949
	54. INH	For curing T.B.	G.F.R.	1951
	55. Hibernal Frenolon, Pipolphen	Tranquillizer	France	1952
	56. Tetran		U.S.A.	1953
	57. Prednisolon		U.S.A.	1955
	58. Oral contraceptives		U.S.A.	1957
Other chemical industry*	59. Duco lacquers		U.S.A.	after 1920
	60. Tetraethyl lead	Increases pressure tolerance of solinega	U.S.A.	1922
	61. Synthetic detergents		Germany	1930
	62. Freon refrigerants	Used in household refrigerators and in air-conditioning equipments	U.S.A.	1931
	63. Synthetic rubber		U.S.A., U.S.S.R.	1932
	64. Cellophane tape		U.S.A.	1933
	65. Methakrylate polymers	Plexi-glass	U.S.A.	1935
	66. Catalytic cracking of petroleum		U.S.A.	after 1936
	67. Nylon		U.S.A.	1939
	68. Polyethylene		G.B.	1939
	69. High-effect synthetic plant protectives		G.B.	1940–1945

* Sources: [5], [32], [44], [59], [119], [166], [181], [183]. Personal information: *János Pap* and *Tibor Vidos*.

Table 20.3 (continued)

Industrion sector (1)	Name of the new product (2)	Short description of new product if not generally known (3)	Country first introducing it (4)	Date of first introduction (5)
	70. DDT		Switzerland	1942
	71. Polyester	Strengthened with glass fibre structural materials also used for coating	U.S.A.	1942
	72. Silicones	Electric insulators, heat resisting lubricating oil, water repellant coating	U.S.A.	1943
	73. Epoxy raisins	Heat resisting and anti-corrosion binding and glueing materials	U.S.A.	1947
	74. PVC-fibre	Used for technical textile goods for its great resistance to heat and chemicals	Germany	1936
	75. Glass fibre	Used for household textiles, strainers and insulators	U.S.A.	1938
	76. Porous plastic (skay)		German Federal Republic	after 1960
	77. Foam rubber		German Federal Republic	after 1960
	78. Quick-freezing		U.S.A.	around 1929
	79. Gasification of coal under surface		U.S.S.R.	1957

* The table was compiled by *Attila Soós. Ilona Antal* assisted in the compilation of the data. One main source to the table was the work by *Jewkes* and co-authors [102]. Further sources are given in the footnotes adjoining each part of the table. We present at the same place the names of the specialists who assisted by providing information informally. We wish to thank them here for their valuable help in the compilation of this table (as well as Tables 20.1 and 20.2).

** Source: [181].

Definition 20.2. The r e v o l u t i o n a r y n e w p r o d u c t s are considerably superior in quality to the products preceding them with similar destination. They tend to transform the behavior of the users thoroughly—in the case of capital goods, the technology of producers; in that of consumer goods the consumption habits and the way of living. They usually create new needs. Their production often gives rise to whole new industrial branches. We shall call r e v o l u t i o n a r y p r o d u c t d e v e l o p m e n t the activity aimed at the invention of revolutionary new products and at the first organization of their production.

The following historical tendency almost appears to be a truism, but it must be pointed out because it is entirely neglected by a considerable portion of traditional economics:

Statement 20.1. The appearance and introduction of revolutionary new products constitutes an element of basic importance in the functioning of the real sphere. The frequency of the appearance of revolutionary new products and the speed of their industrial introduction constitute one of the most important indices of changes in the real sphere.

In this section, we consider only the *first* appearance and the *first* introduction of revolutionary new products. *Mass* acceptance will be treated in a later section.

The second sentence in Statement 20.1 leads to problems of *measurement*. An entire range of Q-indices can be worked out to throw light on the characteristics of the processes connected with the first introduction of revolutionary new products. Let us assume that technical experts, historians of technology and economists engaged in the study of technical progress have selected from among the mass of inventions the products which rightly can be termed revolutionarily new; i.e. that they have constructed a more complete and reliable list than our Table 20.3. In that case, the following calculations could be made.

1. The frequency of the appearance of revolutionary new products, the changes in frequency over time, and the frequency by countries. (We shall revert to this latter problem in the next chapter.)

2. How much time elapses between the invention, the innovator's first initiative, and the first production of industrial dimensions? This indicates the economy's speed of adaptation, or in the present case, the adjustment of production to the development of technical research.

3. The history of the new product's price trend. This is closely related to mass introduction which, as mentioned above, will be discussed later on.

20.5. Gradual Product Development

The revolutionary new product has now entered the scene. At the beginning —in spite of its promising properties—it usually is primitive in character. Think only of the first motor-cars or flying machines— how ridiculous they appear to us now. Or of the low-quality copies which were first produced by the revolutionary new technique of xerography.

Later on, the product which was revolutionarily new at the time of this first appearance is perfected gradually. Its qualitative parameters improve.

Similar work on "minor details" also is carried out in connection with products which must be considered rather old. A case in point is steel. For centuries now mankind has produced steel, but the further possibilities of improving the quality of steel are still the subject of research.

Definition 20.3. The activity aimed at the gradual improvement of the qualitative parameters of some product group beyond the level attained previously at any time and in any place will be called i n i t i a t i n g   g r a d u a l   p r o-d u c t   d e v e l o p m e n t.[6]

It is clear from the definition that gradual product development is as much a pioneer activity as the invention and first introduction of revolutionary new products. (This fact is indicated by the term "initiating".) Catching up with quality parameter values already attained elsewhere and adoption of minor or major innovations or improvements already introduced in some other place is not considered as product development. This should not be interpreted as underrating these activities; they will be discussed in detail later on. It is only in the conceptual sense (and, accordingly, in measurement, too) that we wish to draw a sharp line between following, adoption and imitation on the one hand and the *pioneering* activity described in Definitions 20.2 and 20.3 (i.e. revolutionary and gradual product development) on the other.

There are border-line cases which might belong either to the category of "revolutionary" or to that of "gradual" product development. Once they have appeared, viscose, nylon and polyester fibres revolutionized the quality of textiles; how are we to qualify now the more recent synthetic fibre types? Is automatic transmission, power steering and power braking an invention of revolutionary novelty or does it belong to the sphere of the motor-car's gradual product development? As in the case of all kinds of delimitation, there are border line cases. However, in the majority of cases it is possible to decide

---

[6] In the Anglo-Saxon terminology—although not quite uniformly—the term "invention" is used in connection with revolutionary new products, and "development" refers to already existing products.

unequivocally whether we are dealing with product development belonging to the "revolutionary" or to the "gradual" category.

A variety of Q-indices can be used to characterize gradual product development.

First of all, we can take the most characteristic components of the quality vector of a product group and observe the development of these parameters over time for firms in the vanguard of production.

20.6. Following the World Standard

One of the most important elements in quality improvement is the following of those in the vanguard. Somewhere—in some country, and within that country, in some firm—pioneering work is performed. There, a revolutionary new product is first produced or, at least, some product has first undergone essential gradual development. The pioneers are followed by other firms in the country or by producers in other countries; sooner or later, others start to produce the new product, or they begin to use the same partial improvement or innovation. Following the world standard can be measured with many kinds of indices.

1. One can observe the difference between the standard of an internationally leading quality parameter and the Hungarian standard.

Is the gap between actual quality and the world standard growing or diminishing? At what rate?

2. Let us suppose that the world standard has been reached for a given period. How long it took to overtake those in the vanguard is an observable phenomenon (and it is highly characteristic of the dynamism of the system). It was after 1945 that Japan really began to produce cameras and radios, by the sixties she not only had reached the world standard, but she had become one of the leading countries in the production of these items.

3. The spread of revolutionary new products can be described by national and international statistical data. How the new product drives the rival old products out of use and how the proportion of the new production grows can be represented by time series. Similar time series can be established to illustrate the spread of non-revolutionary, gradual product development.

A few examples have been given in Tables 20.1–20.2.

In our further discussion we shall denote comprehensively revolutionary product development, gradual product development and product development following the world standard by the general term, *product development*.

Here we should like to review briefly two earlier parts of this book. In *Section 11.3*, one of our most important arguments in criticism of the theory of preference ordering was that the goods brought to the market are constantly changing. Revolutionary new products appear while a great number of minor changes takes place in the quality of existing products. Therefore, we cannot speak of constancy over time either of preference orderings or of the set of implementable alternatives. The present chapter (as well as the next) will provide new viewpoints to substantiate this argument.

In *Section 19.8* it was pointed out that we cannot content ourselves with conventional demand functions. The buyers' buying intentions and aspirations must be explained by factors other than prices and incomes. For example, the ideas put forward in this chapter suggest that there should be a relationship between the appearance of a new product and shifts in demand. Demands are shaped also by information of non-price character about the qualitative and technical features of the new products. Empirical investigations already have been made to find out the rate at which new inventions are spreading; this is a phenomena that reflects the spread of new needs.[7]

20.7. Reliability of the Quality

The product development activities all *further* the improvement of the quality of products. A distinction must be drawn between the latter and the activities which ensure that a *quality standard already attained is generally enforced* with respect to each individual product and each individual specimen of a product type.

The world's leading camera producers make efforts to bring out new types from year to year. From time to time, revolutionary new products appear. A recent example is the type which develops and prints instantly within the camera. Meanwhile, the gradual development of favourable qualitative properties has taken place continuously. Individual camera producers have made sudden advances in some fields, but the innovations always have been taken over sooner or later by the others, too.

However, when the consumer makes his decision as to which camera to buy, he considers not only the results of product development but also the reliability of the camera (i.e. that of the producing firm itself). Can he expect the specimen to work perfectly? Camera producers make great efforts to ensure that each specimen fully meets the quality standards they prescribe for it (and usually known also to the informed consumer).

To use a military analogy, product development activities are akin to the *advance* of an army, whereas ensuring reliable quality is similar to the *consolidation and maintenance* of the front-line, once occupied.

In the modern, large firm, these two activities are functionally separated. Development activity is carried out by individual inventors, research institutes,

---

[7] On this subject see *Mansfield's* book [158].

university departments outside the firm, or by specialized research and develop-
·ment departments within the firm. The reliable quality standards of current
production, on the other hand, are ensured by those in operative control of
production, the productive plants and the apparatus of technical and quality
control.

Again, a whole range of Q-indices can be defined to measure the reliability
of products, especially if either national standards, or standards of the company
are known. It is possible to observe the frequency that products deviate from
the standard and the average extent of the deviation. Although producers are
apt to treat such data as business secrets, quality control institutes and associations do exist to safeguard consumers' interests, and these institutes disseminate
information of this character. Repairs under guarantee also allow for inferences
concerning the reliability of quality.

20.8. V-Activities and Q-Activities

On the basis of what has been said up to now in this chapter, the activities
taking place in the real sphere can be divided (on an abstract level) into two
broad categories:

Definition 20.4. The V - a c t i v i t i e s serve to increase the volume of the
product output. They comprise the expansion of production by means of a
better utilization of the existing fixed capital as well as the extension of fixed
capital by means of investment. The Q - a c t i v i t i e s serve to improve the
quality of the product output and the reliability of the quality standards once
attained. This category comprises revolutionary, initiating gradual and world-
standard-following product development as well as the activities aimed at
ensuring reliable quality. The V-activities can be measured by V-indices,
the Q-activities by Q-indices.

Sometimes V-activities take place in a "pure" form, without combination
with Q-activities. This is the case when the firm increases its production exclusively by better utilization of its capacity, without modification of its existing
fixed capital and using the same materials as before. The case occurs also when
investment is confined to adding to the existing fixed capital machinery, equipment and structures of exactly the same quality as that already employed.

However, such "pure" V-activities are a rather rare occurrence. The following
statement is based on well-known empirical facts:

*Statement 20.2. Within the framework of technical progress, V- and Q-activities
are inseparably intertwined. On the input side, new investment increases the
volume of production in part by putting into operation machinery, equipment
and structures of a higher quality than those employed previously. On the output
side, the increase in volume is usually accompanied by an improvement in the*

quality of the product output. At the same time, the improvement in quality tends to bring about an expansion of demand.

The phenomena called in economics "substitution between the factors of production" and "modification of factor combinations" are closely related to the V- and Q-activities. These phenomena refer to the fact that often in a productive plant it becomes possible to put into operation new machinery or equipment of a higher efficiency (i.e. *better quality)* than that formerly employed; this involved a transformation of production technology.

It is true that the dynamics of the V- and Q-activities are not necessarily the same. The one may stagnate for a time in spite of the progress of the other or, if not actually stagnating, it may develop rather slowly as compared to the other. This problem will be discussed in detail in the next chapter. However, to some extent there is joint movement and it is this fact that is summed up in Statement 20.2.

20.9. Comparison

We do not wish to create the impression that our book discovered the problem of quality. First of all, the problem is well known to industrialists and to consumers, and it has engaged the attention of both engineers and housewives. Also it has been treated many times in the economic literature.

Most interesting in this respect are the works on economic history, technical history and economics which analyze expressly the problems of invention and practical application of inventions, innovations and new products. Some of these works are largely descriptive in character, based on case studies and the generalization of their experiences.[8] Other authors additionally employ econometric methods.[9] Unfortunately, these works have remained rather isolated, and their results have not been integrated into the main schools of economic thought.

Economic statisticians, economists engaged in international comparisons and economic historians inevitably are faced with the problem of volume and quality when attempting to measure the dynamics of growth and to make comparisons on the international level. (This has already been mentioned in Section 20.2.) However, they have been interested mainly in the methodological problems of working out V-indices aggregated on the national level; the independent measurement of Q-activities and the international comparison of Q-indices is lost in the process.

There is one element of the Q-activities which has engaged the attention of many representatives of one of the most important schools of modern

---

[8] See, e.g. *Jewkes–Sawers–Stillerman*'s book [102] already mentioned.
[9] An excellent example is *Mansfield*'s work, [158] quoted before.

economic thought, the so-called theory of growth. That is the qualitative improvement of capital goods. Several growth models are known which take into account technical progress "embodied" in machinery, equipment and structures (i.e. their qualitative improvement).[10] Important as this may be, nevertheless, it needs to be pointed out that these studies deal with a *single* element of the ensemble of Q-activities. The models of growth theory leave entirely out of consideration the other elements of the Q-activities. They measure the *results* of economic development exclusively by V-indices. Accordingly, these theories only strengthen the widespread, entirely erroneous and biased view that the economic development of a country should be measured solely by the growth rate of the volume of production and that the improvement in the quality of the product can be wholly disregarded. However, in actual fact the latter is equally important.

Several branches of the theory of growth (especially the so-called neo-classical models) are closely related intellectually with the GE school. At present, the theories of the GE school completely disregard the problem of quality.

It is true that the framework of assumptions and concepts of the GE school does not rule out admission of qualitative differences between the products. As already pointed out in Chapter 3, the GE model's interpretation of the product concept suggests that two related products differing in quality are two different products.[11] Thus, if there is no consumer demand for product 6 (low quality) and rather its substitute, product 7 (good quality), is sought after and the production of the latter is possible, then in the models of the GE school production and consumption will shift towards product 7. *To this extent* there is a room in the GE theory for the problem of the quality of products, in spite of basic assumption 4 concerning the constancy of the set of products.

This, however, represents a rather empty and poor treatment of the subject. In the final analysis, the variables figuring in the models of the GE school are exclusively V-indices. The Q-activities, their motives, the explanations and relationships of their changes are not analyzed. Why does consumer demand shift from the old product to the new one? What is it that prompts the system to produce and introduce new products even if the conversion involves certain costs?

The questions left unanswered by the GE school lead to the subject of the next chapter—the effects of disequilibrium on the increase in volume, the improvement in quality, and also on other processes of the operation of the economic system.

[10] For a comprehensive survey, see the report of *Hahn–Matthews* [78], particularly pp. 58–75, as well as the book by *Andorka–Danyi–Martos* [6].

[11] This interpretation of the product concept has been assumed also in the present book. See Definition 4.10.

21. THE CONSEQUENCES OF DISEQUILIBRIUM

In Chapter 20 we clarified the most important concepts relating to expansion of volume and quality improvement. Now with these concepts in mind, we continue the train of thought of Chapters 17 to 19, the investigation of the operation of the market. This chapter discusses how pressure or suction in the market affects the output of the economic system.[1] Let us number hereunder the components of the output.

21.1. Volume and Inputs

Let us first consider only the short run. In the short run, suction is favourable to the V-activities.

When buyers line up for products, firms endeavour to utilize the available production funds and manpower to the maximum degree possible. It was in the early fifties, when the shortage in materials, energy and consumer goods was greatest in Hungary. This was the heroic age of "shock work", of frantic production. It was in that period that the average number of shifts was highest, with production going on frequently even on Sundays and holidays. These conditions were characteristic not only of mines and factories but also of trade—perhaps even more conspicuously so. Employees of crowded shops attended impatient customers, without rest. This situation is typical of the state of queuing, of the state of suction.

In a state of suction, investment is distributed in a way that it contributes to increasing the volume of production in the shortest possible time. Few resources are allocated to investment projects which do not serve the extensive expansion of production directly.

The volume of production may increase also when the economy is in a state of *pressure*, but here the market situation puts a brake on such increases in the short run. With considerable unused capacity, redundancy and "slack" in most firms, only the cleverest or luckiest firms are able to come close to fully exploiting their capabilities. The rest find it impossible to do so; in this latter case, a considerable part of investment is allotted to purposes which do not directly contribute to increasing the volume of production.

[1] The concept "performance of the system" is explained in Definition 16.2.

Statement 21.1. In the short run suction promotes and pressure brakes the V-activities, the increase of volume. In the former case resources are utilized to the maximum extent, in the latter, there exist unused reserves. In the former case the bulk of investments serve directly to inrease volume; in the latter investment projects are not focused so single-mindedly on volume increase.

In the long run the situation is more complicated. On the one hand, suction strongly stimulates the increase of volume in the long run. But in the long run disadvantages assert themselves; suction puts a brake on technical progress and this also affects growth of volume for the economy as a whole. Conversely, pressure, which promotes technical progress, may in the long run affect the growth of volume favourably.

Whether production involves high or low çosts, with wasteful or economical use of inputs, is affected also by the relation of market forces. However, in this respect the effects of the two types of disequilibrium are not equivocal and not easy to characterize. Both pressure and suction tend to promote wastefulness *as well as* thriftiness.

As pointed out in Statement 21.1, suction results in maximum utilization of the available capacity, resources and manpower. At the same time, the general shortage situation causes frictions in production, which in turn frequently lead to idle hours, to partially unused resources and manpower. It will be shown later that as a consequence of frictions in adaptation surplus stocks may accumulate. Moreover, forced substitution may cause losses, in some cases most grave ones. For example, if a part is missing because some factory has failed to deliver it on time, the user is compelled to improvise. This is more expensive but still less costly than stopping production altogether because of the lack of that part. Another example would be the substitution of an expensive raw material for a cheap one because there is no other alternative.

Statement 21.1 also emphasises the fact that in the case of pressure resources and reserves are not utilized to the full extent possible. This under-utilization— this gap between potential and actual production—frequently may be quite considerable, and this means waste and social loss. On the other hand, the fact that materials, parts and semi-finished products needed for production are available readily facilitates the favorable combination of the factors of production.

21.2. Quality

Let us now examine the Q-activities. The gravest consequence of a continuous state of suction in the almost complete lack of *revolutionary product development*. Let us go back to Table 20.3, presenting the most outstanding products of the past fifty years which were revolutionarily new at the time of their appearance. Obviously, there is a close correlation between the frequency of pioneer introduction of revolutionary new products and the degree of economic development in the introducing country. However, when comparing "suction" economies and "pressure" in economies of about the same degree of development, it must be stated that the power relations of the market also deeply affect the revolutionary production development. It should be noticed that with the exception of a few initiatives no socialist country figures among those introducing the product in question for the first time. Tables 20.1 and 20.2 show a similar situation regarding revolutionary new products. This is a grave truth, and a regrettable one for the adherents of socialism. But because it is the truth, we must face it and ponder over the causes.

This problem was raised in a manner memorable to us, Hungarians, in *András Kovács*'s excellent documentary film "The Difficult Men", dealing with the fate of some Hungarian inventions of outstanding importance. Some of these possibly might have deserved the label "revolutionarily new", or nearly so. However, these inventions were lost; indifference, bureaucracy and conservatism raised innumerable obstacles to their practical introduction. The film made a great impression on its viewers and gave rise to extensive discussions among both economic experts and the non-professional public. Yet, the fate of inventions has not much changed since.

The film revealed the fact that there is no lack in technical talent in Hungary. More than one instance of outstanding Hungarian innovation could be cited from technological history. One of the inventions in Table 20.3, the Biro pen, was invented by a Hungarian; however, its industrial production started abroad. This example shows precisely the essence of the problem. Innovators, engineers, and researchers endowed with talent exist in every country. Moreover, technological history shows us that new techniques do not in every case even come from the most highly qualified and erudite experts. Sometimes an ingenious idea is conceived by an amateur, probaly because he is free of the influence of previously established, traditional technical knowledge, too deeply rooted in the mind of the average specialist. The fact that some outstanding results have been achieved in the socialist countries in the whole range of the natural sciences as well as technical sciences relating to war production shows that the problem is *not* rooted in the lack of talent or in the weakness of scientific and technical culture. The problem is the lack of institutions and organizations which despite all obstacles can effect the *introduction* of revolutionarily new technical innovations, accepting all the risk concomitant with this work, including that of failure, the struggle against conservatism and deep-rooted habit.

Why should an enterprise director[2] accept this risk and take up a struggle when—in a state of chronic suction—he is able *without* such effort to sell the products of his firm easily? With the buyers lining up for the firm's old product, why take upon oneself all the trouble involved in the introduction of a new product?

Certainly, we do not rely on one factor to explain the lack of the introduction of revolutionary new products. A number of problems also may play a role. However, it is our conviction that all other factors are of secondary importance as compared to the cause discussed above; in the "sellers' market", in the state of suction, there is no real incentive to introduce new products.

In the case of pressure the situation is basically different. Given the level of technology the market is divided in some fashion among the firms. Although some shifts take place from year to year, these are not too radical. The *status quo* is upset usually only when some revolutionarily new product bursts into the market. Such a new product is able to conquer the market from its competitors within a comparatively short time and to capture a considerable part of the purchasing power.

Some revolutionarily new products have been introduced by relatively small firms which have gained remarkable growth by means of the invention in question. A classic case in point is the history of Xerox; a comparatively insignificant, small firm launched a new technique of reproduction, growing thereby into a world-wide concern.

In other cases research work relating to some epoch-making new product is controlled by a mammoth firm which then also introduces it. This is what happened in the case of nylon with DuPont. But in such cases, too, the main

---

[2] Let us return to the example of the film "The Difficult Men", discussed previously. The film reporter asks one inventor why there is such a reluctance to introduce invention in this country. The inventor answers: "Well, this is a question difficult to answer, at least to answer in a way that the corresponding part will not have to be cut out from the film... Every technical step, every serious technical step will involve, after all, the taking of certain risks. Now, one would generally risk the difficulties if one has some interest in the matter. If the interests lie in some other direction, risk taking obviously would be more difficult." (From the sound track, [138] , p. 51.)

Two years later, *László Nádasy* published an interview with the innovators in the firm, in the periodical "*Új Írás*" ("New Writing"). He asked them about the destiny of their inventions. The inventors again were full of complaints. To quote one of those interviewed: "for an official, the most important thing is obviously to carry out the instructions of his superior on whom he is, after all, dependent. This is but natural. But for the same reason, it is unnatural that an official should decide—instead of the buyers consuming the product—whether a product is good or bad... The official is dependent on his superiors but the usefulness of the product depends on the user. Let him decide at last, for he is the consumer." ([188], [189], p. 91.)

motive force is that the firm wants to gain an advantage on its competitors in conquering the market.

The effects of the relative strength of market forces show themselves most markedly in revolutionary product development. However, the two types of disequilibrium have a strong effect also on the *other Q-activities*. Suction will brake and pressure will promote both initiative gradual product development and the development work of an initative and adoptive character following the world standard.[3]

Moreover, the market situation also affects the adherence to quality standards. In the case of suction the producers and sellers are not as particular with regard to the reputation and absolute reliability of their products as they are in the case of pressure. The buyer who is forced to queue is not in a position to raise high demands; he is glad to obtain the desired product at all. In the case of pressure the situation is different. It may become widely known that a product is of low and sub-standard quality; this can lead not only to the loss of a single buyer but also to damaging the reputation of the firm, its chances in the market, and its position in the race with its competitors.

Let us sum up the above.

Statement 21.2. Suction hinders, and pressure advances revolutionary product development, and also to a lesser extent the other Q-activities.[4]

It would be drawing a one-sided picture of the situation if we suggested that in the case of suction all products are deficient in quality and no improvement takes place. Equally one-sided would be the converse assertion that in the case of pressure all products are of excellent quality. Undoubtedly there are also tendencies working in the opposite direction. We shall return to this question in Section 21.7, but here we should like to point it out in passing.

First, let us take the case of *suction*. With the economic development of a country—and this implies a growing volume of production, rising level of technical culture and increasing skill of the workers—the quality of products necessarily improves independently of whether there is pressure or suction in

---

[3] In the state of continuous pressure, the choice of products in the U.S.A. is changing extremely rapidly. According to estimates some 6,000 new consumer goods appear in the market every year. In the ten years 1955 to 1965 the list of consumer goods has increased 60 per cent. See *M. Simai*'s article [231].

[4] For illustration of how the lack of goods—the state of suction—hinders quality improvement, we quote here from an article of "Népszabadság" (People's Freedom), the central Hungarian paper, on furniture supply: "The data of the first half-year collected by the Commercial Quality Control Institute demonstrate that the quality— also durability—of furniture have worsened further... In furniture industry the opinion prevails that 'in today's market position anything can be sold' (words quoted from the managing director of one of our biggest factories), and, as a consequence, not much trouble is taken to assure quality and durability" (see *Buzási* [40]).

the market. A rising standard of living is a force acting in the same direction in both cases, because consumers become more exacting. Pressure accelerates this process and suction brakes it; even suction cannot bring it to a halt. *In the case of less strong suction quality falls short of the world standard to a smaller extent, in that of stronger suction to a greater extent—but nevertheless it always follows the world standard.*

Although there is a basic tendency for suction to brake technical progress and hinder the introduction of new products, an opposite tendency also asserts itself. Chronic shortage may provide an incentive to develop products suited to replace those in short supply. It is a generally known fact that shortages in the period of World War II gave a strong impetus to research work in the field of plastics and synthetic fibres as well as to their industrial production.

In a whole range of socialist countries, including Hungary, foreign trade creates a special situation. The firms producing for export are compelled to take into account the strong competition existing in the world market. In foreign markets, especially in hard-currency markets, pressure prevails and therefore, the quality of export products must be up to standard. This is an incentive to improve quality, and the domestic consumer, too, benefits from the results. Unfortunately, the opposite effect also assets itself. Most firms do not produce exclusively for export; they sell a considerable part of their products in the domestic market which is in a state of suction. This acts to lower the exigencies as regards quality: "If our product does not meet the export standard, it can always be sold at home—consumers will buy it here." Thus, the lower exigencies of the domestic market in a state of suction weaken the incentives for improving quality, and ultimately this has an unfavorable effect on the quality of export products also.[5]

Turning now to *pressure*—we must guard against a similar mistake: under no circumstances should we "idealize" the conditions of pressure. Above all, the eternal change of products and improvement of quality should leave a bitter taste in one's mouth when within society there prevail sharp social contrasts and large income inequalities. While, on the other hand huge social inputs are made in order to meet the requirements of higher income strata in an ever more refined manner, on the other hand, the living standards of broad

---

[5] From this point of view, the export markets, where also suction prevails have a similar effect as the domestic market has. On the export markets where Hungarian commodities are in shortage, there will be, obviously, no high-quality requirements made on our products. This is most comfortable for the Hungarian producer—however, it will not provide any incentive to develop the technology and to improve the quality standards.

Perhaps also the notion of *ambition* could be used here. Ambition is—according to Definition 12.4—the difference between present ambition and an earlier result.

strata gravely and unjustly lag behind those of the former. However, we should see that the real evil is rooted in the structure of society, in the ownership relations and the accompanying inequalities in income and in opportunity —and it would be wrong to blame the change of products. *In itself*, this process is obviously progressive; it produces the evolution of productive forces and with it of human culture and civilization, and no one would benefit if it should cease. Beyond this fundamental socio-political problem, the process of improvement in quality is not unequivocal either. While under the effect of pressure there is a steady flow of up-to-date products, in certain fields quality deteriorates decidedly. It is a known fact that some capitalist firms deliberately—usually on the basis of agreement—decrease the durability of products to avoid saturation of the market. A considerable part of goods *must* wear out rapidly so that the need for fresh supply sustains the demand. There are firms specialized in the production of low-quality mass consumer goods and shops selling mainly "shoddy" goods. It is a general concomitant of modern large-scale production that the small plants engaged in the production of individual pieces and masterly handicraft work decline in prominence; the market is flooded with undistinctive mass products—be it furniture, clothing, or decorative articles.

In general, we may conclude that the improvement of quality depends to a large extent on factors independent of the type of disequilibrium; e.g. on education, research and product development, the requirements of defense, etc. Nevertheless, Statement 21.2 is still valid.

Speaking of quality we must also briefly touch upon another question concerning the relationship between product innovation and process innovation. Many economists one-sidedly stress only the latter claiming that if companies are truly interested in diminishing costs they will surely try to promote innovations aiming at decreasing inputs. The efforts of the companies are independent of the disequilibrium of the economy. Moreover, these economists argue that in periods of suction this tendency is reinforced.

There is some truth in this assertion although it gives an oversimplified picture of reality. Upon closer examination of the technical modifications introduced in the production process, we discover that these innovations usually are connected with the introduction of a new *instrument*, such as a new piece of machinery, equipment or measuring apparatus; or, in other cases, they are linked to the use of a new kind of material or to the use of some technical equipment on the administrative level. A change which aims exclusively at a more rational utilization of already existing resources is relatively rare and usually not really significant.

When pressure dominates the economy, the seller and the producer are the ones who are vitally interested in supplying the consumer with newer and better products. For example, it was not the textile industry (the consumer) which perfected nylon but the chemical industry (the producer); or again it was not the printing or reproducing industry (the consumer) which developed xerography but the Xerox Company (the producer) and so on. The producer who develops new products is trying to detect the demands of the potential consumer; in a good portion of the cases, it is the producer and not the consumer on whom the work of developing the new product falls.

In time of suction the incentive on the part of the manufacturer is lacking. In vain the consumer comes to the producer saying, "please give me better products in order to decrease my costs". Why should the producer try to fulfil such a request when he is able to sell his products without any innovation or change.

21.3. Relative Strength and Competition

In the preceding section we have used phrases such as "conquering the market", "gain an advantage on the competitors". Accordingly, we have tacitly accepted the fact that there exists competition and that it is affected by the relation of forces in the market.

The concept of economic competition is given different interpretations by the layman and by the economist, or rather, by the various economic schools. We should like to use the word "competition" in accordance with the interpretation of everyday language, accepting the disadvantage that this may differ widely from the definitions recognized by our discipline.

In the case of equilibrium as interpreted in accordance with Definition 19.4 there is not and cannot be any competition. When the buyer's intention and actual purchase, and as well the seller's intention and actual sale, are exactly identical, there is nothing to compete for. In such circumstances, sellers can peacefully divide among themselves the purchasing power of the buyers, and the latter equally peacefully can divide among themselves the products of the sellers. This is like a race where each participant has been promised a first prize—and actually receives it. The concept of "competitive equilibrium" —although a stock phrase with economists—is actually a complete paradox.

Two types of *genuine* competition are possible. In the one the sellers court the buyers, in the other the buyers court the sellers. Both are genuine forms of competition because there are some who reach the goal and others who do not.

The various forms of monopoly, oligopoly, imperfect and "perfect" competition are extensively treated in the literature. We do not wish to underrate this aspect of the question; however, we believe that it is of secondary importance. The primary question is whether there is pressure or suction in the market.

Let us take first the case of pressure. Let us assume that there is a single monopolistic airline operating in the country. (Foreign companies can operate only in cooperation with it.) This is the situation in many countries. Since pressure exists, the airline will not be able to enjoy its monopoly freely. It will have to compete with the direct substitutes to air transport—the railway and the motor-car. Moreover, there exists indirect competition. One of the principal users of the airline is the tourist trade. The latter, in turn, competes with other pastimes. The family may spend its savings on a pleasant trip, but it also may decide to refurnish the flat or to buy a motor-boat. The air transport competes indirectly with the furniture trade or motor-boat production. It is true that in this case it is practically impossible for passenger transport to sink to nil and for all the cash released thereby to be spent on furniture or motor-boats. But there may be marginal shifts between the various branches and product groups; one may increase 5 per cent while the other increases only 4 or 3 per cent.

Let us now consider the reverse case. If there is general suction prevailing in a country, then the buyers will compete for the seller and the latter will "reign", even in the branches where production and selling are atomized. In every textbook agriculture figures as the classical example of decentralized and atomized production. Yet, in periods of war, when there is a general shortage of food, even the small farmer will be able to dictate as a "monopolist" and the town population will make every sacrifice to obtain food.

Another typical example is the situation of craftsmen and artisans in present-day Hungary. In the majority of capitalist countries, under the conditions of general pressure, the artisan is a modest member of the lower middle-class, placed somewhere below the middle of the income scale, in the neighbourhood of minor clerks and better skilled workers. In Hungary, on the other hand, he has risen to the top of the income scale, benefiting from the shortage in goods and services, from the general suction prevailing in the market of certain goods and services.

The distinguishing characteristic of *competition* is that the aspiration of the organization can be fulfilled only at the expense of the competitors. In the case of pressure, competition exists only among sellers and in that of suction only among buyers; in the case of equilibrium there is no competition.

Definition 21.1. The criterion of c o m p e t i t i o n: the aspiration of the organization can be fulfilled only to the detriment of the competitors.

According to the above definition, in the case of pressure the actual sales of one seller can approach his own aspiration only if one or several other sellers fall short of their own aspirations to an even greater degree. This is the essence of competition. In the case of suction, on the other hand, where the expansion of sales depends only on production, the seller is able to find a buyer for all his products. He does not need to compete.

Similarly, in the case of suction the buyer is able to fulfil his buying aspiration to a higher degree only if this is offset by other buyers' aspirations remaining unfulfilled to an even greater extent; he is competing with the other buyers. In the case of pressure, on the other hand, how much the buyer buys depend exclusively on his own resources and requirements, and he does not drive out the other buyers with his own purchases; he does not compete with them.

The aforesaid can be summed up in the following statement.

Statement 21.3. The character of competition is determined by the relative strength of market forces; the degree of monopoly or atomization in the branch concerned has only a secondary effect on the behaviour of buyers and sellers.

21.4. Adaptation

In every market reciprocal adaptation takes place between production and consumption, between selling and buying intentions. This is self-evident, and there is agreement on this point between all schools of economic thought.

However, the GE school goes further. It suggests that in the adaptation the consumer is predominant, at least in the case of competitive equilibrium. The consumer's preferences are given "from outside" and in the final analysis production adapts to these.

In actual fact the adaptation properties of the system depend very much on the type of disequilibrium, on the relative strength of market forces.

In the case of *suction* production may detach itself to a considerable extent from the consumer's aspirations. As a consequence of the general shortage situation the consumer has to resort to forced substitution. Eventually this may become a habit: the consumer's customary decisions adapt to the structure enforced by the producer.

However, this does not mean that the producer can control production according to his own will. Even if queuing is fairly general, the length of the line is important. If for some products and services the line is too long or too short as related to the customary average, then the disproportion will be corrected within a shorter or longer time and the comparatively backward branch developed at a quicker rate than the others. Thus, here too, there occurs a certain adaptation of production to consumer requirements, if only more convulsively and more slowly than in the case of pressure. In the state of *pressure*, in the short run production will adapt to consumer requirements. As a result of competition among the sellers adaptation is usually quick and elastic. When a new requirement appears somewhere, someone will take the initiative to meet the requirement.

At the same time it would be a mistake to believe that in the case of pressure it is exclusively the producer who adapts to the consumer. Above, it has been explained in detail that in the case of pressure new products constantly are developed and thereby new needs are created. In a sense the effect of production on consumption in such cases is even greater than in the state of suction. Pressure tends continuously and sometimes rather radically to regroup the consumer's requirements and buying intentions.

In connection with adaptation we must mention also the problems of *uncertainty*. Production and consumption could adapt more easily to each other if the intentions of the trading partners in the market were known precisely in advance. However, this is not possible, if for no other reason than because intentions themselves are frequently rather uncertain. Moreover, the trading partners do not reveal their intentions in advance.

The literature of the GE theory raises the following question; how does the market operate under uncertainty; in such conditions, can there be competitive equilibrium? Actually, in our opinion, this is not the essential question. One of the main questions asks how uncertainty can be reduced by use of planning and reciprocal information. (This already has been discussed in Chapter 11, and we shall return to the question once more below.) The other main question asks who should bear the burden of uncertainty? It is here that the effect of pressure and suction enters the picture.

In the case of pressure the burden of uncertainty is carried by the seller. He must be prepared if the buyer gives him preference over his competitors; therefore, he must have considerable reserve capacity and carry reserve stocks. This is the case in industry, but it is particularly conspicuous in trade and services. The buyer does not feel any uncertainty. If he wants to buy, he may rest assured that he will find what he is looking for.

In the case of suction it is the seller who feels secure and the buyer who faces uncertainty. This affects the transaction of his purchases. In a shortage economy the buyer—be it the purchasing agent of a factory or a housewife—buys not when the need becomes acute (when his buying intention, his aspiration takes shape) but when the product in question actually becomes available. Shortage necessarily gives rise to a "hamster" philosophy: "let us provide for more difficult times and buy more—there may be no opportunity later on". This is one reason why in a state of suction, under the conditions of general shortage, redundant stocks tend to accumulate in many places.

In general, when questioning who carries the burden of uncertainty, we must not consider the *material* aspects of the problem exclusively. We must not confine ourselves to the question of who holds reserves. The *psychological* effect also is important. In the case of pressure the seller feels uneasy; will he be able to find a buyer; might he not get into trouble because of selling difficulties? In the case of suction the seller is sure of himself and the buyer worried whether he will be able to obtain what he is desiring. This is not only a psychological problem; it has an effect also on economic behaviour. It is profoundly connected with what has been discussed in Section 21.2, with encouraging or discouraging activities aimed at raising the quality standards of production.

21.5. Selection and Concentration

Economists—and especially the mathematical economists raised in the GE school—have become accustomed to consider the market exclusively as a factor of *coordination*—as an institution which coordinates production and consumption, bringing about a peaceful harmony between the two.

In actual fact the operation of the market in a state of disequilibrium (either pressure or suction) constitutes a *selective* process as well.

First let us consider the case of *pressure*. One seller is able regularly to fulfil his aspirations. However, some other seller has no similar luck; his actual sales fall considerably short of his expectations. The selective effect of the market manifests itself directly in how the global difference of total aspirations and total actual sales is distributed among the individual sellers.

In such cases selection is carried out by the buyer. The less well-informed buyer—especially in the case of an occasional decision of minor importance—chooses *at random* from among the various sellers. However, in the case of repeated decisions the buyer learns from his earlier experiences. Also, if a non-recurrent but basic decision is involved, he strives to obtain extensive information, before making the decision. In such cases, the following are typical selection criteria:

a) The buyer may give preference to the seller whose *momentary* offer is the most advantageous as regards quality, price, service, etc.

b) He may give preference to the seller who has submitted to him the most advantageous offer throughout some *longer period of time*.

c) He may give preference to the seller to whom he has become most *accustomed*.

Anyone of the three criteria may be used, or the three may assert themselves in some combined form. In any case, all three involve the performance of the seller; the novelty of his products, their quality and choice, and also the attention he pays to the buyer. In other words, the selection carried out by the buyer provides an incentive for Q-activities. Moreover, it also provides incentives for economy, for decreasing inputs, because this enables the seller to lower prices, which also is attractive to the buyer.

The distribution of selling tension is not even. The firm where the difference between aspirations and actual sales is frequently rather high eventually begins to stagnate and to decline. On the other hand, where selling aspirations are fulfilled regularly expansion is rapid. This enables the assertion of "economies of scale", the relative economies connected with greater plant size and mass production. It can be seen that in such cases there is favourable interaction between V- and Q-activities on the one hand and economy of inputs on the other.[6]

Under pressure, ultimately selection contributes to the differentiation and concentration of producers.

In the case of *suction* the seller can choose between buyers. Some characteristic selection criteria are:

d) Random selection—the seller is indifferent to who obtains the goods.

e) Higher organs may intervene and allocate the product in short supply according to their own viewpoint. In the most favourable case this rationing is based on carefully weighed principles, taking into account social interests. However, the official engaged in the allocation may be biased.

f) The buyer may attempt to bribe the seller or, in the case of government allocation, the official concerned. This happens—luckily in rare cases only—even with the purchases of firms. More frequently, the situation arises in the case of individual consumers, either in some marked form of bribing or in the form of more harmless "tipping".

g) The seller may give preference to the buyer who is most unpretentious and contents himself with whatever he can obtain.

These criteria—just as those listed above for the case of pressure—also may assert themselves in combined forms.

Criteria d), e) and f) have no unequivocally beneficial or harmful effects. At most, they are harmful inasmuch as they do not produce the favourable incentive of selection as in the case of pressure. Not even the less dynamic, the half-hearted firms fear elimination; suction involves "protectionism".

Criterion g) acts in the same direction. It accustoms the firm to be unambitious and conservative, and to calmly acquiesce in the low quality standard of its products.

Let us sum up the above:

Statement 21.4. In the case of pressure the buyer makes the selection. This stimulates the improvement of quality and the economy of inputs. Selection leads to differentiation and concentration. In the case of suction the favourable stimulating effects of selection fail to come about.

Ever since Walras, the GE school has given great attention to the so-called

---

[6] In *Mansfield*'s work [158] it is pointed out that the firms in the vanguard of technical development in the U.S.A. are growing on the average at double the rate of those neglecting the development of new techniques.

"tâtonnement" process.[7] According to this mode of description, the market "feels" its way towards equilibrium. Sometimes supply is greater than demand and prices decline. Under the effect of declining prices supply diminishes and demand increases. This may bring about a disequilibrium of the opposite sign, with demand surpassing supply. Then the price increases, which acts towards diminishing demand and increasing supply. By "trial and error", the fluctuations of supply, demand, and prices will lead to equilibrium, provided that certain definite conditions exist.

Unfortunately, we do not know of any Walrasian model which defines the selection rules of the "tâtonnement" processes. Let us consider the *Arrow–Hurwicz* paper, which may be considered one of the classics on the subject.[8] The authors describe the functioning of the market by representing its "tâtonnement" processes with the gradient method of mathematical programming. Or, in other words, we can give the gradient algorithm an economic interpretation which is similar to the Walrasian description of the functioning of the market.

We might give a more narrow interpretation to the Arrow–Hurwicz model, regarding it not as a model of the actual *operation* of the market, but only as a model of an *anterior information* process, before contracting. However, what happens if we give a broader interpretation to the model if we consider the *Arrow-Hurwicz* model of the market as one describing the processes of selling, buying, production and consumption *dynamically*, i.e. their operation in the course of time? In this case their paper *fails* to answer, or even to raise, two questions.

—What happens to the unsold products? The algorithm shows what must be done with prices when supply surpasses demand. But what must be done with the surplus products? Are they added to the initial stock of the next period? Or are they destroyed, as was the actual case more than once during the great depressions of the inter-war period? Likewise: what happens to unsatisfied demand? Does it lapse? Or does it accumulate and add to the needs of the next period?

—How is surplus demand or surplus supply distributed among the individual producers or consumers in the course of the "tâtonnement" process, before reaching equilibrium? Evenly? Or does there exist some other selection criteri-

[7] See *Walras* [269], pp. 170 and 520. There is a widespread, misinterpretation of the concept "tâtonnement". Some economists use this term to indicate *anterior* information process between buyers and sellers, before contracting. However Walras, who introduced this concept in economics, gave quite a different interpretation. He called "tâtonnement" the *actual operation* of the market; the "groping" of price, demand and supply towards equilibrium.

[8] See *Arrow–Hurwicz* [15].

on? The algorithm does give the rule of how to reach equilibrium. But what happens before then? As a matter of fact, if the difficulties of selling or buying are not evenly distributed, then some organizations grow stronger, and others weaker, before the market as a whole reaches an equilibrium.

It should be emphasized that we are dealing here not with a special failure of the Arrow-Hurwicz model but with a feature common to the whole GE school. On one occasion, we attempted to reformulate *Oscar Lange*'s famous model of socialism for the purpose of a *simulation* experiment.[9] Lange describes an economic system where prices are regulated by a central office in accordance with Walras' "tâtonnement" rules; it raises prices in the case of shortages and lowers them in that of surpluses. Seemingly, the algorithm of the Lange model is perfectly clear. Nevertheless, it could not be reformulated into a simulation program. As a matter of fact, in a dynamical interpretation rules *must* be inserted to deal with unsold products and unsatisfied needs. What happens to these in the course of time—will they add to supply or demand of the next period? And what happens to them as regards distribution among the organizations of the economic system? Which organization bears the consequences of disequilibrium and to what extent? Unfortunately, O. Lange's study does not offer those engaged in programming the simulation experiment any basis for an answer to these questions. And yet, these are extremely important questions pertaining to the description of actual economic systems.

21.6. Information Relation between Seller and Buyer

In Sections 21.1 to 21.5 we have examined the effects of pressure and suction on the real sphere. First, we analyzed *real processes* exclusively, and in connection with these, the problems of volume, input and quality. Then, we turned also to *control processes*, in connection with competition, adaptation and selection— but again mainly from the point of view of their effects on the real processes.

Now it is worth while to make some remarks also on the *information flows* and the contracting processes taking place in the control sphere independently of the real processes.

Traditional economic theory gives the impression that the information activities of seller and buyer are symmetrical, that sellers and buyers reciprocally submit their offers to each other, bargain and come to an agreement. However, in actual fact the information flows are asymmetrical.[10]

---

[9] See *Lange* [146]. We shall revert to the Lange model below.

[10] The asymmetry of the information processes of buying and selling is pointed out in *Heflebower* [86]. However, he describes only the asymmetric state under pressure and does not discuss the information flow taking place under the conditions of suction.

To a certain extent, concentration of information tasks on part of the sellers is unavoidable, since a seller is confronted by many buyers, while a buyer usually buys a definite product only from a single seller. But, beyond the "natural" division of informative tasks between seller and buyer, there are also other asymmetries.

In the case of *pressure* a disproportionate part of the information tasks falls to the seller. The seller endeavors to seek out the buyer. In the fields where this activity is possible at all, the seller sends his representatives to call on the buyer. Enormous amounts are spent on publicity. This is one of the most detrimental consequences of pressure. Sellers manipulate buyers. Through a flood of advertisements, efforts are made to convince buyers that a product offers something new—even if no real innovation is involved or at most a slight alteration has been made.

In the case of *suction* a considerable part of the information tasks falls upon the buyer. There is much less publicity (which is to some extent a healthy phenomenon). Of course there are advertisements, but characteristically it is often the buyer who advertises, informing the potential sellers of his needs.

The purchasing agent again and again calls on the supplier, urging the shipment of his orders. The individual consumer, the housewife, repeatedly looks in at the shop to see whether the long-awaited goods have arrived.

This one-sided burdening of the buyer with the information tasks is partly modified when central institutions intervene and distribute goods in short supply by administrative methods. In that case, the center asks for information to be sent in by the producers and users and issues instructions to both. The tasks connected with the preparation, release, and processing of information are thus distributed among producers, users and the central administrative organs.

To sum up, we can make the following statement.

Statement 21.5. The market has no generally valid information structure; its information structure is dependent of the market power relations.

21.7. Survey of the Effects—Counter-Tendencies

In *Table 21.1* we present a summary survey of the effects of pressure and suction, on the performance of the system for the hypothetical case when the economy functions exclusively in disequilibrium.

In Chapter 15 we decried the spell of the number "one", and we do not wish now to explain all achievements or failures of the economy with one single reason. The type of disequilibrium has a strong effect on the achievements of the economic system—but this effect may be strengthened or weakened by other factors.

TABLE 21.1

The "pure" effect of pressure and suction

Domain where the effect asserts itself	Pressure	Suction
Volume	In the short run brakes the increase in volume	In the short run stimulates the increase in volume
Inputs	Partial idleness of resources Free combination of inputs	Tight utilization of resources Forced substitutions of inputs
Quality	Stimulates introduction of revolutionary new product	Does not stimulate introduction of revolutionary new products
	Stimulates improvement of quality and ensures reliable quality	Does not stimulate improvement of quality or ensure reliable quality
Competition!	Sellers compete for buyer	Buyers compete for seller
	Even the monopolist behaves "like a competitor"	Even the seller of the atomized branch behaves "like a monopolist"
Adaptation	Producer adapts to consumer in the short run	Consumer adapts to producer in the short run
	New products transform consumer needs	
Uncertainty	Burden of uncertainty carried by seller	Burden of uncertainty carried by buyer
Selection	Selection is made by buyer	Selection is made by seller or central administrative organ
	Generally progressive selection criteria	Generally indifferent or counter-productive selection criteria
Information flow	Generally the seller informs the buyer	Generally the buyer seeks to obtain information about buying possibilities

In the highly centralized socialist planned economy, the detrimental effects of *suction* are mitigated by two main factors. One is the regular intervention of government and social-political institutions, directed against the negative phenomena. Various strict regulations are enforced to prevent the debasing of quality. The government maintains quality control organs, and in many fields quality standard are prescribed centrally. Technical development is assisted by government funds and there is an extensive network of research institutes. There is a whole range of material and moral incentives to improving quality, reducing costs, meeting in the most flexible manner the needs of the consumers, and providing the buyers with matter-of-fact information.

The other counterbalancing factor is the conscience of both managers and lower rank personnel. Most people like to work honestly; they are ashamed of turning out faulty products. For engineers technical development is a natural ambition. Managers find queuing embarassing and strive to bring the composition of production into harmony with demand.

The two factors are interrelated. Moral and material incentives appeal, first of all and in many cases most successfully to the conscience of workers and employees.

In connection with suction we have emphasized the *positive* counterbalances to the basic detrimental effects on the output of the system. In connection with pressure, on the other hand, we must discuss the *negative* counterbalances to the basic favourable effect. The competition of sellers for the buyer brings abut not only new products but also numerous undesirable aspects of the competitive struggle: greedy speculation, ruthlessness to competitors, cheating the buyer, a flow of false and wasteful advertisement.

21.8. Tension

The positive and negative effects also depend on the *strength* of the pressure or suction in the market. So far we have considered only the *sign* of the disequilibrium; are the sellers or the buyers predominant? However, the *degree* of predominance is not immaterial.

Let us consider first the case of *pressure*.

The concept of *tension* of aspiration is based on the difference between the seller's aspiration and actual sales.[11] As a first approximation we may say that pressure on the market will be the greater, the greater the seller's aspiration tension.

The phenomenon is analogous to the operation of a hydroelectric power plant; the water is able to perform work because there are two water levels differing from each other; the hydro-electric generators are driven by the water rushing down from the higher level to the lower one. The greater the difference between the two levels, the larger the generators the water can drive and the greater the amount of current that can be generated. True, there is a tendency for the water to reach a state of equilibrium where the two water levels are equalized in accordance with the law of communicating vessels. However, should this state be attained, although a fine state of equilibrium would be brought about, the production of current would cease altogether. The water is

---

[11] See the *general* definition of tension of aspiration 12.3, as well as its specification for the tension of selling and buying aspirations in Chapter 19.

able to drive the generators only as long as there is a difference between the two water levels.

There is a similar phenomenon also in connection with electricity. Electric current flows only where there is electrical tension (i.e. a difference between the two poles in electric potential). Here, too, there is a tendency towards equilibrium, towards equalizing the electrical potentials. But if this equilibrium is realized in the sense that the difference between the potentials is not continually reproduced, then the electric current will cease at the same time.

In our case the analogous phenomenon is the tension, the difference between aspiration and result. A whole range of processes (improvement of quality, etc.) are driven forward by this tension—just as the difference in water levels drives the generator and the difference in potentials the machines operated with electric current. When the tension ceases, the driving force of the process in question disappears.

21.9. Intensity

Tension by itself does not give a full explanation. Let us imagine two large-scale plants of equal size, say, two shoe factories. Both strive at selling one million pairs of shoes per year and both succeed in selling 800 thousand pairs only. The tension of selling aspiration is thus equal in both factories. Nevertheless, there may exist some fundamental differences. Let us assume that the first factory is in Hungary in the fifties. Here, the manager and the managerial staff are quite indifferent as to whether the *selling* aspiration can be fulfilled or not. Both bonuses and honors depend exclusively on the fulfilment of the *production* plan, irrespective of whether or not the products are sold.

The second factory is located in a country where it is of vital importance whether or not its products can be sold, and what profit results from the sale. Production without selling would mean a net loss. If our firm has been struggling with selling difficulties for a long time and if it cannot sell off its stock it may go bankrupt. Therefore, in such a case there would be hectic activity to promote sales; salesmen would be sent out to call on the shoe dealers and products would be advertised widely. Also, managers would try to keep abreast with the latest fashion in shoes or even to launch a new fashion since this, too, may contribute to successful selling.

In the terminology introduced in Chapter 12, in the second firm there is a much higher *intensity* of selling aspiration than in the first one. The example indicates that *the intensity of selling is highly dependent on the interestedness of the sellers*. If the seller is not very interested in the selling results, he will not work intensively. The more important the results are to him, the more they represent a question of existence, the more intensive is the aspiration.

On what does intensity depend? Without trying to rank factors by importance, we list some relevant factors:

1. How long have stocks of the commodity destined for sale been held? Within a period that the firm considers "normal", it does not worry about the presence of stocks; however, after expiration of this period the unsold commodity begins to cause ever greater fretfulness. Storage involves costs; the state of the commodity deteriorates; it becomes obsolete; and so on.

2. What are prospects for the future? If sales prospects promise to be favourable, it is not so urgent to sell. But if they are bad, it is important to sell, and the sooner the better.

3. What is the business situation of the firm, its monetary reserves, its creditworthiness? Can normal business be carried on even if it does not sell now, or will it get into trouble if the sale does not succeed?

4. What are competitors doing? Are they making aggressive efforts to conquer the market or are they not striving to win over the usual trading partners of the rival firm?

5. Finally, there is the most essential point; how important is the fate of the firm to those who decide on selling? To what extent are they concerned with the momentary position and the future of the firm? How important is it to them whether the firm stagnates, grows or is ruined? And additionally, is there a real chance of failure or does the state or a bank guarantee survival? Can such an organization interfere and prevent the bankruptcy of the firm? To summarize, to what extent are the decision-makers identified with the "interests of the firm", and more precisely, with its sales interests?

In Chapter 12 we emphasized that intensity cannot be measured directly. It finds its expression only in the *promoting* activities carried out to ensure the attainment of the aspiration.

In the case of pressure, attainment of the selling aspiration level is promoted by three types of activities:

A. Improving the quality, the various kinds of Q-activities.

B. Informative processes: personal visits to buyers by agents, various forms of advertisement, including both factual information and deception of buyers.

C. Relative price reductions.

In relation to tension and intensity of aspiration, we should like to call attention to an important phenomenon, the interrelation among the various promoting processes.

Improvement in quality, advertisement and price changes usually appear *together*. New products are announced by advertising and even within a general pressure situation "excess demand", suction, appears in relation to the new product. This enables the seller to charge a high initial price for the new product in demand. At the same time, the old obsolete product is offered for "sale", and this is also accompanied by a certain advertisement and price reduction.

The problem is not only that the various sales promoting processes appear together, but additionally, that each process *acts* on the other one. Consider the effect of the price on the improvement of quality. The phenomenon is well known. In the case of pressure, there are two kinds of motives prompting the producer to improve the product. The first (which is not stressed sufficiently by the GE school) is that the firm strives to expand or, at least, to preserve its share in the total turnover of the market. The other important motive (which justly receives sufficient emphasis in the GE theories) is the expectation of high profits. While the new product is scarce, a high price may be charged for it. However, the process is somewhat more complex than is recognized in the standard market theories describing it. Not all new products stand the test and earn abundant profits. But since great successes are not infrequent, the *hope* in itself mobilizes response. In fact, the circumstance that every hundredth product brings huge additional profits for the enterprise provides greater impetus to improve quality and increases the *intensity* of the Q-processes to a greater extent than would the circumstance of the additional profits being evenly distributed among the hundred new products.

To sum up:

Statement 21.6. Pressure on the market increases as the tension and intensity of selling aspiration increases. The intensity is indicated by the extent and growth of activities promoting the attainment of selling aspiration, i.e. A) product development and quality control as well as B) advertising—the process of informing and convincing the buyers and C) price reductions.

The formalization of Statement 21.6 constitutes an important problem for research. It would be necessary to specify functions for the individual promoting activities in order to illustrate the effects of pressure. For the sake of illustration only, we mention that functions of the following type could be obtained:

$$\frac{dQ^{(j)}}{dt} = (\alpha^{(S)} - \omega^{(S)})w^{(S)} + \ldots, \tag{21.1}$$

where $Q^{(j)}$ is the jth Q index, i.e. one of the indices measuring the quality improving activities, and $w^{(S)}$ the intensity of selling aspirations. The three points at the end of the formula indicates that the growth rate of the Q index depends not only on tension and intensity but also on other factors (which have been discussed in the preceding sections). These also must be taken into account as independent variables.

The form of formula (21.1) has been borrowed from the natural sciences. It is customary to define the velocity of various physical and chemical processes as a linear function of the deviation from some equilibrium value (in our example, from the selling aspiration).

It should be emphasized once more that formula (21.1) does not even pretend to be a final formulation; rather it serves to provoke thought. It is impossible to decide *a priori* the form of the functions defined in Statement 21.6. This requires further theoretical work and particularly empirical observation. In the final outcome, genuine real-science *laws of motion* of the economy must be discovered in this way; these laws relate the *motive forces*, the tensions, the deviations between the equilibrium value and the actual value of some variable to various social and economic processes (the increase in volume, the improvement of quality, technical progress).

Suction, too, can be treated in a manner analogous to the above description of pressure. The intensity of suction depends on the extent of the tension, i.e. the unsatisfaction of the buyer. The higher it is, the more impatient is the buyer. Of course, impatience also depends on the harm caused to the buyer by the shortage; how intensively does he strive to fulfil his aspiration? Let us recall our motor-car purchaser in Chapter 19. If his buying intention is very intensive, he does everything in his power to obtain a car. Again and again he calls on the selling organization and urges them to sell him a car. He tries to "pull strings" and he may even attempt to bribe the seller. If, on the other hand, the matter is not very important to him, he waits patiently until his turn comes.

The determinants of intensity under suction may be expressed in a more general form as follows:

1. How long has the buyer waited to satisfy his needs? (Provided that he can wait and the need does not become obsolete.) The buyer who has waited longer is more impatient; his demand is more intensive.

2. What are future prospects? If a shortage is expected, it is worth while to make efforts to purchase and accumulate stocks. If supply is abundant and continuous, procurement is not so pressing. If a price increase is expected, it is worth while to buy now. If a price reduction is imminent, purchasing is not urgent.

3. Are there any reserves and stocks, and of what size? If reserves and stocks are abundant, procurement is not urgent; if they are small, procurement becomes more important.

4. What is the buying behaviour of competing firms? Do they aggressively push the firm away from favourable sources of procurement or do they respect established patterns?

5. What is the importance of the product which the firm intends to buy from the point of view of the production of the firm? It is possible that no other product can be substituted, and thus that the product is indispensable; or the

only substitutes may be poorer or more expensive goods. Perhaps, substitution is easily done. This leads us to factor 6:

6. What is the effect of success or failure of procurement on the activity of the firm? In an extreme case, if procurement does not succeed, the firm might be compelled to stop production. Has it monetary reserves or credit for this case so that a temporary shut-down would not mean complete bankruptcy? Alternatively the failure of procurement may cause only a partial disturbance in the production of the firm and in its financial situation? How great is this disturbance; what financial, organizational, and other difficulties are involved?

7. We have saved the deepest effect for last. To what extent are those on whom procurement actually depends and who decide on the conditions of demand, identified with the business success or failure of the firm, its continuous functioning, survival or extinction?

Statement 21.7. Suction in the market is a monotonic increasing function of the tension and intensity of buying aspiration. The intensity is indicated by the extent and growth of activities promoting the attainment of buying aspirations, thus, among other things, by the urging of sellers (or the distributing central institutions) and probably by bribes, forced substitutions, and the buyer's information collecting activity.

21.10. Normative Standpoint

We have endeavoured to compare pressure and suction objectively, from the viewpoint of the descriptive-explanatory real-science. However, although we have pointed out both the favourable consequences of suction and the unfavourable ones of pressure, it should already be clear from what already has been said that in our opinion from the viewpoint of the performance of the system, pressure has more advantages and less disadvantages than suction.

Therefore, *economic policy should endeavour to make general pressure prevail in the market, with three qualifications:*

1. *There should be tension between selling aspiration and actual selling—but not too great tension. Tension should be sufficient to make the seller worry about selling but it should not be accompanied by too great idleness of resources.*

2. *The intensity of aspiration should be strong. Sellers should be highly interested in the success of selling.*

3. *There should operate counter-forces and counter-processes to neutralize or at least mitigate the detrimental effects of pressure: speculation, ruthlessness towards competitors, cheating the buyers, wasteful advertising.*

21.11. Comparison

At the end of each preceding chapter, we have compared our theory with the GE school, mainly on the basis of how well each theory describes and explains. Now, let us turn to the comparison of normative approach. Having presented in the forgoing section our own standpoint, let us survey the economic-political suggestions of other theories.

We begin with the GE school. The theoretical works belonging to this school do not generally formulate economic-political suggestions. Our remarks below could be dismissed easily by the GE economist with the following argument; "We have never suggested that economic policy-makers *should* bring about competitive equilibrium. We would only say that under such and such conditions this equilibrium could be brought about, and that if brought about, it offers such and such benefits."

This argument is only partly justified. The truth is that an economic theory will suggest normative viewpoints merely by the selection of the questions to be answered. When a school continues to scrutinize the conditions necessary for equilibrium to come about, the conditions necessary for that equilibrium to be stable, and the conditions under which it meets various optimality criteria, then it inevitably suggests that equilibrium is desirable. If a hundred economists brought up in the GE school were asked whether it is good or bad for an economic system to be in equilibrium, ninety-nine certainly will answer that it is good.

Here, we have come to one of the most essential criticisms of the GE school— probably the absolutely most essential one. Our book is profoundly opposed to the viewpoint that equilibrium is good. As a matter of fact, in our view—and we have endeavoured to clarify this in the present chapter—what is desirable is not that demand and supply should be in equilibrium but that the aspirations of both sellers and buyers should be intensive and that one type of disequilibrium, namely pressure, should assert itself at this high degree of intensity. The marriage between an impotent man and a frigid woman may be considered as some kind of "equilibrium"—nevertheless, it can hardly be considered the ideal form of relationship between the two sexes. An intensive clash between opposed forces—a passionate one, we might say—is highly preferable to a halfhearted "state of equilibrium" in the economy.

In our first description of the theory of the GE school (Section 3.6) and again in our discussion of preference ordering (Section 11.9), we have emphasized that it is one of the merits of the theory under criticism to have brought into the fore the concept of consumer *sovereignty*. Every humanist will subscribe to the idea that the production of material goods should serve the satisfaction of human wants; production should adapt to men's needs, and man's needs should not be forced to conform to any pattern of production. However, the fulfilment of this humanistic norm is not served by an economy in a strict state of equilibrium. In Section 21.4 of the present chapter we already pointed out that *it is only in the case of pressure that the seller is compelled to adapt to the buyer's requirements; it is only in that case that the consumer is "sovereign"*. (And even in this case he is sovereign only in the short run. In actual fact his needs are influenced by technical progress and by the new products appearing in the market.)

We do not wish to create the impression that our book is the first to take a stand for the requirements to be summed up in Section 21.10. Authors with a different theoretical background already have reached similar viewpoints.

1. The idea appeared in the works of *Karl Marx*. Although Marx generally abstained from normative suggestions concerning the reasonable organization of the economic system, in his theory of reproduction he presented some remarkable ideas in this connection. He poses the question of how the economy should be organized "once the capitalist form of reproduction has been eliminated" and refers to the fact that in some years there is more fixed capital being used up and in others less. In the interest of its replacement "the total product of the means of production ought to increase in some years and decrease in others. This can be accomplished only by continuous relative overproduction. On the one hand, more constant capital must be produced than that directly necessary and, on the other hand, stocks must be built up of raw materials, etc. in excess of the direct annual needs. (This applies especially to subsistence goods.) This type of overproduction means that society has taken under control the material means of its own reproduction".[12] As can be seen, Marx is speaking not simply of reserves and stocks but of *continuous relative overproduction*.

2. Researchers engaged in the comparison of economic systems introduced the term "buyers' market" versus "sellers' market", terms which we have used repeatedly. They have revealed many of the effects of the two types of market situation.

The intellectual precursors of the Hungarian reform of economic administration and management also have adopted this terminology.[13]

While most economists are well acquainted with the terms "buyers' market"

---

[12] *Marx Capital*, Vol. II, in Hungarian language [168], pp. 476–477. In German-Vol. II [167], p. 473. We have used this quotation previously in our book [126] published in 1957 and dealing with the overcentralization of economic administration
[13] Thus, e.g. in *György Péter*'s articles [201] quoted above.

and "sellers' market", these concepts usually have not been integrated with other theoretical knowledge. In another compartment of the mind the conceptual framework of GE theory continues to exist. In one compartment there is the normative requirement that the desirable state is the buyers' market. In the other compartment there is the norm of equilibrium. It is time to recognize the fact that *"buyers' market" means continuous disequilibrium.*

3. Similarly, a number of economists, both in socialist and capitalist countries consider it desirable for the economy to retain some "slack" (surplus, unused capacity), because some "slack" will foster adaptation and even stimulate progress. However, it has not yet entered the general consciousness of the economic profession that *continuous "slack", the less-than-complete utilization of resources, means continuous disequilibrium.*

4. There is also mention—especially in Western literature—of a gap existing between the economy's potential and actual capacity.[14] This concept is based on the partial idleness of the resources of the economy, first of all of manpower. In the case of complete utilization of resources, the units of the economy would be able both separately and collectively to show even higher performance. *Continuous gap between the actual and potential performance of the economy means continuous disequilibrium.* (Here we recall the concept of "potential production increment" introduced in Chapter 19.) Economists usually point out the negative consequences of this disequilibrium—unemployment, slow growth rate—and these undoubtedly exist when the gap is unnecessarily wide. Less frequently, economists perceive (although sometimes they mention it) that the gap, if it is not too wide, may also have positive efforts. It may stimulate technical progress and the flow of new products; it may improve adaptation and selection, and so forth.

5. In connection with the less developed countries it has been suggested that growth occurring in a state of disequilibrium—"unbalanced growth"—is more advantageous than growth in an equilibrium context.[15] Several authors are of the opinion that—in the terminology of the present book—continuous suction and the appearance of bottle-necks may give an impetus to the country's development.

The economic growth of backward countries involves special problems the analysis of which would exceed the scope of this book. In any case, we are under the impression that those advocating development under suction appreciate only the beneficial effect on the increase in volume but underestimate the detrimental effect on quality and technical progress. In the light of Hungarian

[14] See *Okun*'s article [195].
[15] See the studies by *Hirschman* [89], *P. Streeten* [244] and *S. Mason* [172].

experiences the economic policy of suction-type "unbalanced growth" would appear one-sided and in many respects harmful.

Finally, as at several points in this book, we would like to emphasize the problem of theoretical *integration*. All the ideas and statements dealing with the appearance and the effects of disequilibrium should be amalgamated into one homogeneous theory.

22. REPRODUCTION OF TENSIONS

22.1. Delimitation of the Subject

In the following we shall expound some ideas about the *causes* of disequilibrium. What is the explanation for the fact that economic systems do not achieve a state of equilibrium but operate for long periods in the state of suction or pressure? How are tensions continuously reproduced, maintaining lasting discrepancy between production and consumption, between selling and buying intentions and aspiration? Why is it that the central tendency of production and consumption, in selling and buying intentions is not the equilibrium? (This was illustrated in Figures 19.2 and 19.3.)

An answer to these questions still requires much research. Though often we have mentioned that our ideas still are very immature, we should like to stress that fact particularly at the present moment. In this chapter we deal with a fundamental problem of economic history and economic systems theory which has not been investigated sufficiently thoroughly by science.

Those who are politically biassed against socialism or those who are more objective yet superficial easily arrive at the conclusion that suction is a necessary consequence, or at least a concomitant of socialism, and pressure that of capitalism. It is not easy to argue with this viewpoint since —as has been stressed in Statements 19.4—19.5—in most of the socialist countries there prevails mainly suction and in most of the capitalist countries, pressure. Still, as we have anticipated in Section 19.10, we decidedly discard this "explanation". It is our conviction that the socialist system also may exist in the state of pressure and that also general suction may prevail in the market of the capitalist system.

Our conviction is to a certain extent consistent with the facts of *economic history*. On the one hand, according to the available information, the market of the Yugoslav economy in the middle of the sixties functioned in a state of pressure at least for some time in the wake of economic reforms. This change was not the result of decentralization or to the introduction of workers' self-management, etc.,—suction could have continued for years even with these. Rather it was the changes made in investment policy, in price formation and wages, and in credit and financial policies that tilted the relation of market power.

On the other hand, in times of war, pressure gives way to suction in most capitalist countries—even where no changes are made in capitalist ownership relations, where there is no nationalization.[1]

A further example is the commodity market of several backward African, South-American and Asian countries, where suction prevails despite the fact that the power and ownership relations of the system can by no means be considered socialist.

The examples listed weaken the correlation between suction and socialism or between pressure and capitalism. But considering only the facts of economic history, the correlation is still rather strong. Therefore, we believe that logical arguments are more important. We should like to delineate the main factors *immediately* responsible for suction and pressure. Having surveyed these, we can investigate the chain of causes further, looking for the cause of the causes. We shall attempt to show that the immediately responsible causes do not necessarily appear under the conditions of socialism or capitalism, respectively, but are themselves consequences of a definite *economic policy* and may be changed *while maintaining* the given ownership relations.

22.2. Suction: Consumers' Intentions

The causes immediately responsible for the state of suction may be classified into three major groups:

1. Factors affecting the trade in consumer goods.
2. Requirements imposed upon firms; structural disproportions.
3. Factors affecting the trade in investment goods.

Let us start with the first group of factors. The phenomenon may be summarized as follows:

Statement 22.1. The first immediate cause of suction is slow, repressed inflation in the trade of the consumer goods. The consumer cannot completely satisfy his buying aspirations but regularly performs forced substitution, including forced savings.

Ex post, the self-explanatory equilibrium conditions always prevail.

—Either the consumer spends his disposable income on the buying of consumer goods or he saves it.

—All consumers combined cannot buy more than the stock of products available. That is, in the last resort, considering a longer period, production sets a limit on consumption.

---

[1] This phenomenon is effectively described in the article on the American war economy by *Galbraith* [68a], entitled: "The disequilibrium system."

The *ex ante* situation is different. In the case of suction there are grave disproportions; to a considerable extent buyers' aspirations remain unsatisfied.[2]

There are two kinds of (interrelated) reasons. One is the fact that disposable income grows in an inflationary manner. The inflationary process is pushed forward by the increase of employment and wage rates, in summary, by the growth in disposable incomes; prices rise but the rise is retarded by various measures, such as government price fixing. Owing to this fact, total disposable income is greater than the sum of the previous purchases and the previous saving intentions. This "superfluous" disposable income stimulates buying aspirations. The other factor of this process involves the efforts made by the consumers to reduce risk.

The state of suction can be furthered also by the rigidities of the price system. More flexible prices could contribute to a short-term adaptation: they would reduce demand for shortage goods and could render more attractive the buying of such articles as are now bought under necessity, because of the shortage. Simultaneously the producer would be induced to increase production of the articles in shortage and to reduce production of those articles which are consumed as a result of forced substitution. Nevertheless, however important the effect of *relative* prices i.e. price proportions, may be, a still more important part is played by the proportions of the *general* price and wage level, the total commodity supply, and the total purchasing power.

The buyer knows that he cannot fulfil all of his aspirations. His aspirations might include a car, a flat, new furniture and a travel abroad. He knows that if—by some miracle—all four aspirations were met, he would not have the money to afford these simultaneously. But there are no miracles, and he fares well if some of his aspirations can be met: either the flat, or the car, and so on.

The feeling of non-satisfaction due to suction strengthens the efforts of the buyer at safety, the multiplication of his claims as described above. Thus, there is a peculiar *suction spiral* and the feeling of non-satisfaction grows. Suction prevails in the control sphere and produces all the effects on the producer and consumer described in the preceding chapters—in spite of the fact that in the real sphere there is "equilibrium" between purchase and stocks, consumption and production, as described before.

An additional remark is warranted. The above process—the repressed inflation in the trade of consumer goods, the suction and the feeling of non-satisfaction—is fully compatible with the growth of real consumption over time and thus with a rising standard of living. Total consumption measured in constant prices may increase over time while the disproportions due to suction prevail. This actually happens in most years in most socialist countries; the

---

[2] The term "unfulfilment ratio" was defined in Definition 19.2.

standard of living regularly rises under conditions of suction. It should be added that the consumers believe that this rise in the standard of living is smaller than it is in reality, precisely because it is coupled with an enduring feeling of non-satisfaction.

22.3. Suction: "Tense" Production Plans, Structural Disproportions

A second group of the factors immediately responsible for suction are the requirements imposed on enterprises.

National economic policy in many socialist countries endeavoured over long periods of historical development to prompt plants to "maximum performance". Recall the period 1949–1953 in Hungarian economic history, the period of "tense production plans". Firms were expected to produce the maximum volume of output with their scarce material, energy and labour resources available. Every financial stimulation (e.g. the system of bonuses and wage-payment), every moral reward was coupled to success in increasing the volume of production; a poor utilization of capacities might have brought moral or even legal sanctions. Therefore every plant thirstily sucked in material, energy, and manpower.

The practice of "tense plans" hindered the formation and reproduction of reserve capacities, surplus resources, and "slacks", which are indispensable for the maintenance of the state of pressure.[3]

Structural disproportions between entire productive branches are a phenomenon similar to that discussed in the previous section. As in the situation prevailing in consumption—as expounded in Section 22.2—the phenomenon can be understood only if we separate the *ex ante* disproportion prevailing in the control sphere from the *ex post* equilibrium of the real sphere, that is, aspiration and intention on the one hand and realization on the other hand.

In the real sphere, obviously the following equilibrium condition prevails *ex post*:

The using productive firms cannot purchase more products than are made available to them by the supplying productive enterprises.

The *ex ante* situation is different; here we turn to the processes of the control sphere, the intentions and aspirations. In the case of suction, the following disproportions prevail:

In order to fulfil and overfulfil the tense plans expected from them, productive

---

[3] For the sake of completeness the pursuit of quantity must be mentioned among the factors contributing to the state of suction. Since the introduction of the reform of economic management this factor has ceased to exist in Hungary.

firms demand more material, energy, in general, more means of production than are really available.

In this sense, there exist structural disproportions. In the literature of the socialist countries, frequently even in the official statements of leading statesmen, there are phrases such as: "Production of basic materials has lagged behind the requirements of the manufacturing industry"; "energy production has lagged behind industrial development", "production of machine tools has not reached the level required by the engineering industry", and so on. These expressions all reflect the disproportion described above.

Statement 22.2. The second immediate cause of suction is disproportion among the productive branches; the aggregate real output of firms supplying a given product usually lags behind the aggregate purchasing and utilizing intentions of the firms using the particular product, if the latter make efforts to fulfil and over-fulfil their tense plans.

22.4. Suction: Investment

Finally, we turn to the third immediate cause of the state of suction which is perhaps the most important one: the disproportions related to investment.

A large part of investment decisions are made centrally in socialist countries. The five-year plans of many countries contain the most important, concrete investment decisions: where, when and what kind of important new plants should be established, what new roads should be built, how many and what kinds of machines should be given to agriculture, and so on.

In addition, there are also decentralized investment decisions which are made either by local, state administrations (e.g. county or town councils), by institutions (e.g. universities or hospitals) or by productive firms.

To make the problem of investments more palpable, we introduce some concepts and notation.

Let us denote by k the elementary investment decision. This can be made either by a government agency, by a firm or by any other institution.

Suppose that a decision was made in period t_0; its implementation could begin at the earliest in period $(t_0 + 1)$ and it could be concluded in period t_1. e.g. in 1963 the government decided to build a new factory; work began in 1965 and was completed in 1969. In this case t_0 is 1963 and t_1 1969.

Definition 22.1. The e l e m e n t a r y i n v e s t m e n t d e c i s i o n k is v a l i d in period (t_0, t_1) which includes the time from decision-making to the completion of the investment activity or the annulment of the decision. The period (t_0, t_1) is called the v a l i d i t y p e r i o d of the decision. Its duration is the period $T = (t_1 - t_0 + 1)$.

22 **KORNAI:** Anti-Equilibrium

Definition 22.2*. The set $\mathcal{K}(t)$ is the set of all elementary decisions k which are valid in period t. The set $\mathcal{K}(t)$ is called i n v e s t m e n t   i n t e n t i o n.

To implement the investment decision, investment goods (machinery, equipment, structures) are required. In the economy there are n kinds of investment goods in all.

Whenever a decision is made about an investment project, the decision includes not only the final result of the action (e.g. what factory should be built; what and how much it should produce after starting its activity), but also the schedule of realization and the necessary inputs. Whether this is stated expressly in the investment decision or not, we may assume that to each investment decision there belongs a *time-series* of claims on investment goods. For the k investment decision ($k \in \mathcal{K}(t)$ let us denote the claims on input pre-estimated at the time of the decision-making as follows:

$$d_k(t_0+1) = d_{1k}(t_0+1), \quad d_{2k}(t_0+1), \quad \ldots, d_{mk}(t_0+1)$$
$$d_k(t_0+2) = d_{1k}(t_0+2), \quad d_{2k}(t_0+2), \quad \ldots, d_{mk}(t_0+2) \qquad (22.1)$$

$$d_k(t_1) = d_{1k}(t_1), \qquad d_{2k}(t_1), \quad \ldots, d_{mk}(t_1)$$

It is possible that all components of the first two (or even more) rows of the vectors are zero, provided that some periods elapse from decision-taking until the actual commencement of the investment activity.

With the lapse of time and the advancing implementation of the investment, only the claim-vectors of the remaining periods are of importance. And the original estimates may be modified. During the implementation of the decision it may turn out that more or less of some investment goods is needed; some partial action should take place later or sooner; and so on. From the point of view of our further dicussions, we are interested exclusively in the current claims on inputs.

Definition 22.3*. Let us call c l a i m   o n   i n v e s t m e n t   i n p u t arising in period (t) and denote by $d(t)$ the vector of n components which determines the quantities of the n-kind of investment goods necessary in order to implement the set $\mathcal{K}(t)$, the investment intention, the valid investment decisions.

$$d_j(t) = \sum_{k \in \mathcal{K}(t)} d_{jk}(t) \qquad j = 1, \ldots, n \qquad (22.2)$$

The definition shows that the vector $d(t)$ is a variable of the control sphere; it is an aspiration or intention in a double sense. Partly it expresses only the intention of a purchase, not the actual purchase. Partly, this intention itself derives from another intention, the investment decision.

We have to consider two further vectors of n components related to the n

kinds of investment goods. The first is the *vector of the investment goods actually purchased in the period t in order to carry out the investment activities*, to be denoted by $x(t)$. The second is explained in a definition.

Definition 22.4*. Let us call the vector of the investment goods physically available in the period t the i n v e s t m e n t   p o t e n t i a l   and denote it by $y(t)$.

The components of the vectors $d(t)$, $x(t)$ and $y(t)$ are measured in physical units of measurement.

As in Sections 22.2 and 22.3, we begin with the *ex post* equilibrium condition which, of course, exists in the real sphere:

$$x(t) \leqq y(t) \tag{22.3}$$

The actual acquisition of investment goods cannot exceed the available investment potentials.

In the case of suction there exists an *ex ante* disproportion in the control sphere.

$$d(t) > y(t). \tag{22.4}$$

The disproportion described in (22.4) manifests itself in well known phenomena. The press of the socialist countries frequently criticize the planners, the economic decision-makers, for beginning too many investments, because this causes realization to become too protracted. The smooth realization of investment is frequently hindered by shortages; the investor "queues" for building capacity, building materials, imported machinery, instruments, equipment.

Statement 22.3. The third immediate cause responsible for suction is the over-ambitious character of investment intentions in comparison to real material and technical conditions of carrying out the investments,[4] to the investment potential.

22.5. Suction: A Combined Survey

Having surveyed the three immediate causes separately, now we may make a comprehensive statement.

On learning of the phenomena of suction, most Western economists are inclined to say that these phenomena depend exclusively on prices. These economists would explain the *whole* group of phenomena with repressed inflation; wages grow while the growth of prices is restricted. Therefore, if either

[4] The phenomenon described here is related to the problems that were examined by the Yugoslav economist *Branko Horvat* in his work [94]. Horvat points out that every economy has an absorptive capacity in respect to investment and if investment plans are higher than that tensions and losses result.

22*

the wage inflation is stopped or rapid rise in prices is restrained, the suction will cease.

But our analysis suggests that only the first category of the causes is directly related to monetary processes, to the relation between the purchasing power of the population and the price level of consumer goods. However, the factors under 2. and 3., exert their influence *not* through prices, credit or money; these disproportions appear in "physical" forms. (E.g. when formalizing the disproportions (22.4) we used exclusively variables expressed in physical units of measurement.) The problem is not that in socialist countries "too much money" is allocated to investments but that the investment decisions (which also can be described in engineering terms by estimated real inputs and real outputs measured in physical units of measurement) generate in their aggregate too high investment intentions relative to the realistic possibilities available, to the investment potential. After the decision has been made, the money to be used for the purchase of investment goods can be created; whenever the action has started, the purchasing intentions necessarily will appear. The investor begins to put in his claims on building capacity, imported machinery, equipment; his intentions already exist, almost independently of whether the money necessary for the purchase is at hand or not.

Additionally there is the fact that in the socialist countries—particularly before the reforms, but to a certain extent even after them—some real processes are not controlled by variables measured in money terms (e.g. investment credits) but by direct decisions, instructions, suggestions expectations). ("Start to build the new factory with an annual capacity of 10,000 tons!")

In Sections 22.2–22.4 we described separately three groups of immediate causes. In fact, these are not independent of each other, but they mutually influence and strengthen each other. There is a particularly strong connection between the factors under 3. and 1. By simplifying the explanation somewhat, we can say the following:

Overambitious investment intentions lead delays in the realization of investments. Those working on the realization of the investments obtain incomes while the additional production due to their activity—the additional supply—appears only with a considerable time-lag. This is one of the factors promoting wage-inflation induced from the side of purchasing power.

The unfavourable effects of suction—the "queueing" and the dissatisfactions involved—prompt the economic leadership to restore equilibrium by increasing production. But this again leads to the effects mentioned under 2. and 3., to tense plans and an excessive volume of investments. Thus, it does not abolish but rather increases the tensions due to suction.

In the final analysis, it seems that all three immediate causes can be reduced to a common source:

Statement 22.4. The reproduction of suction is ultimately related to the impatient chasing of economic growth, the forcing of the acceleration of the growth rate.[5]

The rigid separation of cause and consequence can no longer be continued at this point: the chase after volume, after higher growth rates, is both a cause and a consequence of the state of suction. We may also put it in another way; there is a close positive correlation between the chase after growth and suction. This formulation is a fortunate one because it leaves room for exceptions. There are situations in which there still is strong suction although the rate of growth is slow.

The summary Statement 22.4 is supported by economic history. Yet, it cannot be considered proven; it may be regarded rather as a working hypothesis. Further research is necessary in order to clarify fully the cause of tensions accompanying suction.

22.6. Reproduction of Pressure

Let us turn to the explanation of pressure, considering mainly the forms that appear at present in the developed capitalist countries. We must keep in mind the fact that the regulatory variables expressed in monetary terms (—purchasing power, credit, monetary savings and prices—) have a much wider scope in a capitalist economy than they do in a socialist economy. This was especially true before the reforms in the latter countries.

1. As a first approximation, as in the discussion of suction, we investigate the relation between prices and wages. It is known that in most capitalist countries inflationary processes have been going on for decades, sometimes at a slower, sometimes at a quicker rate. We do not wish to deal here with the historical beginnings of these processes; nor with the manner in which the original state of pressure came about. Let us be content to recognize that the process is already going on and pressure prevails. Then the rise in both prices and incomes can continue; that is the value of money diminishes while the state of pressure is maintained steadily. The only requirement is that the rise in incomes must not precede that of prices but at most keep pace with it. The increases of prices and wages are, of course, interrelated but, in a capitalist pressure economy price-inflation has a leading role. This ensures that, in spite of the steady rise in prices and wages, at no time can the purchasing power of the consuming population absorb the mass of products offered for consumption. Under capitalist market conditions this disproportion reverberates affect-

[5] Our train of thought here is related to the comments of *F. Jánossy* in his article [101].

ing production through the monetary and credit connections among enterprises. In the final analysis solvent buying intentions lag behind the selling intentions backed by real stocks of goods or potential production possibilities.

This phenomenon is related also to the influence of the big concerns on price structure. Even if selling meets with difficulties, the firm will introduce new products, instead of adapting itself to the given market situation by reducing prices.

As in the case of enduring suction, enduring pressure does not preclude the growth of real consumption over time.

The description is only a makeshift outline. Much further research will be required before we are able to describe more precisely slow "controlled" inflation in the case of "pressure" and that in the case of "suction". Yet today, every government makes efforts to control inflationary processes and to this end they try to slow down the rise both in prices and wages. The difference between such control in the cases of suction and pressure may be summed up as follows.

Statement 22.5. In order to slow down and control inflationary processes, in economies with suction it is the price rises and in those under pressure, the wage rises which are held back more efficiently.

2. As we have pointed out in the case of pressure there is unused potential productive capacity, a slack, in the whole of the economy and within that in almost every firm taken separately. How does this come about?

Every enterprise decision-maker knows that considering the whole of the economy there are and there will be unused capacities, big stocks of products and idle resources. But he hopes that perhaps it will be *his* enterprise that will be able to utilize capacity fully. "Maybe fortune will favour me. Should the confidence of the buyer turn towards me, I do not want to turn him away because in that event he would not look for me the next time."

We have here a vicious circle. Owing to the pressure—as has been pointed out in Chapter 21—the buyer may choose from among the sellers. Knowing this, every seller tries to prepare for the case when the confidence of the buyer turns towards him. Thus, he makes efforts to create surplus capacity—which again increases pressure by increasing the selection possibilities of the buyer. In the final analysis, this reduces the probability—from the point of view of the individual seller—that the buyer will choose precisely him.

The same idea may be formulated in another manner. The market is characterized by uncertainty. To reduce uncertainty of sales, every seller creates separate reserve capacities. The size of this reserve is as great as required by his *own individual security;* he does not want to miss potential buyers. However, considering *global social* needs, the sum of individual reserve capacities is much greater than would be necessary for a safe supply for all buyers. (This is why planned economies are tempted to abolish this "exaggerated" reserve.) A side effect of this exaggerated excess capacity, is stimulation of important progressive technical processes and qualitative improvement.

3. How "slack" arises is related to technical progress, the appearance of new products and the improving quality of the old ones. When producers supply the users on some given level of quality, there can be equilibrium in the narrow static sense. But a new product might appear and begin to draw away the buyers. Soon, the producers of the old product will develop excess capacity which persists until they follow the initiative and shift to production of the new product.

Since the initiative, the first introduction of a new product, entails great advantages strong firms make efforts to prepare for that in advance. For example, large chemical firms and pharmaceutical plants reserve some additional capacity for this eventuality; should a new invention appear, they may quickly shift to its production.

4. Finally, in the case of pressure as in that of suction, an explanation is found in the proportions of investment. The Keynesian school frequently deals with the equilibrium and disequilibrium between savings and investment. This is related to another, much less discussed aspect of the problem; the relation between the investment *intention* and the *real investment goods* available for the implementation of this intention. With the notation introduced in Section 22.4, this can be described for the case of pressure as follows:

$$d(t) < y(t), \tag{22.5}$$

that is, the investment intention is smaller than the real investment potential available for the investment activities (machinery, equipment, building activities, etc.).

The inequality is precisely the reverse of the relation (22.4) describing suction; there the investment intention was larger than the realistic possibilities of implementation.

This lower level of investment intention may be explained by many factors. Money savings, which may serve as a financial basis of decentralized investments, are low; the decision-makers are too cautious; the government or the banking system pursues a strong restrictive credit policy; public investments are low. These factors (and maybe others) may appear separately, but it is more frequent that several or all of them exert their effects simultaneously.

In the final analysis, the core of the matter is that in an economy under pressure there always exists idle, unused real capital which can be used for real investment. In the financial world this is reflected by the existence of mobile money capital which quickly seizes upon investment possibilities that promise to be profitable.

The idleness of resources is obviously a loss. On the other hand, it has some advantageous effects (particularly if it is not too great); it allows investment decisions to be more flexible, it facilitates the investments necessary for the rapid introduction of new products and new inventions.

We should like to sum up that which has been said about investment in a statement, the truth of which we can not yet prove satisfactorily; however, the analyses hitherto performed seem to justify it:

Statement 22.6. The main regulator of the state of suction and pressure is the control of the ratio of investment intentions and investment potential, the production of real investment goods.

Both sides of the inequalities (22.4) and (22.5) can be controlled by purposeful measures; both the capacity of the industries turning out investment goods, and the investment intentions, the ensemble of investment decisions. The former involve longer real processes; the latter involve informative, decision and control process which are influenced even more easily.

Having surveyed the main factors shaping the state of pressure, we propose the following summary statement.

Statement 22.7. The main immediate causes of the state of pressure are: 1. strong breaks on the increase in purchasing power in slow inflationary processes: 2. surplus capacities created in firms in order to meet smoothly the purchasing intentions presenting themselves to the firm: 3. selling difficulties of the firms turning out old products when new or improved products gain ground on the market, and 4. the lagging behind of investment intentions in comparison to the real potential available for the implementation of investments.

In the final analysis, we find here, too, a strong correlation between the growth rate and pressure (or rather, the extent of pressure). *A very slow growth is usually accompanied by strong pressure. Increase of the growth rate in most cases diminishes pressure. If the growth rate is strongly accelerated, the pressure will turn into suction.*

22.7. Transition from Pressure to Suction and Vice Versa

Up to now, we have considered only systems in a state of pure pressure or those in a state of pure suction. What happens when a system changes from one state to the other?

Let us take first the case of pressure turning into suction. This has happened in capitalist countries when they changed over from peace economy into war economy. To a certain extent it is a similar phenomenon when the capitalist economy turns from the state of depression into the state of "Hochkonjunktur" or "boom".

This situation also occurs when the capitalist system is supplanted by the socialist one in a revolutionary way—not after a war but under non-warlike, peaceful conditions. Such a transition took place at the end of the fifties in Cuba.

Statement 22.8. In the period of transition from a state of pressure into that of suction the growth rate of the volume of real output increases greatly. The transition entails transitory acceleration.

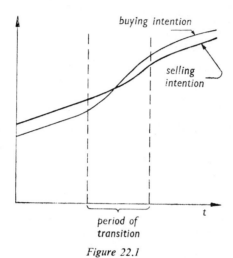

Figure 22.1

Transition from pressure into suction

The transition is shown in *Figure 22.1*, where the time of transition is represented as distance between the two broken vertical lines. As has been clarified earlier, it is always the lower line, the lower intention, that predominates since only this amount can be realized. The higher intention will not be fulfilled. Before the transition it is the purchasing intention which predominates; after it the selling intention. But the selling intention suddenly grows, during transition when the chase after the volume of production, the utilization of the hitherto idle "slacks", begins. It is as if the system obtained "free of charge" as a gift, additional fixed assets. Obviously, this induces a sudden rise in the volume of production. But this surplus can be acquired only once. If, from a former state of relatively unutilized capacities the system arrived at a state where resources are fully utilized, further growth can take place only as a function of expanding fixed assets, technical development, and rises in productivity and efficiency.

What would happen if a system operating in the state of suction should shift into the state of pressure? Since there is insufficient historical evidence, we depend on logical analysis to establish the following hypothesis:

Statement 22.9. In the period of transition from the state of suction into that of pressure, the growth rate of the volume of real output diminishes to an outstanding extent. The transition involves a transitory slowing down of development.

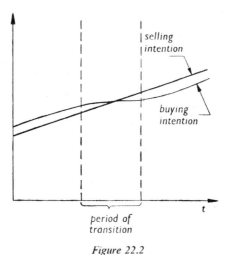

Figure 22.2

Transition from suction into pressure

The transition is shown in *Figure 22.2*. It is advisable to put a brake on the increase of purchasing intentions during the time of transition and to let the selling intentions (the production potential) develop. The "slack", the reserves of resources and capacities, as well as more differentiated stocks of products must be created. In the final analysis, this is a particular kind of major investment that does not yield any additional production at the given moment but which affects the growth processes later—by accelerating technical progress and improving the adaptive and selective properties of the system.

It is understandable that the socialist countries are hesitant to take such a step. The transition would lead partly to a transitory slowing down of the growth rate (the importance of which is undoubtedly overestimated, fetishized, as it were). On the other hand, it would probably involve also transitory sacrifices as regards the raising of consumption. Regardless of these facts we believe that the problems and difficulties described in Chapters 20–21 sooner or later will compel socialist countries to organize the transition from the state of suction into that of pressure. As was explained in Chapter 16, the performance of the system is composed of many different kinds of effects. Their comprehensive consideration will sooner or later force us to transition from suction to pressure.

22.8. A Degression: The Reform of the Hungarian Economic Control and Administration

As we have said in the foreword to this book, in 1968 there was a deep reform of economic control and aministration in Hungary.[6] The authority for decision making was delegated to a considerable extent to the firms. Partial decentralization was carried out with respect to pricing as well as the control of investment. The right of decision making was delegated almost entirely to the firms with respect to short term decisions on production and inputs. Firms (mainly the executives) become much more interested in raising their profits.

Using the terminology of this book we can say that as a result of the reforms, there has been a profound change in the information structure of the economic system together with the response functions and decision algorithms of its economic institutions. Likewise the decision indicators, the aspirations and their intensity have changed.

Prior to the reform suction prevailed in the Hungarian economy. It is true that suction was not uniformly strong in every field. Still, in general, the state of suction was dominant and pressure was an exception. In this respect, the reform has brought about hardly any change whatever. There has been some reallocation of resources. In some fields (perhaps in the food industry and in some branches of the light industry) power relations have changed and pressure developed (or, at least) suction became lighter. Still though it is true that suction dominates even today. Strong tension due to suction is felt in important branches such as the market of meat, of cars, of furniture and of housing, and the market of investment goods.

Thus, an ambiguous situation has developed. Partial decentralization of decision rights and the increase in interest in profitability has not yet provided sufficient incentive for technical development or for a more flexible adaptation to the buyers' demands—since there still prevails a sellers' market and not conversely. This should be obvious, since, as we have seen in Chapters 20–21, the adaptive properties of the system depend not so much on the institutional system but rather more on the type and degree of disequilibrium, on the relative strength of the buyer and of the seller.

[6] From the rich literature on the subject see, e.g., the book by *Rezső Nyers* [194], works of *Tamás Nagy* [186] and *István Friss* [67], as well as the collection of studies [68] in English.

The initiators of the Hungarian reform of economic control emphatically called attention to this fact.[7] However, in the final stage of preparing the reform, this aspect of the problem was pushed to the background.

Let us face wholly sincerely and openly some basic ideas of the Hungarian reform which usually are not discussed publicly although everyone is aware of them. Some Hungarian economists and economic leaders thought along following lines; "The advanced capitalist countries have attained very remarkable results in technical development, in improving the quality of the products, in attentively meeting the buyers' demand, in flexible adaptation. Therefore, let us borrow some important institutional features of modern capitalist economy—a greater decentralization of decisions, a greater freedom for price movements, profit incentives—without renouncing the fundamentally socialist features of our system: socialist ownership relations, planning and central control."

But, in this form, the idea is only a half truth. All the results achieved by modern capitalist economy in technical progress, in improving quality, in adaptation, can be explained only partly by decentralized decisions and the profit motive. The presence of the state of pressure is an explanation of at least equal if not higher rank. Therefore, if Hungary carries out the former institutional changes but remains in a state of suction, either the expected, favourable results will not appear or they will assert themselves only to a lesser extent.

With some exaggeration, we could add a touch of humour by saying that some of the economists preparing the reform of the Hungarian economy became victims of their credulity towards the GE school. The real economic life of capitalism does not pay much heed to what is being taught in the universities under the title of general equilibrium theory. In reality, we find a highly concentrated economy in which there are not many traces of atomized, self-regulating perfect competition; rather, capitalist economies operate with a complex information structure, on the basis of complex motivation and under the effect of diversified government intervention. Their economic successes may be attributed at least, partly to the lasting state of pressure. But, even if capitalist economies do not function according to the suggestions of the GE school, some groups of reform-economists in the socialist countries were "taken in" by it. Practical men in New York, London, Paris or Amsterdam surely do not take the theoretical ideas of exclusive profit-maximization, market price mechanism and the "invisible hand" governing the economic processes as seriously as do some men in Budapest. The reform-economists do not know closely the

---

[7] See primarily the articles by *György Péter* already quoted [201] as well as Chapters IV and VI of the book of the present author, published in 1957 [126].

mathematical models of the GE school, with all of the fine details. But the basic ideas of the school, its "suggestions", have reached their minds. Illusions have developed that a partial decentralization of the scope of decision, profit incentives, a freer movement of prices—coupled with socialist planning and central government control—would be *sufficient* to secure economic efficiency.

In our view, these are really *necessary but not sufficient.* An indispensable condition of efficiency, technical progress, improvement of quality and more flexible adaptation is a certain amount of pressure—not too strong, but palpable. The GE school is silent on this subject and all those who took the one-sided teachings of the school too seriously, are "victims" of this reticence.

All that we have said is not a critique of the reform measures hitherto taken. It is our conviction that the reforms were fundamentally correct and necessary, they should be confirmed. It is desirable to develop further the economic reform, so that production becomes more efficient, the growth of firms depends more on their own results, and their interest in adaptation and reduction of costs increases. We only wish to stress that sooner or later they must be complemented with changes in market power relations. It is not the task of the present book to take a stand on when and at what rate this can be carried out. Precisely because it requires extensive preparation and perhaps considerable sacrifice, the transition must be thoroughly worked out by weighing all social, political and economic consequences. It would not be wise to press impatiently for a change. However, we are justified to illuminate from the theoretical point of view what would result from such a change, from transition to pressure—and what would be the consequences of remaining in a state of lasting suction.

22.9. Comparison: Excess Supply and Excess Demand

One might raise the question of whether it really was useful to introduce a new concept, pressure, in the previous chapters (Chapters 17 to 22). Would it not be more simple to say that pressure is nothing else but what in the GE school and in the closely related neo-classical price theory is called "excess supply", while suction is the same as "excess demand"?

There is little use in quibbling over *mere terms.* The two pairs of concepts, pressure—suction and excess supply—excess demand obviously are related. Yet there were several considerations which lead us to introduce new terms.

1. In the neo-classical price-theory both supply and demand are functions of prices. In our system both selling and buying intentions do not *only* or necessarily depend on prices. We wanted to introduce a universally valid concept which could be used in describing both the centrally organized economies of the socialist and that of the capitalist contries. However, in the socialist

economic systems capital investments, for example, have very little to do with prices (see Section 22.4).

As has become clear earlier in this chapter, the state of suction in the socialist economy cannot be explained simply by the phenomena of the *financial* sphere: by the disproportion between demand and supply, or between purchasing power and the quantity of goods offered at the given prices. In that form, and without special restrictions, this can be said only about the consumer's market. On the market of the means of production the main problem is not that there is—using a simplified expression—"too much money" in the hands of the buyer. To cover the requirements of its current production, for a normal or maybe tense utilization of capacity, the firm demands more material, more spare parts, more machines and manpower than it has at its disposal. Similarly: for the realization of the agreed investment decisions there would be need for more machines, more building activity and more foreign currency than can be afforded by the investment potential of the country.[8] In this case therefore it is not a disproportion in the financial sphere we are faced with, but a gap between the *actual real* requirement aroused by plans and decisions, and the *actual real* possibilities.

2. The usual concepts of supply and demand concern decisions immediately previous to selling and buying. In contrast we wanted to study intentions and aspirations formulated *much earlier*. We wanted to observe the measurement of satisfaction or dissatisfaction of intentions and aspirations.

3. The use of any term which has had a long history creates some usual *association* of ideas. If one utters the word 'excess supply' before one hundred economists brought up in the neo-classical tradition nighty-nine of them will continue with the following sequence of ideas; "If price decreases, the supply decreases and demand grows until finally the equilibrium price clears the market."

On the other hand, we would like to associate our notion of pressure with the concept of an adaptive mechanism consisting of at least four simultaneous processes:

a) The decrease in prices of already manufactured products (that is, the process well-known from neo-classical price theory).

b) Adaptation based on non-price information. The seller and the buyer

---

[8] Aspiration is no wishful thinking but a *serious* intention. In the case of the buyer the seriousness of the intention is proved by the fact that he is prepared to spend his money for the buying. As regards however the centrally approved investment decisions of the socialist economy, it is more appropriate to apply a different "criterion of seriousness": the intention or aspiration is serious if action is prescribed in a valid decision.

realize, on the basis of direct information and of stock-changes, that there is excess supply and they decrease production without waiting for actual price changes to occur.

c) Introduction of a new product by a producer.

d) Advertising and other activities which the producer uses to increase the sale of both of his old and new products.

All these processes are extremely important but, from the historical perspective, process *c*) is the most significant.

There are also other grounds for criticizing the neo-classical chain of arguments associated with the notion of excess supply. Pressure (or if you prefer 'excess supply') *does not* necessarily lead to a decrease in price,

—because there is always faster or slower inflation, the rate of which is not so much influenced by the price setting but rather by the financial and credit policies of the government and of the leading companies.

—because many prices are determined by the oligopolist companies which are quite capable of fixing prices.

—and finally because the choice among products takes place under ever-changing conditions and therefore, the change in the price level cannot be measured accurately over a long period of time.

4. Finally, the last argument against the use of the terms excess supply and excess demand is that they carry certain normative overtones.

The word 'excess' suggests that it would be better to have less supply than 'excess' supply. However, in the course of this work we have tried to point out that from the point of view of the healthy development of the national economy it indeed is beneficial that some pressure, some "excess", be present.

22.10. Comparison: Macro- and Microeconomics

Another important difference emerges *in the relation between macro- and microeconomics*. It was Samuelson's famous textbook[9] which introduced the concept of "neo-classical synthesis". By this he meant the integration of Keynesian macroeconomics with the GE microeconomics following Walras. We fear that it is not a real synthesis if macroeconomics is treated in Part 2 of Samuelson's book and microeconomics, in Parts 3 and 4; and there is no discussion of their interrelation, organic connections and interaction.

The GE school gives the impression that the rules of behaviour in microeconomics are completely independent of the state of the macroeconomic situation. Whatever characterizes *the whole* of the economy, the behaviour of *the parts* —it is thought—still can be described satisfactorily in a way that

[9] See *Samuelson* [213].

the firm maximizes profits and the consumer maximizes utility. We have deviated sharply from this idea in Chapters 17–22.

Statement 22.10. Many of the regularities in the behaviour of economic organizations, (of firms, households and control institutions), as well as in the information structure of the system depend to a great extent—on the general state of the system, on the type and measure of disequilibrium, on the relative strength of the buyer and of the seller.

Of course, Chapters 17–22 have not exhausted the analysis of interrelations between macro- and microeconomics; at most they have offered some basic ideas. However, it is certain that instead of a pseudo-synthesis, a *real synthesis* is needed, and this must be based on the observation and comparison of different economic systems. It is only in this manner that generally valid statements —of basic importance from the point of view of the economic systems theory— can be derived concerning the effect of the macro-state of the economy on micro-economic behaviour and vice versa.

22.11. Further Research Projects

The previous chapters (from Chapter 17 to 22) have outlined some problems concerning the disequilibrium theory. The author himself is far from considering the theory presented here to be complete. A whole series of possible research projects connected with the topic could be undertaken. Here we would like to mention only a few:

1. It would be helpful to examine more closely the conditions of *observation* and *measurement*. First, however, we must make more exact and more operative a great number of concepts, such as buying and selling intentions, dissatisfaction, potential production increment, intensity and so forth.

2. Several works have dealt with the theory of flexible prices, that is, with prices which themselves can regulate supply and demand. However, the theory of *relatively more inflexible prices* still is waiting elaboration. That is, the theory of such prices which together with other regulating and informational mechanisms can regulate the economic processes.

3. The present book mainly deals with two "pure" cases: the condition of general pressure and general suction. However, there are also typical "mixed" cases; whole spheres of the same economy might be in a state of permanent pressure while others are in a state of permanent suction. This is the case, for example, in the economies of most of the developing countries. Permanent "coexistence" of different types of disequilibria does deserve further study.

4. We have emphasized the role of inflationary processes in developing pressure and suction. Yet, market disequilibrium can exist side by side with

relatively stable currency. The relationship between inflation, the stability of currency and permanent market disequilibrium is still far from clear.

5. Further studies should be made in order to clarify the relation between the problems, raised in this book concerning the general disequilibrium of the economy and certain problems raised by Keynes and his followers, especially in theories of investment, saving, and employment.

One thing is clear; our knowledge concerning the actual functioning of different economic systems can be broadened considerably in the future if we spend at least as much mental energy on the analysis of the reasons for and consequences of different types of disequilibria as we have spent in the past on determination of the abstract conditions of equilibrium.

23. MARKET AND PLANNING

23.1. Two Extreme Views

Now that we have come to the end of our discussion of the market, we should attempt to summarize the role of the market in the functioning of economic systems.

There are two extreme views, one proposes a "pure" market economy; the other, "pure" central planning.

According to the former the market is in itself capable of controlling the economic system. In so far as the price system meets the theoretically established requirements of optimality, it in itself will provide the basic information necessary for control. Any interference with the economic processes is superfluous.

According to the latter view, the economy must be centrally regulated according to plan. If planning is of a sufficiently high standard, exact and reliable, there is no need for any other control; in particular, it would be superfluous to expose the economy to fluctuations and frictions due to the market.

A justification of either view could be proven only with extremely strong abstractions foreign to reality. It is paradoxical but true that we would have to make essentially *the same* unrealistic assumptions to "prove" the justification of either of the extreme views. Among other things, we would accept the following:

1. There is strict rationality; the "homo oeconomicus" predominates in the economic system, whether on the lower level (according to GE theory), or on the higher (according to the theory of perfect planning). There are no unavoidable conflicts. Undisturbed harmony can be brought about. Both views believe with naive optimism in the certain absolute victory of rationality.

2. There is no uncertainty in the economic system. The consequences of every decision can be foreseen.

3. In addition, the mathematical models which formulate the "perfect market" or the "perfect central planning" are compelled to apply strong restricting assumptions about the real sphere (e.g. elimination of increasing returns, convexity of the set of production alternatives, etc.).

The problem is that none of the above assumptions is acceptable. (Each has been treated in this book several times and thus there is no need for refutation in this place.) To ask, "Planning or market"—is to ask the wrong question. Rather, what we must deal with are two, *complementary* control subsystems of the complicated and complex economic system.

23.2. Comparison of the Two Sub-Systems

In Section 5.5 we introduced the notion of control *sub-system* to denote the relatively separate parts within the control sphere,[1] and we listed five kinds of them in all. From among the five we deal here with two, the market and national economic planning.

To make the survey easier we present the comparison in tabular form. *Table 23.1* compares the information flows in the two sub-systems. Both subsystems fulfil a useful role, but neither are fully reliable.

From the table, it can be seen that we interpret the concept of market broadly indeed. This already was made clear in Definition 17.3 as well as Chapter 17–22. There are markets where the buyer and the seller freely agree on the price of the product. But we also use the word "market" for control sub-systems regulating the market where the prices emerge in another manner, e.g., where they are set by official authority or dictated by monopolistic firms. The market differs from other control sub-systems in that the seller and the buyer are in immediate informative contact with each other and agree to the transactions with a "horizontal" information flow. In this sense there is no modern economy without a market and thus there was a market in socialist industry even at the times of highest centralization. The question is not *whether there is* a market or there is not (there must be one), but *what* kind of a market? What is the algorithm of price formation? Or, a question of equal importance: does pressure or suction prevail in the market? And what is the relation between the market and other control sub-systems; what is its relative weight in the whole system of control?

Using Table 23.1 let us now compare the two sub-systems: the market and national economic planning.

The market works with fresh, actual information, but it is short-sighted. Planning looks far ahead, but, accordingly, its data base is highly uncertain even with the most careful data collection.

The advantage of the market is that since the buyer pays money and the seller receives money, both of them thoroughly consider their offers and whether they should enter into contract. In this sense the information supply is responsible. The responsibility of the persons taking part in planning is several times removed. Their actions do not directly affect "their pockets". Therefore, irresponsible information is not infrequent. On the other hand, they are less biased, more objective. Buyers and sellers are necessarily egoistic; they consider primarily their own momentary interests. The two control sub-systems differ from each other in their adaptive qualities.

[1] See Definition 5.9.

23*

TABLE 23.1

Information flow in the market and the planning sub-systems

Characteristics	Market	Planning
Major information types	Offers and counteroffers	Statistical reports
	Advertising	Plan proposal and counter-proposal
	Order	Plan-bargaining, critique of the plan-proposal
	Confirmation	
	Contract	Plan-decision
	Modification of contract	Instruction or recommendation to those implementing the plan
	Payment for fulfilment of contract	
Character of reflection	The information directly reflects the real action	The information reflects the real action perhaps through several transmissions
Time lag between anterior information and real action	Preceding it only slightly, almost simultaneous	Preceding it by considerable time (1–5–15–20 years)
Role of memory	Small; short-time horizon	Large; long-time horizon
Measure	Variables measured in both physical and value terms: price information has the outstanding role	Variables measured in both physical and value terms: the former have the outstanding role
Is the flow of goods accompanied by money flow?	Yes	No
Vertical *versus* horizontal flow of information	Both, but the horizontal flows are dominant	Both, but the vertical flows are dominant

The market is an adaptive, learning system where participants continuously grow wiser from their own earlier failures. For example, if a productive firm produces too much, it has trouble selling it all; if it has turned out too little, it forgoes good selling possibilities. There is a high cost of such failures: in the final analysis it is blamed not only by the firm but also by society. Planning is cheaper. Plans are formulated by applying "trial and error" methods; the repeated trials take place only on the paper or in "plan-bargains", not in the fluctuations of real processes. The adaptation cost that can be saved by planning is a gain for society. With careful planning it is possible to prevent disproportions which could be eliminated by the market only ulteriorly, at the cost of adaptation sacrifices resulting from fluctuations in real processes.

In the final analysis, we can draw the following general conclusions:

Statement 23.1. Neither the market nor planning can reliably control the modern, complex economic system alone. In itself, either one is a regulator which functions in less than a completely reliable way. Therefore, on the basis of the principle of multiplying information,[2] the combined activity of the two is necessary for a satisfactory control of the system, for the improvement of its performance.

23.3. Factors Determining the Combination of Market and Planning

Statement 23.1 is of descriptive, real-science character; it establishes the existence of a historical tendency toward mixed economies. Every modern economic system is "mixed" in the sense that both subsystems appear in it.

But the statement says nothing concrete about the actual combination realized in some system, which depends on many factors. In the following we try to group the factors according to their major criteria.

1. The political, power and ownership relations of the system. Under socialist order, social ownership relations play the dominant role and this promotes planning. On the other hand under capitalist order private ownership plays the dominant role and this hinders planning. It is a historical fact that economy-wide planning began in the Soviet Union.

Private ownership is accompanied by competition, trade secrets, etc., and these render general exchange of information difficult. A private firm does not like central interference in its own affairs. Of course, this attitude immensely hinders central planning. This attitude generates the illusion, spread partly under the effect of theoretical economists, that there is no need for central planning since the market can solve every control problem.

Analogous illusions are found within the socialist countries: illusions about the omnipotence of planning, about the unfailing foresight of economic processes and the undisturbed possibility of their rational control.

Thus, it was not only the actual power and ownership relations but also

[2] See Statement 5.2.

illusions and misbeliefs that developed under each of the two systems that acted to produce a one-sided emphasis on planning in the socialist countries and on the market in the capitalist ones.

One can recognize an entire economic system as capitalist or socialist according to the political, power and ownership relations *dominating* in it. But within an economy, the ownership relations are usually not uniform. In the capitalist countries there exist state-owned firms as well as firms owned by social institutions; and there are means of production under private ownership in the socialist countries. Accordingly, the concrete proportions between public and private property also influence the combination between planning and market. A broadening of the scope of public property will act in the direction of increasing the role of the planning sub-system.

2. General economic policy and, within that, the policy of raising the volume of production and the standard of living. The more general economic policy is directed toward increasing the volume of real processes, raising the growth rate, the more does planning come to the fore. Planning promotes a high concentration of resources in those actions which immediately promote the growth of volume. This happened in the Soviet Union and in other socialist countries. But also this is the situation in the developing African and Asian countries which set themselves the task of rapidly liquidating backwardness. A highly centralized planning has appeared also in these countries.

When the consumption by the population is on a low level and stagnates on that level (or rises only slowly in comparison to that level), it is relatively easy to plan centrally the production of consumer goods. The more the planning of consumers' needs comes into the foreground the more it becomes necessary for production to adapt to consumers' needs and for the production of consumer goods to become increasingly differentiated. Increasingly it is difficult to plan everything centrally. This explains the division of emphasis over the last fifteen years in the socialist countries between raising the level of living more quickly and subjecting the methods of economic administration to reform.

3. Pressure or suction. Pressure strengthens the role of the market, and suction that of planning (although either general state of the market may exist under either subsystem). In the case of suction "rationing" is necessarily introduced. On the other hand, in the case of pressure the control of most real processes can be left to a direct agreement between interested sellers and buyers. This is one of its major advantages.

Factors 1–3 hitherto listed affect the *whole* of the economic system; they influence *in general* the relative "weight" or "proportion" of each of the two subsystems. But we must not believe that plan and market will necessarily combine in a uniform, even manner within a given economic system. In this respect there may be essential differences, first *by* industries and then by type of *decision problem*. Let us now review the factors 4–6 which explain these deviations between industries and according to type of decision problem.

4. "Indivisibility", increasing returns, standard versus fundamental decision.[3] The nature of the decisions are different in the cases of a textile mill deciding to produce 100,000 or 105,000 metres of fabrics next week and an electric company deciding whether to create a new hydroelectric plant. The first decision is largely controlled by the market. In making the latter decision, however, it is not customary to reason that "the price of electricity has risen, so let us build a new power plant" or that "the price of electricity has fallen, so there is no need for a new investment in power plants". Instead, efforts are made to assess the future development and structure of the economy and to analyse the future demand for energy. This example contrasts two simple cases and shows that the two differ from each other in a whole series of characteristic features. These different features are surveyed in a tabular form in *Table 23.2*.

5. Uncertainty. The more complex the decision problem and the longer the period which must be anticipated the greater is the importance of the reliability of the forecast. The decision-maker, as has been pointed out in this book earlier, makes efforts to reduce uncertainty mainly by collecting information. National economic planning is a tool serving this end.[4]

Factor 5 is related closely to the fourth one. The risk involved in the decision is great if it is connected with some large-size indivisible unit (in our former example: the building of a big power plant), and at the same time the decision must be taken with deficient information about the expectable consequences. The greater the risk, the more the decision-maker will feel the necessity of collecting thorough and many-sided information.

6. Effects measurable and non-measurable with price. In every society there are inputs and results, favourable and unfavourable, which are not evaluated directly in terms of money and therefore have no price.

Western literature deals with this problem under two headings. One is "externalities", which are contrasted to "internal" effects appearing in the calculations of the profit maximizing firm or of the household drawing up its budget (i.e. with all inputs and outputs whose effects are measurable with prices). The firm pays higher wages to its workers, if working conditions are polluted, but it pays nothing to the population if the factory pollutes the air and that air settles down on houses and gardens in the surrounding area.

The other heading is public goods. Here belong urbanization, water regula-

---

[3] For the definition of standard and fundamental decisions see Section 9.3.
[4] In France the managers of 371 firms were asked about the advantages and disadvantages of planning. They thought that the most important advantage was that the plan provided them with a vast amount of information free of charge. See the article by *Sckoelhammer* [223].

TABLE 23.2

Effect of the type of decision on the combination of market and planning

	Factors which favour the market sub-system	Factors which favour the planning sub-system
Degree of concentration	Little concentration, atomized market	High concentration
Character of decision	Standard decision (smaller modification in real variables against earlier situation)	Fundamental decision (major modification in real variables against earlier situation)
Indivisibility	The real variable controlled by the decision is continuous; there is no essential indivisibility	The real variable controlled by the decision is not continuous; there are essential indivisibilities
Character of the input-output function	There are no increasing returns	Increasing returns assert themselves
Time horizon	Decision can be taken with a short time horizon	Decision can be taken only on the basis of a longer time horizon

tion, the preservation of works of art, and so on. Nobody pays for enjoying the beauty of a city or for warding off a flood; yet everybody benefits from them.

One of the fundamental problems of the capitalist economy is the neglect of externalities and public goods. The extent of this neglect differs by countries; perhaps the most extreme examples can be found in the USA. Sociologists, economists and politicians are much concerned about the grave troubles caused by pollution of rivers, lakes, and the air, by increasing noise, and by decreasing public safety. One way out of these difficulties is to internalize the externalities; firms should pay for effects which were not priced in the past. It is in the USA where one most frequently finds tolls on roads and bridges; it is in the USA that instead of general health insurance, the health service is based most on costs and there is no institutional care for old people. A "dollarization" of every human activity increases efficiency somewhat and forces economizing on resources; however, on the other hand it leads to many anti-humanitarian, rigidly materialistic and even outright irrational and wasteful phenomena.

In the socialist economies there is the opposite problem. Activities were not accounted for in money terms where it would have been expedient to do so. The effect was reduced to income incentives for working. Economists frequently have pointed out that the share of benefits above wages paid, "outside the envelope", is too great in comparison to the consumer goods available for purchase.

Although there is no delimitation that would be generally valid between the "internal" and the "external", between the spheres accounted for and not accounted for in prices, it is certain that there always will exist an external sphere. And with the growing wealth of society this sphere is bound to grow; this will become a historical tendency. It is possible (and in certain cases justified to some extent) to internalize some of the external effects, to account for them through prices, in terms of money. But in reality this can be done with most externalities only in forced manner. Their true regulator must be planning; the natural form of their description is information of non-price character. One may argue whether tolls should be assessed on large highways. However, it is certain that one cannot assess a toll at the corner of each side-street. Therefore, the development of the road network must be planned as an integrated whole.

One of the advantages of socialist economy is that the dominating role of social ownership relations allows increased possibilities for planning.

Let us sum up what has been said.

Statement 23.2. There exists no uniform combination of the control sub-systems of the market and national economic planning independent of age and endowments. Their relative weights depend primarily on the political power and ownership relations. In addition, as regards the entirety of the system, the weights are greatly influenced by economic policy (growth rate and living standards) and by the general state of the market (pressure or suction). Within a given system planning is given a greater role in more highly concentrated industries, in fundamental decisions involving greater risks and relating to indivisible major units, and in the control of effects not measurable with prices.

A common characteristic of the factors listed hitherto is that they express circumstances *objectively* existing at a given historical moment. However, in addition there also are *subjective* factors at work, operating through intentions of those controlling the economy. Today the role of planning is greater in the Netherlands than in Belgium and this cannot be explained by differences in the situation of the two countries; much more important are the differing opinions of Dutch and Belgian politicians and economists. Similarly, if the role of the market is different in Hungary than it is in Poland, this fact should be explained not so much by objective differences between the two systems as by differences between the opinions formulated in Budapest and in Warsaw.

23.4. Comparison

Now that we have commented on the relation between market and planning, let us return to discussion of the GE school.

The *partial* market models of the GE school relying on the idea of "perfect competition" may be accepted as an approximate presentation of the interaction between supply, demand and prices, valid *for a special, narrow class of markets*. This special, narrow class is characterized, among other things, by the following properties:

a) Both the demand side and the supply side consist of a great number of atomized organizations.

b) There exists no lasting tendency for the reproduction of market tensions, for constant disequilibrium, and for the preponderance of either side of the market.

c) There are no increasing returns. There are no large indivisible units. Production (and with it, sales) can adapt continuously to small changes in needs, and conversely.

d) Prices may develop freely, according to agreements between sellers and buyers.

For example, in many countries the well known trade of pigs and corn belong to this narrow class. This is a well-known case of the "cobweb problem". Time series are available and these show the interaction of prices, supply and demand, according to the basic ideas of the GE model.

However, the problem is that GE economists want to squeeze the whole world into a theoretical construction which is suited only for the theoretical description of a narrow class of phenomena. This attempt is to no avail, as the model cannot accomodate the world. The partial market model is capable of describing and explaining the functioning of a definite kind of partial market. But its extension into general equilibrium theory is unacceptable.

In modern economic systems, there are markets that fit the GE model and also others for which the above assumptions a)—d) are not characteristic. In addition, *combined* functioning of all partial markets raises further problems.

Modern economic systems (as was stressed in Statement 23.1) are controlled by a complex control sphere, and the market is only one regulator of it. The GE school suggests that planning is a superfluous luxury, that an "optimal price system" could be found which in itself is sufficient for control.

The task of science is not to seek the "optimal" *simple* regulator of an unrealistic Walras-world but to describe, explain and improve the *complex* control system of the real economy.

In discussions of earlier drafts of our book the following idea frequently was raised.

True, the theories of the GE school cannot be accepted as an accurate description or explanation of a *real* economy. But it should be accepted as a plan for designing a *new* world. Should the leaders of a country find themselves in a position in which they are able to develop the functional mechanism of their system themselves—as it happened in the drafting of the Hungarian reform—the system should be formulated according to the models of the GE school.

It is true that the descriptive and the normative approach can be separated to a certain extent. The world can be changed. But not even the boldest change can entirely neglect the realistic possibilities. We cannot arrange the control system of an economy in a way that does not reckon with the real motivations of human activities, the finite limits of human intellect and capacity, the complexity and intricacy of the system, and so on. Therefore, before putting forward normative proposals, careful descriptive explanatory analysis is needed. *Marx* mobilized the men of science not only to explain the world but to change it.[5] His warning is topical today. But its opposite is no less topical; it is not enough to change the world—it must also be explained. What actually exists must be understood before we take a stand on what *should be*. For example, if we find that in the Soviet Union, the United States, Albania, Yugoslavia and the Netherlands there simultaneously exist autonomous and higher control processes, price-type and non-price type informations, market and planning, we may assume safely that their simultaneous existence is an objective necessity. Therefore, we should not propose an arrangement which stresses one-sidedly only one half of each pair of phenomena.

Before making proposals, we have to study very carefully everything that has developed in different economic systems. It is not that these are necessarily good and no changes should be made. The reason that such studies should be made is that in everything that comes about "in a natural way" there must be something that is necessary for the survival, functioning and development of the system which called it into being. These elements of the system may be improved, but they cannot be neglected because obviously they have a task to perform.

In the language of mathematical decision theory, the problem of "descriptive versus normative theory" may be formulated as follows.

The task of descriptive theory is to explore and describe the set of *feasible systems;* normative theory desires to select the best element of that set. The GE school, if considered normative, proposes to choose an "optimal system", but in searching for that system it reaches *outside* the set of feasible alternatives.

[5] See *Marx* [171]; "Theses on Feuerbach," p. 367.

PART IV

RETROSPECTION AND A LOOK FORWARD

24. ANTECEDENTS IN THE HISTORY OF THEORY AND RELATED TRENDS

24.1. A Survey of the Discussion

In Part I of the book we presented a comprehensive review of general equilibrium theory; in each chapter of Parts II and III we criticized that theory from a different point of view. As we stressed in Chapter 3, our critique was addressed primarily to the models of the early fifties, the renaissance of the GE school which began with the works of Arrow and Debreu.[1]

In the last, fourth part of the book we shall examine what *preceded* and what *followed* this era of the GE school. First, in Chapter 24 we shall look back in the history of theory, back even to Adam Smith. Also we shall discuss the works of some contemporaries.

Then, in Chapter 25, we shall present a survey of further developments. Partly, we shall discuss research intended to reform and improve the equilibrium theory of the original Arrow-Debreu models. Partly, we shall discuss trends independent of the GE school and others critically opposed to it; the direction of the latter investigations is similar to that of the present book.

We should like to stress that it is not our intention to present a full history of economic doctrine, where each author would figure according to his true importance. Our selection is arbitrary. The criterion of selection was the following. We mention only those works which are worthy of attention from the point of view of our subject, considering either the critique of the GE school or the positive formulation of economic systems theory. And in discusssing these works we restrict our comment to those aspects which are related to our subject. Thus, we do not pretend to evaluate the activity of the authors in their entirety. In this sense, we treat one-sidedly the activity of such authors as Marx, Lange, Kantorovich, Keynes, Arrow—to mention only a few names. We do not write a Dogmenhistorie, but only offer *illustrations* from the history of theory relevant to the subject of "Anti-Equilibrium".

Finally, after having presented a survey of the works of our "allies" in Chapter 25, of the trends and efforts which we believe to be similar to our own, in Chapter 26 we shall try to sum up the most important remaining tasks of research.

---

[1] More precisely, not only the articles of Arrow and Debreu but all that which has become a "school", which has spread in a wide circle, which students are taught in the university textbooks, etc.

24.2. Summing up the Antecedents

The first mathematical formulation of *general* equilibrium theory is that of *Walras*, published in the years 1874–77.[2] Thus, GE theory is barely a hundred years old.

"Modernization" of the general equilibrium theory is associated with the names of *Arrow, Debreu, Gale, Koopmans, McKenzie, Uzawa, Wald*: their first works in this field were published in the fifties.[3]

Let us first glance at the period preceding Walras.

1. We should notice the great number of threads by which Walras is related to the English classics, to *Smith* and *Ricardo*, (interpreting the relation in the sense of Definition 3.1). It is incorrect to assume that the general equilibrium theory of Walras radically turns its back on these antecedents in every respect.

a) The existence of the "homo oeconomicus", the man who manages his affairs according to strict rationality, is a basic psychological assumption of the English classics. The exclusive motivation of economic action is egoism, the maximum assertion of own interests.

b) The classics are interested largely in the phenomenon of price. They consider price to be the only economic information worth analysis in the functioning of the economic system. Accordingly, their continual subject is the equilibration of demand and supply by price.

c) The "black box" view of the market appears in the concept of the "invisible hand" of Smith; this in itself connects and harmonizes the producers and users.

2. A fundamental antecedent is the appearance of *Gossen*, and with him, of the idea of utility function. The representatives of the Austrian and the English school of marginal utility, contemporaries of Walras but working independently, improved this field of ideas.

3. A further important antecedent is *Cournot* who was the first to describe the relations between economic phenomena in the form of a function (demand curve, etc.) and first to examine the conditions of equilibrium.

Thus, the activity of Walras had an integrating character; he put many ideas which had appeared independently of each other into a unified system.

In the works of Walras we find the whole system of the assumptions, concepts and questions of the equilibrium theory. *In comparison, modern theory has introduced neither essentially new assumptions, essentially new concepts, nor essentially new questions. It was exclusively the precision of describing the theory*

---

[2] *Walras* [269].

[3] Their most important works are: *Arrow* [10], *Arrow–Debreu* [14], *Arrow–Hurwicz* [15], *Debreu* [50], *Koopmans* [124], *McKenzie* [175], *Wald* [268].

that was developed after Walras; assumptions were formulated with precision and the theorems proved in a faultless, up-to-date manner.

The modern equilibrium theory is nothing else than a mathematically exact formulation of Smith's "invisible hand" which harmonizes the interests of egoistic individuals in an optimal manner. At the time of Smith—in the era of atomistic competition when producers and consumers were connected exclusively by prices and markets—this description of the functioning of a capitalist economy was not unrealistic (though not exact either). More than a hundred years were required for Smith's intuition to be expressed in a faultlessly exact form; by the time this was achieved, it became utterly anachronistic. The present-day capitalist economy differs essentially from that in the times of Smith, to say nothing of the appearance of the socialist economic system.

That the formulation was perfected a century too late is rather depressing. Let us hope that current hunches about the actual capitalist and socialist economies of our days will attain precise formulation in less than a century.

24.3. Consumption, Production and Market Theories

As we have seen already, Walras drew together contemporary ideas about consumption, production and the market. Similarly, modern equilibrium theory can be considered a synthesis of the most developed forms of "consumption theory", "production theory" and "market theory".

As in the case of the integrating equilibrium theory, the three partial theories have been formulated much more precisely since the time of Walras; their mathematical apparatus was developed further, their conceptual framework was refined, and some of the assumptions were made less restrictive. But everything that we have said about the equilibrium theory in a comprehensive form applies also to these partial theories.

We already have made several remarks on the theory of consumption in Chapters 10–11 of this book; we do not wish to return to them now.

However, now we shall make some remarks on the theory of production. Earlier that theory dealt exclusively with the firm turning out a single product under the condition of diminishing returns. Now the model represents a firm with many products, choosing from among alternative technologies, whose input-output functions may be linear.[4] But this does not mean that present theory has relinquished the really restrictive assumptions which are remote from reality. The book by *Dorfman, Samuelson* and *Solow*[5] justly points out

[4] See e.g. the book by *Dorfman* [53].
[5] See *Dorfman–Samuelson–Solow* [54].

that the conceptual framework and the theorems of traditional production theory reappear in the modern linear programming model of the firm.

The weakness of modern mathematical market theories is rooted in their fundamental assumptions.[6] In the modern model it still is assumed that the economic organizations optimize, whether they be atomistic small units within the environment of perfect competition, or oligopolies, or monopolies. Further it is assumed that *price* is the *exclusive* type of information linking economic units.

It is the model of perfect competition that fits best in equilibrium theory since a common assumption of both is that price is formed independently of the individual economic decision-makers; it is given externally for them. However, this is the market theory most foreign to reality. The other mathematical market theories reflect the actual degrees of concentration more realistically, but they are not much better as regards the other two assumptions (optimization and the role of prices).

24.4. Barone's and Lange's Models of Socialism

Let us turn to the more distant relatives.

In 1908 a disciple of *Pareto*, *Barone*, investigated the possibility of rational functioning of a centrally administered socialist economy.[7] He proved this possibility by applying the general equilibrium theory of Walras.

In the later debate the following argument was made; it cannot be assumed that the socialist state would be capable of solving millions of equations in order to secure equilibrium, and the rational allocation of resources.[8] However *Lange* attacked that argument in the thirties.[9] Lange showed how a socialist economy could use the market price formation process for central control. A central price control office could replace the atomistic market, setting prices independently of the decisions of the firms. By changing prices with "trial and error" methods, they could be shaped into equilibrium prices. Simultaneously, the volume produced by the profit-maximizing firms would conform to the level necessary for equilibrium.

The similarity of the Lange-model and the equilibrium theory is conspicuous—profit-maximizing firms; equilibrium as requirement; price as exclusive information.

When Lange was asked to help in the actual establishment of a control

---

[6] See e.g. *Schneider* [222].
[7] See *Barone* [25].
[8] See e.g. the study by *Hayek* in volume [84].
[9] See *Lange* [146], and *Morva's* exposition [184] in Hungarian language.

system for the Polish economy, he never recommended that his model of the thirties be implemented. He himself must have thought it impossible that an economy could be controlled exclusively by means of equilibrium prices.

24.5. Welfare Function, Optimization on a Global Social Scale

In the models of the GE school each economic unit has its own autonomous preference system; still, the economy may reach a common Pareto-optimum.

But in the eyes of many people this is unsatisfactory: they would like the economy to attain a genuine *optimum optimorum*. However, for this purpose there must exist a "welfare function" expressing the joint interests of the society.[10] Although this idea emerged first among Western economists, in the so-called welfare economics, one can find related views also in Soviet economics. Many mathematical economists have made this idea the focus of their investigations of economic systems theory.[11]

We have touched upon the problem of optimization on the national economic level in several places of our book (e.g., in Chapters 11 and 23). To avoid repetition we revert to the problem only briefly.

A common basic assumption, more general than basic assumptions 7 and 8 of the GE school, might be phrased as follows: *all economic decision-makers behave in a strictly rational way*. In other words, they have a preference ordering; they solve a conditional extreme value problem; they optimize. This assumption holds equally for the individual consumer, the productive firm and the government or the planning office.

We do not want to deal with the problem of how a national economic objective function can be used for planning purposes.[12] We examine it here only from the viewpoint of a single problem. Can we assume the existence of an objective function summarizing the joint interests of society in a real science theory *describing and explaining* the real functioning of the economy? In my opinion we cannot.

Many ideas that we expounded in Chapter 7 about conflict and compromise within the firm are valid for the entirety of all societies, among them the socialist society. Society has its classes, strata, groups of interests—with their conflicts and communities of interest. The examples seem to be superfluous—the common interests and the conflicts between town and countryside, young and

---

[10] A classical book on "welfare economics" *Pigou's* work [203]. For welfare functions see *Boulding* [37].

[11] A characteristic example of this school is the book by *Katsenelinboigen* and associates: [106], [107] and [108].

[12] We have dealt with this in Chapter 27 of our book [128].

old, white-collar and manual workers, etc., are well known, exposed in socio-logical investigations and also from everyday experience. In reality, the highest decision-makers seek compromises which are acceptable to each group as far as possible. Thus, for the descriptive-explanatory purposes *instead* of the mathematical formalism of a "preference ordering on the global social scale", a mathematical apparatus similar to the one outlined in this book for the modelling of conflict and compromise within the firm is needed.

24.6. The Economy Controlled with the Aid of Shadow Prices

A new form of the theoretical ideas outlined in Sections 24.4–24.5, and particularly, of the Lange-model of socialism, is the utopia of an economy controlled with shadow prices.

This idea, which is widespread mainly among those adhering to the Soviet school of Kantorovich,[13] retains the original Lange-model in many respects. Here too, the main task of the centre is to establish prices for the profit-maximizing firms. The difference is that the centre no longer needs to "feel out" the equilibrium price; rather, the equilibrium price can be *computed*. For this purpose the centre can employ a linear programming model and prescribe its dual solution, the optimal shadow prices, as actual prices to be used obligatorily in financial accounts.

The idea has several variants. Some authors would hold the centre responsible not only with establishing prices but *also* with direct control of some real processes. Others recommend, instead of a single central model, a multi-level system of models.

The school has contributed many valuable elements to the elaboration of a formalized theory of the socialist economic system, many of which also can be used in practice for improving both planning and the state control of prices. However in our commentary we shall not discuss the merits of the school but rather what we believe to be erroneous in its line of reasoning.

---

[13] In the wake of the important work by *Kantorovich* [111] a whole series of Soviet authors discussed the use of shadow prices. An interesting mathematical formulation of the equilibrium theory may be found in the article by *Polterovich* [204]. The articles by *Katsenelinboigen* and associates already mentioned, [106], [107], and [108], are related to this conceptual framework. The most important Hungarian echo of the school may be found in the works of *György Simon and György Kondor* [234] and [235].

Western readers may obtain an instructive survey of the Soviet discussions and the idea of an economy controlled with the aid of shadow prices from the articles of *A. Zauberman* [279] and [280].

It is obvious that the concept of an economy controlled with shadow prices is related to the GE school. It is built upon the same basic assumptions: optimization, preference ordering, exclusiveness of price informations, simple information structure, deterministic character, convexity of the set of production alternatives, and so on. Thus, all that we have said in this book is valid to this theory, also. Still, we should complement our previous arguments with some further points.

One point of view of a critique regards how the duality of the real and control spheres is reflected in the mathematical models. The linear programming models of national economic planning, the models of the Kantorovich-type are[14] fundamentally models of the *real sphere*. Of course, owing to several simplifying assumptions (linearity, continuity, etc.) they do not faultlessly represent even the real sphere; however, they approximate the solution of this problem to an acceptable extent. Yet they reflect almost nothing of the *control sphere*. The response functions of the control units, the information flows, etc. are missing.

As a matter of fact, price is a variable of the control sphere. We cannot draw valid conclusions for prices from a model which considers prices outside their natural medium, independently of other types of information and from the functioning of the control sphere.

We believe that in our critique of the Kantorovich-school we have come to a deeper and more general problem; what economic interpretation should (and may) be given to the duality theorems of mathematical programming? The usual interpretation is based on the assumption that the equations and inequalities of the model describe the properties of the real sphere. Everything that is essential in the behaviour of the *living* participants of the economic system, of individual persons and of organizations, comes to expression in the preference orderings, the objective functions. Only *if this were true* would the dual of a suitable model of the real sphere really yield the information necessary for control. But, in our opinion, *this is not true*. Chapters 10 and 11, and beyond these our entire book, have attempted to convince the reader how complex is the behaviour of living people and of the organizations and institutions that they comprise; this behaviour cannot be formalized satisfactorily in the exclusive form of utility functions. And if this is true, the dual cannot yield directly all information necessary for control. In that case the effect of the price system can be studied only with a model reflecting not only the real but also the control sphere.

We characterize the behaviour of those who wish to derive a price theory

---

[14] Here also belong the national economic programming calculations performed in Hungary under the direction of the author. (See: [128] and [133a].)

from an "activity analysis" model, from a mathematical programming model planning choice among real activities, with a simile. Their behaviour is similar to that of a student who, at a *biology* lesson, is asked to relate chapter six of the biology textbook but who responds by reciting correctly chapter six from the textbook of *physics*.[15]

We do not want to oppose a dualistic approach to the functioning of economic systems. On the contrary, as we have stated in Chapter 4, this approach is very necessary; perhaps, though, the concept should be given a wider interpretation.

We may summarize our conjecture as follows:

Statement 24.1.

There exists a dual correspondence between the real sphere, on the one hand, and the control sphere, on the other. Definite real spheres can be operated only by definite classes of the control spheres (and, within that, of price systems).

E.g. a modern, concentrated economy cannot be operated by a control sphere having a single-channel, exclusively price-type signal system, but only with a combination of multi-channel, multi-stage and multi-type signal systems.

The rules of a duality thus interpreted have not yet been worked out. Once duality theorems in the above sense appear—duality rules constituting a true real-science theory, a verified, formalized, hierarchical system of ideas—the duality theorems known today will figure only as percursors in the history of economic theory. Today, we cannot know what present theorems will remain valid as real-science theory and what will be regarded only as interesting items in the historical development of economic ideas. However, it seems that the duality theorems of today do not constitute a starting point for a genuine theory of prices. The germs of a future theory of prices must not be sought

---

[15] When we challenge some of the efforts of the shadow price school, we wish to stress the following: we believe that it is grotesque that some economists—both in Hungary and in other socialist countries—desire to enter into polemics with the marginalism-equilibrium-shadowprice-conception on a "Marxian" basis. Their "Marxism" consists in repeating that the price should be based not on marginal costs but average costs; that the price should have a "value centre" or it should be of "production-price-type", etc.

To continue the simile from the text, the student now responds to teacher of biology by fluently explaining Chapter 6 from the textbook of *chemistry*. Kantorovich, Walras or Debreu have not worked out the models of the control sphere in the sense required above. Nor did Marx. Perhaps one may disagree because Walras and Debreu made their theories appear as true models of the control sphere. But Marx never even wanted to undertake such a task. Marx's subject is not how to control the real sphere; competition and the phenomena of the "surface" were barely mentioned by him. It was only his overambitious disciples who wanted to find an answer from his work to a question differing from the one to which he had sought an answer.

around the Kuhn–Tucker theorem, but in a systematic description of price as *one* of the components of the *complex* control sphere.

The second part of our critical remarks against control by shadow prices concerns the division of labour between planning and the market. To see the problem, let us turn back to Table 23.1. Mathematical programming may be a useful tool *in one* of the control subsystems, in planning. But the market functions appear in a *separate* subsystem. The data basis, the information structure and the motivation of the participants are different in the two subsystems. The price actually used in financial accounts is an element of the functioning of the *market* subsystems, it cannot be deduced from a subsidiary tool of the other subsystem, *planning*, and thus from the model of mathematical programming.

24.7. Neo-Liberal Ideas

In the preceding three sections we have spoken of ramifications of the GE school which are related to the *central control* of the economy; those sections analysed the possibilities and methods of planning. But the equilibrium theory points in the opposite direction as well; in the direction of *absolute decentralization*.

To the equilibrium theory there belongs a market where prices are developed independently of the individual economic organizations; they are given from outside for the economic organizations. Regulation of prices may occur by government price authority (as in the Lange-model) or by means of perfect competition. The latter is an atomistic competition where the individual units are too small to influence price in themselves.

Thus, the Walras–Arrow–Debreu model may be interpreted also as the model of perfect competition. In this way it is related to every school which considers the atomistic competition not as an abstract intellectual experiment but as a system deserving practical implementation.

One such interpretation is that of the neo-liberal school, represented by *Röpke*[16] and others. They sharply refute concentration, planning, state intervention in any form; they regard these not as necessary concomitants of economic development, but as unequivocally harmful distortions which should be eliminated. Though the argumentation of these economists is literary—their system of ideas and argumentation is extremely close to those of the equilibrium theory.

The Western neo-liberals show a particular intellectual relation with the

[16] See e.g. *Röpke* [212].

"naive" group of the reformers of socialist economy. Particularly in the early stages of maturity of the reform ideas, there appeared theories definitely refuting all forms of central interference and planning. Advocates of this proposed that the functioning of the economy should be exclusively left to the market, the equilibrium prices, the movement of demand and supply, the interest in profits.

However, in the practical preparation of the reform the naivity of these views became more and more conspicuous. We do not believe there remains a single Hungarian economist who still would take this view.

24.8. Production Prices

In Section 24.2 we have pointed out that Walras partly diverges from the path of the English classics but, at the same time, he retains certain of their ideas. Some of the ideas and concepts common to both the English classics and Walras appear also in *Marx's* Capital.[17]

Demand, supply, the market, equilibrium—all these phenomena were in the focus of interest of classical economists. And in this respect Marx remained entirely within the bounds of his age. He differed radically from other authors when he spoke about production relations, class relations, surplus value, and, indeed, these were the fundamental subjects of his work. But when he touched upon secondary subjects—market price formation, market value, competition, capital flow, etc.,—he adopted the conceptual apparatus and reasoning of his age. He believed it was self-evident that the capitalist wants to maximize profits, that this motivation controls the flow of capital from one industry to another. He believed that the notions of demand and supply were self-evident and also that these two forces caused price fluctuation.

Marx's theory of production prices is based on this line of reasoning. Here, Marx assumed that profit-maximizing capital would be continuously redistributed among industries, and, as a result of these efforts there would be a tendency toward levelling out the rates of profit.

However startling it may seem to some people, the relation of the Marxian ideas concerning competition, the market, and the flow of capital to those of the equilibrium theory is evident. In our opinion these details of the Marxian work are not sufficiently elaborated; they describe the real movement of prices and capital in a highly simplified manner.

It is true that Marx did not care particularly to examine the problem. He indicated several times that it would be desirable to work out the analysis of

---

[17] The most characteristic part of this is Chapter 10 in Volume III; the equalization of the general profit rate as a result of competition; market prices and market values; extra profit. See *Marx* [167].

competition (in our own terminology we would say the theory of the control sphere of capitalist economy), but he never undertook this himself. He concentrated his attention on *political* economy. It is no wonder that in this respect he simply accepted without thorough criticism the current ideas of his time— summarized mainly by Smith and Ricardo.

We have indicated already in Section 24.2 that in Marx's time these ideas were not so foreign to reality as they are today. However, it is incorrect to assume that this simplified description of the control sphere of the capitalist system truly reflects the functioning of the *modern* capitalism.

Later Marxian economists generally were content to repeat Marx's description of the "surface".

As a result, many contemporary Marxian economists in the West extremely simplify the motivation and the behavioural rules of capitalist firms in a way completely identical with that of the representatives of the GE school. They see nothing else than the maximization of profits.[18]

There are works of political economy dealing with monopoly capitalism, treating monopolistic prices, but a really thorough analysis has not been undertaken as yet.[19]

Just as the followers of the neo-classical school have neglected verification of their own theory of prices, the Marxian economists have neglected to check empirically their theorems on price formation. To this day no Marxian work has been written that would prove factually the hypotheses relating to prices, or, should reality not support them, would suggest that they be modified or discarded. Yet the spirit of Marxian research required precisely this approach.

24.9. Equilibrium Theory and Politics

Three questions should be raised about the relation between a social science theory and politics.

1. What is the political content of the theory? Has it internal political content at all or is it only in interpretation that it is desired to reveal such? To inves-

---

[18] E.g. *Baran* and *Sweezy*, the well-known American Marxists write the following: "The profit becomes the immediate, exclusive, unifying, quantitative aim of corporate policies." (See [24] pp. 39–40.)

[19] As mentioned before, *Lenin, Hilferding, Luxemberg* and other Marxists recognized early the importance of concentration and as a result the diminishing influence of atomistic competition, the role of oligopolies and monopolies, etc. But even these Marxists primarily elaborated the political economic and political sociological aspects. Understandably, they were not interested in the effect of these processes on the control of the capitalist economy. Unfortunately, this problem was never studied by other Marxian economists.

tigate this problem the *theory itself must be analysed* independently of the "milieu" in which it was born and in which it is advocated.

2. What political interpretations can be given to the theory? Here we are concerned with the *ideological role* of the theory.

3. Finally, what political objectives motivate the creators and improvers of the theory? This leads to a history of the origin of the theory, investigating the behaviour of the research workers from *political-sociological* aspects.

Question 1: in our view, equilibrium theory is politically a completely indifferent, sterile theory. This is testified precisely by its strictly axiomatic formulation. The 12 basic assumptions described in Section 3.2 may be good or bad, but they are politically indifferent. If they were acceptable at all, they could be used equally in Greece, Sweden, Albania, and Yugoslavia.

Question 2: equilibrium theory can be given many kinds of political interpretation. It may serve the ideology of a strictly centralized socialist economy (Barone, Lange), and also it may serve as an ideology for a completely decentralized capitalist economy (Röpke). The circumstance that each interpretation *justly* considers the equilibrium theory as a proof of its own ideology supports the conclusion contained in the answer to question 1 that the theory is politically sterile.

Question 3: the political motivation of the creators and appliers is heterogeneous. Undoubtedly, marginalism emerged in opposition to the English classics and to Marx. Several of the later representatives of the GE school also coupled the neo-classical, equilibrium theory with pro-capitalist, anti-socialist interpretations. The theory of marginal productivity was used, for instance, as a moral justification for the incomes of capitalists and landlords; the theory of perfect competition was used against socialist planning, and so on.

However, several other representatives of marginalism, of equilibrium theory, used the conceptual framework in the interest of political ideas differing essentially from the former ones, that is either to justify "moderate" reforms of capitalism (welfare economics, etc.) or expressly for socialist purposes (Lange).

At the same time, several representatives of the school have performed their economic activities in quite an apolitical, neutral manner.

These economists may testify that the theories of the GE school cannot be termed as necessarily anti-socialist (independently of the fact whether conscious antisocialism, pro-capitalism played any role in its conception). The 12 basic assumptions, the conceptual framework and the group of questions to be answered are acceptable from the political point of view of socialists and non-socialists, Marxists and non-Marxists alike. (As we pointed out in the preceding chapter, some of the ideas of the equilibrium theory were not foreign to Marx.) A whole series of scientists have proved in a convincing manner, using various approaches and different logical and mathematical tools, that the conceptual frameworks of Marxian political economy and the GE school, may be compatible to a certain extent.[20]

In summary, *the equilibrium theory must be discarded not because it is a "bourgeois" theory; as a matter of fact, it is not "bourgeois" but rather politically indifferent. It should be discarded because—owing to the weaknesses of its basic assumptions, concepts and question formulation—it is useless as a real-science theory.*

24.10. How Can the Hardening of the Deficiencies Be Explained?

A historical survey shows how old and far-reaching is the theoretical school that we are criticising. It is instructive to enumerate some factors that contributed to the emergence and the conservation of the deficiencies of the theory.

Obviously, no objective economist should assume that the adherents of the equilibrium theory are blind or do not want to face the problematic aspects of the theory. It is well known that the leading representatives of the school of equilibrium theory are distinguished personalities among the economists of the world, with outstanding intellectual power, great erudition and superior mathematical knowledge. They themselves know best that increasing returns, uncertainty, etc., do exist.[21]

What, then, is the reason that excellent minds (let alone the legion of their followers) have devoted so much energy to working out and conserving a theory which, as a matter of fact, is hopelessly incapable of functioning?

1. One of the motives is an intellectual impatience with the "maturing" of economics. We have already mentioned this problem in Section 2.6. The main spring may have been the following idea: "Let us have a bad model rather than none." However, the truth—as exemplified by the GE theories—is that a long series of bad models does not converge to a good one. On the contrary, the emphasis on bad models hinders the patient preparation of really good models.

2. A factor in this impatience may be the painful feeling that economics lags behind the natural sciences. The lag is particularly depressing when economics is compared with physics. Driven by impatience the economist

---

[20] See, among others, the Political Economy by *Lange*, [147] the articles by *Johansen* [103] and *R. Frisch* [65] as well as the book by *Bródy* [38].

[21] E.g. *Koopmans*, in the same collection of essays where he gives a classic summary of modern equilibrium theory, treats in details the complications owing to increasing returns and uncertainty. See [124] pp. 150–165. *Samuelson* takes a similarly critical attitude against the GE theory in one of his articles. (See [216].)

forgot that the formalizing physicist worked on the basis of empirical physics, relying on thousands of observations and experiments. This is what justified a formal description of hypotheses which could be immediately subjected to verification.

Physics has acted in a pressing way not only with the example of its maturity but also directly, with the formulation of its questions, and its mathematical apparatus. At the time when *Cournot, Walras* and *Pareto* were active, classical mechanics was the brilliant star of natural sciences: it was from classical mechanics that economics took over a whole series of notions and formulation, as the equilibrium of opposing forces, stability, static and dynamic equilibrium and so on. Economics also borrowed from the same place the whole mathematical formalism of the differential and integral calculus.

Since then other formalized branches of physics have adopted classica mechanics (partly amalgamating classical Newtonian mechanics); also other formalized natural sciences have appeared. But the GE school has not been able as yet to free itself from the spell of classical mechanics.

3. The narrow mathematical apparatus used in economics is also related to the above. It is true that economics applies not only calculus but also linear algebra, theory of sets, and probability theory. Last but not least, it was precisely at the urging of economists that mathematical programming, and the theory of games and decision theory have developed into more or less autonomous branches of mathematics. But even if taken together these are but a few fields from the much broader science of mathematics. Application of other mathematical disciplines is entirely sporadic.

It is worthwhile to reflect on the example of *John von Neumann*. His case is unique—but instructive precisely on this account. Neumann was no economist; he was interested in economics as a mathematician, but he was at least equally interested in quantum physics or the computer. His type is a rarity even among great mathematicians as he knew mathematics comprehensively and his work was at the forefront of and advanced ingeniously the most diverse fields of mathematics, in disciplines far apart from each other. It was no coincidence that von Neumann made such a great contribution to free economic reasoning from the spell of differential calculus giving economists a *new* mathematical apparatus.

Economics must break away from the narrow framework of its present mathematical apparatus. It seems this will be achieved only with the cooperation of *professional mathematicians* who are well versed in many branches of mathematics. But a further step forward is rendered difficult by the fact that many great mathematicians do not follow Neumann's example. This is deplorable because, unfortunately, even the best prepared mathematical *economists* are usually narrow and their mathematical knowledge is limited; they are versed on a higher level only in the mathematical disciplines traditionally employed in economics.

4. An extremely narrowing factor is habit. Ever newer generations of economists are "raised" on the ready-made conceptual schemes of the GE school. And just as the railway cannot leave the track, this framework of ideas sets a way to be followed. Additionally the closed character of the system, its "beauty" is attractive.

The conceptual framework of the GE school and its formulation of questions has an extremely conservative effect on economics. Although two parties might have different points of view, if both of them wish to answer the same question within that framework they must take a common road. "What is the centre around which demand and supply make prices fluctuate?" If this is the question, the notions of "demand" and "supply" already attract the discussants into a mode of discussion within the same tradition. (Whatever answer they give to the question they have taken a new road.)

5. In our book we have mentioned repeatedly the disintegrated state of economics as a brake on cognition. Here we wish to call attention to another phenomenon indicating disintegration—the sharp separation between the studies of capitalist and socialist economic systems.

Expressly or tacitly when constructing models the Western followers of the GE school have in mind the *capitalist* economy; however, they stress (or distort) some features of the capitalist market. Beside the theoretical economists, in capitalist countries a legion of empirical economists are engaged in describing the functioning of capitalist economy. Both groups—both the theoretical and the empirical economists—know but very little about socialism. Similarly there are also economists in the socialist countries who specialize in the problems of capitalism; usually without sufficiently knowing their own economy, and without a deep, prior analysis of socialism.

Of course the *socialist* system is studied by many in the socialist countries. Besides, it is studied in the West by the so-called "Sovietologists", specialists in socialism, some of whom aim at objectivity but others who work in a biased manner. Here we have the reverse of the situation mentioned above; the majority of those investigating the socialist system know capitalism but superficially.

Most so-called "comparative" economists try to prove the superiority of one side. We can find hardly any theoretical works which describe, classify and explain the general and the specific, the common and the distinctive features of the different systems.[22]

---

[22] Only recently has early experimentation begun in working out formal models that would be suited for the comparison of different systems. This effort is still in its initial stages. (See e.g. [278].)

However, without an international comparison, particularly without a dynamic comparison from the point of view of economic history, we cannot even pose the question in a correct manner. Only by comparing the different ways of development can we obtain an answer on the merits of such concepts as the development of "pressure" and "suction"; price-type and non-price-type information; various combinations of profit and nonprofit motives, and so on.

6. The most important source of the conservation of errors is *the absence of strong claims based on practical experience.*

The shocks of capitalist crises *compelled* Western economists to work out a real-science theory that could serve as a basis of anticyclical policy, of state interference. This led to the emergence of the Keynesian school.

The necessity of anticyclical state interference, the pressure of economic problems deeply transformed Western *macro*-economics but, as we have pointed out previously, it is separated from *micro*-economics by an abyss. Micro-economics, or more correctly, the economic systems theory dealing with the behaviour of economic units and their inter-relations, was not questioned about its relevance to everyday operation of capitalist economy. The control sphere works well or badly. Yet it functions in a satisfactory way and secures the current, operative, short-term control of the economy without the contribution or scientific help of the theoretical economists. It is immaterial to actual practice what the theoretical economist says about the behaviour of the firm, about prices, or the consumer; the economy goes its own way. Lacking pressure of practical matters, economists at the universities could tinker safely with their theories; nobody would reproach them if the theories were not workable.

In some capitalist countries small or major changes do occur in the control sphere. However, these are mostly slow, gradual, and for the most part spontaneous and they are effectuated without deep theoretical foundations.

When in the socialist countries (among them, in Hungary) the necessity of a reform emerged, there also emerged *the need for the creation of a true, workable systems theory.* In these countries there is an urgent need for true real-science theoretical statements about the functioning of the control sphere.

Let us hope that this appeal from the practical would provide our discipline with new incentive.

25. REFORM OF THE EQUILIBRIUM THEORY AND THE NEW TRENDS

25.1. Reform versus Breaking Away

In discussions concerning the present state of economics, frequently we have been confronted with the following view:

Our discipline is now in a state similar to that of physics at the turn of the century. We already have a "classical physics": the general equilibrium theory. Now we must create the "modern physics" of economics, which would have more general validity but would comprise—as a special case—our classical physics, the GE theory.

We believe that the simile does not hold and that therefore it gives rise to unjustified self-satisfaction.

The laws of classical Newtonian physics are a very good approximation of *an extremely wide field* of phenomena in the material world—for bodies moving considerably slower than light and consisting of many atoms. True, the new physics, the beginnings of which are present in the elaboration of Einstein's special theory of relativity, covers a much more general field; but— within its own scope—Newtonian physics is still valid.

We could be quite satisfied if we had attained the situation in our discipline that had been attained in physics before 1905. But the situation is different. In our book we have endeavored to show either that the theories of the GE school cannot be accepted as verified real-science theory, or if it has any real science content, it covers but a small field. Each of its assumptions is very special, and combined they can explain a class of phenomena comprising an even smaller, narrower field.

Many economists feel that the narrow scope of GE theory is oppressing and that it should be expanded. With some arbitrariness, the efforts aimed at an expansion can be divided into two main currents; The one is "reformist"; these efforts are directed toward *improving* equilibrium theory, without discarding its bases. This work is being performed by the initiators, pioneers of the modern general equilibrium theory and their disciples. Their intention is to retain as many of the results of the equilibrium school as is possible, thus maintaining its traditions and authority; on the other hand, in order to render it more efficient— they try to relax the strong assumptions and exchange the unrealistic assumptions for more realistic ones. The other main current is "revolutionary", it either *discards* the equilibrium theory, sharply criticizing some of its features, or simply *neglects* equilibrium theory without argument, simply pushing it aside, and starting novel investigations quite independent of the equilibrium theory.

In our book, there is scarcely a criticism of or argument against the traditional theory that has no precedent in some other, earlier work. As announced in Chapter 3, the sources of our critical remarks may be found largely in the research which will be reviewed in Sections 25.3–25.4.[1] The present book is intended to make a step forward not so much in asserting original, independent criticisms as in collecting the fragmented arguments from various works in a single place and systematizing them. Perhaps this synthesis of the critique of the GE school may increase the effect exerted by the hitherto mostly scattered remarks.

25.2. On the Attempts to Relax the Basic Orthodox Assumptions

In the following section we deal with the first current of criticism, the attempts to improve the GE theory. The survey will be somewhat like a mosaic; various works will be quoted without an order of importance. Each of them is characterized by modification of one, two, or three basic assumptions described in Section 3.2, taking over all other assumptions from traditional GE theory. It is precisely the *partial* change that indicates the intention of these theoretical development activities; they strive to reform, not to create something revolutionarily new.

1. Dynamic equilibrium models have appeared. Above all, we emphasize the works of *Koopmans, Arrow* and *Hurwicz*.[2] These works and similar ones utilize the apparatus of dynamic programming, control theory (within that, the so-called Pontryagin methods) to investigate the properties of equilibrium growth paths. Maintaining most of the basic assumption of equilibrium theory they relax the assumption 1.B about the stationary character of development.

2. So-called decomposition, "multi-level" models have been worked out, with algorithms suited for their solution. Such models may be interpreted as descriptions of an economic system consisting of sub- and superordinated units, and the solution algorithms as descriptions of the decision preparation processes.

---

[1] The literature reviewed in Sections 25.3–25.4 has greatly helped us in formulating several ideas of our book (both the critique of the GE school and the positive ideas). Since these are mostly indirect inspirations, coming through several transmissions, it was impossible to provide concrete references in all places of our book. Instead, we want to stress here in general terms how much we owe in the elaboration of the ideas of "Anti-Equilibrium" to the works mentioned in the following sections.

[2] See e.g. *Arrow* [11], *Kurz* [144], and *Koopmans* [123].

Similar interpretation has been given to the model and algorithm of multi-level planning by E. *Malinvaud* and by several works of the author.[3]

As we have pointed out in Section 6.5, this modification relaxes partly assumption 3: beside the productive and consumer units there appears a special administrative unit, the "centre". Additionally, this model relaxes assumption 10; information of price character is not exclusive; non-price, quantity information appears also.

Even with the above modifications, many theorems of the equilibrium theory remain valid.

3. Several attempts have been made to relax the assumptions about perfect divisibility, the continuity of variables.

Baumol and *Gomory* investigated the problem of what dual price system belongs to the primal activity program if the variables or part of them are not continuous.[4] *Vietorisz* studied the functioning of a multi-level economy in the case of indivisibilities.[5]

It turns out that in the latter case the main theses of the equilibrium theory no longer hold.

4. Some pioneering works have been published to do away with basic assumption 6D, to investigate the effect of increasing returns, maintaining, however, the general framework of the equilibrium model. This was the approach of M. *Aoki*.[6] These studies, as well as those mentioned under 3, turn far away from the original conclusions of Walrasian theory and thus, from theoretically supporting the automatic equilibrating role of perfect competition. They showed precisely the opposite, namely, that in the case of increasing returns some kind of central interference is indispensable (e.g., levying taxes, or allocating investment funds).

5. Attempts have been made to relax simultaneously the assumptions relating to the convexity of the production and consumption sets and to the concavity of the preference functions (basic assumptions 6-7-8). An example is the study by *Shapley* and *Shubik*.[7]

Their results lead one far from the original theorems of the GE theory. Prices turn out to be insufficient for regulating the system; stability cannot always be secured.

6. Attempts have been made to model the antagonistic interests, conflicts and coalitions between the participants of the economic system using game theoretic

See e.g. *Malinvaud* [157] Chapter 7, and *Kornai* [128], chapter 25, as well as the study [127]. A few further studies on the subject: *Hogan* [93], *T. Marschak* [165], *Gy. Simon* [232], *Weitzman* [273] and *Waelbroeck* [267].

[4] See *Baumol–Gomory* [28].
[5] See *Vietorisz* [265].
[6] See *Aoki* [7], [8].
[7] See *Shapley–Shubik* [227].

tools. The works by *H. Scarf* must be stressed especially.[8] A central notion of these studies is the "core" of an N-person game. The core is a particular arrangement of coalitions, each being efficient in the sense that no other arrangement can improve the situation of any partner in the coalition without making worse the position of the other partners. This is related to the concept of the Pareto-optimum. The concept of the "core" allows a much looser interpretation of equilibrium; conditions of its existence may be investigated under less restrictive assumptions than in the case of traditional equilibrium theory.

7. *Arrow* and *Hurwicz* examined the process of price formation; this amounts to a certain relaxation of assumption 5.[9] *Kondor* attempted to improve this approach by accounting for the effect of regular lags.[10]

These models maintain the original conclusions of equilibrium theory. However, neither the Arrow–Hurwicz study nor that of Kondor abandon the static (or stationary) character of the original GE theory model; they perform only a certain dynamization of price formation. Besides, they also maintain the basic assumptions relating to convexity and to the absence of uncertainty.

As we mentioned earlier, these models do not consider the problem of stocks. Though prices are modified on the basis of unsatisfied demand or oversupply, the material results of oversupply, the stocks, "disappear". The oversupply manifest at some date does not appear at the next date as an initial stock. Therefore, the Arrow–Hurwicz algorithms cannot be considered as formalization of the *functioning* of a market; at most they can be considered as a dynamic model of a single decision preparation process.

8. It was *Debreu* who first made an attempt to relax the assumption relating to uncertainty; this appears in the last chapter of his classical book.[11]

Later, *Radner* examined in more general terms the extent to which the main theorems of the general equilibrium theory can be maintained basic assumption 12 is relaxed or weakened, that is, if uncertainty is partially or wholly taken into account.[12] However, these studies make it clear that the more realism that is inserted in our assumptions about uncertainty, the fewer the number of the original Arrow-Debreu theorems that can be maintained.

The survey presented in the eight paragraphs above is incomplete. We have only mentioned the attempts at a partial or total relaxation of individual assumptions; we have not evaluated them thoroughly.

However, we should like to add a few general comments on the attempts mentioned above.

[8] See *Scarf* [219], [220].
[9] See *Arrow–Hurwicz* [15].
[10] See *Kondor* [137].
[11] See *Debreu* [50].
[12] See *Radner* [207] and [208].

Each of the reformers makes a small dent in the foundations of the GE model. He hopes that the impact of his attack will destroy a dilapidated wing of the building, but he supposes that the other parts will remain intact. However, if all attacks on the foundation were made simultaneously, the entire building would collapse.

Let us pass now from the simile to the scientific problem; the main theorems of the GE theories remain valid in their entirety or with only small modification if the system consisting of 12 basic assumptions is modified only in few points— and in the non-sensitive ones. If we make corrections simultaneously at many points, and particularly, if we make them at the most important, most sensitive ones (convexity, uncertainty), the theory will collapse. It is worth noting that none of the attempts at reform touched upon the heart of the GE theory, the basic assumptions about preference ordering. A synthesis of the careful attempts to improve the equilibrium theory may turn the "reform" into a "revolution", into discarding and transcending the orthodox theory.

In this connection, it is worthwhile to remember what one of the greatest scientists of our age, *Heisenberg*, said about the closed systems of axioms of modern physics:

"It is characteristic of closed systems that at their beginning appears a subtly defined system of axioms. ... There is, however, a point of great importance which is often neglected, unfortunately not only by the non-professional but also by the physicist, which is that a closed system of this kind cannot be further improved and perfected. ... We ought to say even that such a closed theory cannot be corrected at all since, as a result of its system of axioms, it has become really a mathematical crystal, some rigid thing which may be correct or incorrect, but without an intermediate case."[13]

It seems that the axiomatic equilibrium theory is also a "mathematical crystal". It cannot be improved (or only in some relatively unessential points). In my opinion, the proper task of economics is not to polish this mathematical crystal further. It is much more important that economists should seek a theory suited to describe the economic systems by *opening new approaches*.

25.3. New Economic Currents: Formalized Models

Now, we turn to review the second main current, comprising work on the problems of economic systems theory *outside* the GE school. As a matter of fact, the expression "main current" is not precise since there is no one uniform, coherent school, but several ones more or less distinctly separated from each

---

[13] See *Heisenberg* [88], pp. 231–232.

other. Let us take in turn the most important ones, first of all the works using formal models.

1. The borderline case between the *internal* reform of the GE school and a complete break with the orthodox school is the literature dealing with imperfect competition, oligopolies, monopolies. The initiative is coupled with the names of *J. Robinson* and *Chamberlin*.[14] They and their followers retain many of the basic assumptions of the equilibrium school, but they deviate sharply in others.

In this literature many valuable contributions may be found on the conflicts between competing enterprises within the capitalist system. Instead of the naive picture of harmony in the market, here the real fight appears.

Unfortunately, this branch of literature has not become really integrated with the other parts of economics (and, within it, mathematical economics).[15]

2. The starting points suited for the mathematical formalization of conflicts existing in society are found in the theory of games of *von Neumann* and *Morgenstern*. Though initially this was still coupled with the traditional economics, the connecting roots are being cut one after the other. Mathematical models of "bargaining," "threat" and other conflict phenomena have appeared.[16]

3. A highly important theoretical achievement of the last decade has been the "team-theory" of *Jacob Marschak* and *Roy Radner*.[17] This theory formulates the problem as follows. Consider an economic system, consisting of units separated from each other. The system (in the terminology of the authors, the "team") has a common objective function. We may prescribe definite rules of decision defining the behavior for the units. There is a flow of information among the units. There are various information structures possible (e.g. different degrees of centralization; continuous or occasional, etc.). The following questions arise. What are the characteristic properties of the rules of decision and behavior and of the various types of related information structures? How do these affect the common objective function? Considering the latter, what is the "value" of the different information structures, their effect on the objective function?

Some assumptions of the theory are strong; for example, it assumes the existence of a common objective function of the system, its measurability, and even some of its special mathematical properties. Still, perhaps it is this theory that has reached the highest point in theoretically modelling economic systems.

[14] See *Robinson* [210] and *Chamberlin* [43].

[15] A good survey of the isolation of the theory of imperfect competition may be obtained from several studies in Volume [141].

[16] See e.g., *Harsanyi* [83].

[17] See *Marschak* and *Radner* [160], [161], [162], [163], [206].

4. The work that is closest to the subject of my book, to its problem formulation, is found in an article by *L. Hurwicz*. We have drawn on this article primarily in the elaboration of Chapter 4 of this book.

Hurwicz' line of reasoning is as follows.

There is an "environment". In this environment economic activities take place which can be described by a matrix of resource flows. We define certain formal criteria to evaluate the flow of resources, stating when it is efficient, non-wasteful, pareto-optimal, etc.

In Hurwicz' model the disaggregation of the economy into units (interpreting the concept of unit according to the terminology of our general model) is initially given. This, too, is a component of the environment.[18]

The problem is the following; we seek the best "adjustment process" that may be assigned to a given environment. The adjustment process may be described by prescribing the "language" of the economic units, as well as the response function expressing the reactions to arriving information.[19]

According to Hurwicz, the *datum* of the problem is the environment, the *unknown* to be determined is the adjustment process to be assigned to it.

To evaluate the adjustment process, Hurwicz similarly introduces several criteria, among other things, the notion of "information efficiency". With some simplification, this can be explained as follows. From two adjustment processes the one has greater information efficiency which attains the same result with less detailed information.

In conclusion, the study reviews two adjustment processes. The first is the "competitive process"; it essentially corresponds to a dynamic version of the Arrow–Debreu model. The second he calls "greed process." The "language" of the latter does not contain prices; in the framework of an iterative process each unit adapts to the immediate offers of the other units. Hurwicz establishes that the information efficiency of the competitive process is greater—but it can function only in an environment where there are no increasing returns. The other process (without prices) has less information efficiency, but it may operate also with increasing returns.

In our opinion, what deserves the greatest attention in Hurwicz' study is not the above-mentioned conclusions, but the formulation of the problem itself, the way of approaching the problem as well as the introduction of some important concepts.

In our view, we should depart from the way indicated by Hurwicz only in

---

[18] As can be seen, the interpretation of "environment" by Hurwicz differs essentially from what is called "outside world" in the general model. (See [97].)

[19] Hurwicz adopted the notion of "response function" from cybernetics, from the mathematical models of the nervous system. However, in all of economic literature, we found this term only in his work.

respect to a single feature. We need not insist that for a description of adaptation optimization models should be used.

Unfortunately, to date there have been but few followers of the study by Hurwicz.

5. Some works appeared dealing with the formal analysis of the advantages and disadvantages, conditions and consequences of centralization and decentralization. In this relation the work of *T. Marschak* must be stressed.[20]

6. The work of *T. C. Koopmans* and *J. M. Montias* was aimed at a formal description of the hierarchies emerging in economic systems.[21] This, already referred in the chapter on multi-level control, is also an important initiative. The paper also represents a significant step forward in other respects, mainly in the foundations of a conceptual framework for the theory of comparative economic systems.

7. Some attempts have been made to apply cybernetics and general systems theory to economics. Of particular importance is the activity of some Soviet researchers, primarily the interesting work of *E. Z. Maiminas* on the information aspects of the economic planning processes.[22] In Poland *Greniewski* and his collaborates intensively dealt with the general social, economic and planning interpretation of cybernetics and mathematical systems theory.[23] Their activity acted as an incentive in the preparation of this book (particularly of Chapter 4).

8. Some parts of economic systems have been modelled in simulation experiments. A beautiful example is the model of *Balderstone–Hoggatt*,[24] already mentioned in this book. A few further (unfortunately, not too many) examples could also be quoted.[25]

Undoubtedly, these have promoted knowledge of some partial fields of the economic system. However, it is conspicuous that the authors refrain from drawing far-reaching general conclusions. They do not compare their own experiences with the theses deduced from the deductive models; they do not criticize the latter. Thus the statements arrived at through simulation "peacefully coexist" for the time being with those deduced from the deductive models, although frequently they differ substantially.

9. Experimental games[26] aimed at the examination of particularly complex

[20] See *T. Marschak* [164], [165].

[21] See *Koopmans–Montias* [125].

[22] See *Maiminas* [156].

[23] See *Greniewski* [73], [74].

[24] See *Balderston–Hoggatt* [23].

[25] See works by *Albach* [3], *Bonini* [35], *Forrester* [62], *Frigyes* [64], *Orcutt* and co-authors [198], and *Schmidbauer* [221]. General information on the simulation methods can be found in the works by *Guetzkow* [77], *Naylor* and co-authors [187], and *Shubik* [228], and [229].

[26] See, e.g., the writings of *R. Selten* and *H. Sauerman* [217], pp. 1–168.

decision problems, series of decisions and conflict situations promise to be interesting. In these, the participants of the experiment behave as if they were confronted with a real decision problem, e.g., as if they were managers. The experimenting researcher observes the complex situation that emerges.

10. The simultaneous equation systems of econometrics often are called upon to prepare government decisions, plans, economic policy. However, it is worthwhile to examine these also with the eyes of the researcher of economic systems theory; what do they show about the behavioural rules of the economic units, about the functioning of the control sphere, about the actual effects of the various information variables on the real sphere?

25.4. New Economic Currents: Non-Mathematical Works

Let us turn now to works not employing (or making little use of) formal models.

1. The work of *Keynes*[27] began a new stage of Western economic theory. The subject of his investigations were the macroeconomic relations between national income, employment, investments and savings. This subject is not covered in the present book; it belongs largely outside the scope of economic systems theory. However, there is a contact at a single point; the phenomena of pressure and suction are related to the macroeconomic disproportions, as has been indicated already in this book.

Keynes had an immense effect on Western literature; however—as we have mentioned before—to this date there has been no organic integration between Keynesian macroeconomics and the traditional Western microeconomics (nor could there be).

2. There are many interesting ideas on the functioning of the capitalist system in several works in economic sociology, or political sociology. Particularly deserving attention are those which analyze the actual behaviour of economic units (mainly industrial enterprises, commercial firms, banks), the characteristic regularities in their behaviour, and the conflicts within organizatons and the solutions of those conflicts. Some authors engaged in the "theory of formal organizations" and in "behavioural theory" are sharply critical of many fundamental ideas of the theory of the firm and of the market as taught by the orthodox GE school.[28]

3. Some important works have appeared dealing with the bordering fields of

[27] See *Keynes* 118], further its most important comprehensive commentary in *Hansen* [82].

[28] An example of such polemic writing is an article by *H. A. Simon* [237], a leading personality of the behaviourist school. Some further characteristic works by *Simon* and *March* are: [159], [236], [238].

economics and psychology. We have already referred to the work of *G. Katona*.[29] This direction of research is very important, since one of the great weaknesses of traditional economics is that its psychological foundations are only rough and superficial.

4. In recent years there has emerged in several Western universities a new field called "industrial organization."[30] Its purpose is to describe realistically and to classify the actual production and market structure of various industries as well as their production, selling and price policies.

5. A few works have been published which stress the fundamental importance of intervention on the part of the capitalist state and of planning experiments. These works attempt to identify new phenomena which distinguish the present-day capitalist economic system from that of some decades ago (and from unrealistic theories). *J. K. Galbraith*[31] is representative of this trend. It is true that the works belonging to this category often are not very deep; therefore they are open to attack both by those who criticize them with some political purpose in mind and by those who wish to defend scientific standards. Still, this school describes present-day capitalist economy more realistically than do the elegant but useless models of perfect competition.

6. For the sake of emphasis, we have left to last the literature emerging in the last fifteen years in the socialist countries.

The discussions during the preparations for the reforms in the socialist countries raised almost every important, practical problem of the economic systems theory. It is true that the answers were not the result of theoretical generalization. They rely on sound practical considerations rather than scientific proofs. For the time being the abstract theory of the mechanism lags far behind the continuously changing practice, which is feeling out the ways of improvement and development. Life has pressed on, not being able to wait for theoreticians to work out the theory of controlling the economy.

In any case, a thorough study of the literature of these discussions may be important in inspiring further theoretical research.

25.5. Fragmentation of Economic Systems Theory

The currents reviewed in the previous Sections 25.2–25.4 may be justly regarded as future building stones of the new economic systems theory. It would be desirable if an increasing number of them became reliable elements to be fitted to the others.

---

[29] See *Katona*, [113], [114], [115].
[30] An outstanding representative of the school is *Bain* [22]. See further [42].
[31] See *Galbraith* [69], [70].

We should like to stress again that our survey of the schools and currents is incomplete; we reviewed them only for the sake of illustration. The heterogeneous nature of the camp of "allies" should be evident from our survey. (We mentioned in all $8+10+6 = 24$ groups, schools or currents.) A common feature of all is that they more or less break away from traditional GE theory, from the market and price models, and approach the problems of economic systems in other ways. However, it is also a common characteristic that each of them investigates only a sphere or partial field of the system; thus none can be considered as a ready and comprehensive theory of economic systems. In other respects, they highly differ from each other in the subject and method of analysis, the scientific world outlook and political intention.

This latter fact is what justifies the following conclusion to our chapter on Dogmenhistorie. We should once more emphasize the fragmented, disintegrated state of the theory dealing with economic systems. It would be better if each current were aware of the existence of the others.[32] More frequently they entirely neglect each other.

Macroeconomics and microeconomics have developed almost hermetically separated from each other. The same is true of formalized and literary economics, the economics of socialism and capitalism, the Walras-school and the behaviorists, econometrics and mathematical programming; these examples certainly do not exhaust the list of non-intersecting sets of sciences in the multidimensional space of economics.

The time is ripe for a broader synthesis.

---

[32] A leading representative of the behaviorist school said to me with some self-irony that they are now sufficiently "in fashion" to be referred to at least in footnote in every work dealing with the theory of the firm. This suggests a belittling of their results; authors adhering to the traditional theory of the firm do not attempt to revise their own models and theories even in the light of results of investigations that are better founded empirically.

26. EPILOGUE

In the first sentence of the Foreword we called our work a "semi-finished product". We should like to explain why we consider our work a semi-finished product; additionally we shall proceed to explain why we still decided to publish it.

In dealing with epistemological and scientific-methodological problems, Chapter 2 set a very high standard. A finished, mature economic real-science theory is expected to answer essential questions, to work with precise concepts and to verify its statements with empirical observations. It whould have formal models and an hierarchical theoretical structure, that is one with general laws and also others narrower in scope yet in harmony with the former.

The most this book can give is far from meeting the requirements we ourselves have set for a mature economic real-science theory. We have gone no further than to raise some questions that we deem important, to outline the conceptual framework of economic systems theory, and to make some statements or, rather, hypotheses or hunches, which have not yet been rigorously verified. We do not have finished mathematical models; at most we indicate in some places how the problem could be formalized. Accordingly, our book does not present any mathematically proven theorems.

Without trying to be complete we shall indicate some important tasks for research, by title only.

1. Axiomatic foundation of economic systems theory, with axioms of more general force and, therefore, more realistic than those of the GE school.

2. Empirical observation of the information structure of modern economic systems, classification and typology of the structures, formal modelling of control based on the multiplication of information.

3. Observation and modelling of the multi-level control of economic systems.

4. Observation and modelling of conflicts and compromises taking place within institutions.

5. Observation, typology and computer simulation of decision algorithms.

6. Development of a methodology for the observation and measurement of aspiration level and other extensive and intensive indices.

7. Observation and more precise delimitation of autonomous and higher funtions, and perhaps a distinction of more phases. Mathematical modelling of autonomous functions.

8. Observation of the adaptive properties of economic systems, modelling of adaptive functioning.

9. Observation of the selective regularities of economic systems, establishing the regularities of selection.

10. Development of a methodology for the observation and measurement of buying and selling aspirations and intentions. Observation and computer simulation of algorithms of market processes.

11. Observation of quality improving processes and their measurement. Stimulators and brakes of quality improvement.

12. Regular observation of the phenomena and consequences of pressure and suction. Detailed elaboration and formalization of a theory explaining the continuous reproduction of tensions.

The above list is only a sample survey of research tasks; the necessity of research into these and as well several other subjects emerge from this book. As a matter of fact our work was intended not so much to publish finished results as to prompt new investigations. We believe that after ten or twenty years, when substantial progress will have been attained in the research of the problems outlined here, the original forms of the concepts, classifications and statements found in the present book will not be retained. Some of them will be discarded, and others will be modified. It is not the point of the author that *these* concepts in particular and these classifications should spread, that particularly *these* statements should be proven exactly. Our point is exclusively that research should go on as efficiently as possible *in the direction* outlined.

Thus, the book wishes to outline a *program for work*, to be complemented, of course, with the ideas of many other research workers; we want to recruit research workers who have a mind to participate in realizing this program. If we simply survey the camp of researcher-economists, mathematical economists included, we are glad to find many tens of thousands of them. But at the same time, we also feel some heartache that so many devote their precious intellectual energies to investigate barren problems. If only a half or a quarter of the group of workers who now pick up the grains left by the "great men" on the already harvested field of the general equilibrium theory, began to investigate the really relevant problems of economic systems theory, results could be achieved perhaps in a few years.

The map of our discipline is full of blank spaces. We hardly have looked into the internal structure of the functioning and control of economic systems. We know very little about the actual processes of planning, government control, price formation and the decision-making of the firm. We have few mathematical models that formalize these control processes in ways conforming with reality. It is such a task for which much greater forces need to be mobilized.

Perhaps it is unusual that a work claiming scientific standards does not itself solve problems but urges others to do so. Still, my book chose this unusual way, the more so since it would be hopeless to expect success in solving the problem without the joint efforts of many research workers.

As have the other sciences of our days, economics has become very specialized, divided into several fields separated from each other. Since we wanted to integrate economics, we had to scan a whole series of specialized fields within economics, let alone the borderlands of other sciences (sociology, psychology, cybernetics, etc.), into whose domain we also ventured. It is to be expected that the specialists of all these trades will regard as amateurish if not the whole book, at least the part bungling in their special field. In addition, those who themselves have made efforts at a synthesis relying on a more orthodox conceptual framework may brand my efforts at integration as "eclecticism".

Taking upon ourselves these real weaknesses and the accompanying justified criticism, (and even the expected unjustified ones), we thought it was justifiable to publish our book in its present state. As we indicated at the end of the preceding chapter, *time is ripe for a synthesis of economic systems theory.* This synthesis cannot succeed at the first attempt. Even a first attempt—with half or rather a quarter of a success—may be useful; it may promote the development of more mature, more rounded, more elaborated integration, and synthesis. But at any rate, now is the time to begin the synthesis on bases that are broader than general equilibrium theory and more in agreement with facts.

To conclude, let us make a general remark. We have sharply criticized works representing the *peak* of intellectual efforts over the course of a century. Knowledge of this fact obliges the critic to great modesty and caution. Still, we would rather take upon ourselves the odium of immodesty than diplomatically to circumvent our confessed belief that a *radical* turn is necessary. The peak reached by the equilibrium theory is extremely impressive, and, perhaps, its present-day adherents are capable of building a look-out tower on this peak. Still, we think that we should descend from the peak to the plains and begin again from a much lower level to climb another, steeper and higher peak.

REFERENCES

The bibliography below does not pretend to completeness, listing as it does only a minor part of the literature on the book's subject. The list includes exclusively the titles of works to which reference is made in the text.

[1] ABADIE, J. M.–WILLIAMS, A. C.: "Dual and parametric methods in decomposition", [72], pp. 149–158.

[2] ADELMAN, I.–MORRIS, C. T.: "An econometric model of socio-economic and political change in underdeveloped countries", *American Economic Review*, Vol. 58. (1968) pp. 1184–1218.

[3] ALBACH, H.: "Simulation models of firm growth", *The German Economic Review*, Vol. 5 (1967) pp. 1–26.

[4] ALCHIAN, A. A.: "Uncertainty, evolution and economic theory", *Journal of Political Economy*, Vol. 57 (1950), pp. 211–221.

[5] AMOS, A. J.–BILLINGTON, A. E. etc. (ed): *Food Industries Manual*, London: Hill, 1962.

[6] ANDORKA, R.–DANYI, D.–MARTOS, B.: *Dinamikus népgazdasági modellek* (Dynamic models of the national economy), Budapest: Közgazdasági és Jogi Könyvkiadó, 1967.

[7] AOKI, M.: *Increasing Returns to Scale and Market Mechanisms* (mimeographed), Stanford: Stanford University, 1967.

[8] AOKI, M.: *Dynamic Processes and Social Planning under Increasing Returns* (mimeographed), Stanford: Stanford University, 1968.

[9] ARROW, K. J.: "Alternative approaches to the theory of choice in risk-taking situations", *Econometrica*, Vol. 19 (1951) pp. 404–437.

[10] ARROW, K. J.: "An extension of the basic theorems of classical welfare economics", [193], pp. 507–532.

[11] ARROW, K. J.: *Applications of Control Theory to Economic Growth* (mimeographed), Stanford: Stanford University, 1967.

[12] ARROW, K. J.: "Utility and expectation in economic behaviour", [121], pp. 724–752.

[13] ARROW, K. J.–ENTHOVEN, A. C.: "Quasi-Concave Programming", *Econometrica*, Vol. 29 (1961) pp. 779–800.

[14] ARROW, K. J.–DEBREU, G.: "Existence of an equilibrium for a competitive economy", *Econometrica*, Vol. 22. (1954) pp. 265–290.

[15] ARROW, K. J.–HURWICZ, L.: "Decentralization and computation in resource allocation", *Essays in Economics and Econometrics*, Chapel Hill: University of North Carolina Press, 1960.

[16] ARROW, K. J.–HURWICZ, L.–UZAWA, H.: "Constraint qualifications in maximization problems", *Naval Research Logistics Quarterly*, Vol. 8. (1961) pp. 175–191.

[17] ARROW, K. J.–KARLIN, S.–SCARF, H.: *Studies in the Mathematical Theory of Inventory and Production*, Stanford: Stanford University Press, 1958.

[18] ARROW, K. J.–KARLIN, S.–SUPPES, P.: *Mathematical Methods in the Social Sciences*, Stanford: Stanford University Press, 1960.

[19] AUGUSTINOVICS, M.: "Egy pénzforgalmi modell" (A model of money circulation), *Közgazdasági Szemle* (Economic Review), Vol. 12 (1965) pp. 189–203.

[20] AUGUSTINOVICS, M.: "A model of money-circulation", *Economics of Planning*, Vol. 5. (1965) pp. 44–57.

[21] BAIN, S.: "Chamberlin's impact on microeconomic theory", [141] pp. 147–176.

[22] BAIN, J. S.: *Industrial Organization*, New York: Wiley, 1968.

[23] BALDERSTON, F. E.–HOGGATT, A. C.: *Simulation of Market Processes* (mimeographed), Berkeley: Institute of Business and Economic Research, 1962.

[24] BARAN, P. A.–SWEEZY, P. M.: *Monopoly Capital*, New York–London: Monthly Review Press, 1966.

[25] BARONE, E.: "The ministry of production in the collectivist state", [84], pp. 245–290.

[26] BAUMOL, W. J.: *Business Behavior, Value and Growth*, New York: Macmillan, 1959.

[27] BAUMOL, W. J.: *Economic Theory and Operations Analysis*, Englewood Cliffs: Prentice-Hall, 1961. in Hungarian: *Közgazdaságtan és operáció-analízis*, Budapest: Közgazdasági és Jogi Könyvkiadó, 1968.

[28] BAUMOL, W. J.—GOMORY, P. E.: "Integer programming and pricing", *Econometrica*, Vol. 28 (1960) pp. 521–558.

[29] BEER, S.: *Kybernetik und Management* (Cybernetics and management), Frankfurt a. M.: S. Fischer, 1962.

[30] BELLMAN, R.: *Adaptive Control Processes: A Guided Tour*, Princeton: Princeton University Press, 1961.

[31] BENSON, P. H.: "A model for the analysis of consumer preference and an exploratory test", *Journal of Applied Psychology*, Vol. 39 (1955) pp. 375–381.

[32] BEREI, A. (ed): *Új magyar lexikon* (New Hungarian Encyclopaedia), Budapest: Akadémiai Kiadó, 1960–1962.

[33] BERGE, C.: *Topological Spaces*, Edinburgh: Oliva, Boyd, 1963.

[34] BLAU, P. M.–SCOTT, W. R.: *Formal Organizations*, San Francisco: Chandler, 1962.

[35] BONINI, C.: *Simulation of Information and Decision Systems in the Firm*, Englewood Cliffs: Prentice-Hall, 1962.

[36] BORCH, C.–MOSSIN, J. (ed): *Risk and Uncertainty (Proceedings of an International Economic Association Conference)*, New York: St Martin's Press. 1968.

[37] BOULDING, K. E.: "Welfare economics", [80], pp. 1–38.

[38] BRÓDY, A.: *Érték és újratermelés* (Value and reproduction), Budapest: Közgazdasági és Jogi Könyvkiadó, 1969.

[39] BRUS, W.: *A szocialista gazdaság működésének általános problémái* (General problems of the functioning of the socialist economy), Budapest: Közgazdasági és Jogi Könyvkiadó, 1966.

[40] BUZÁSI, J.: "Elveszett esztendők — avagy: kertelés nélkül a bútorproblémákról" (Lost years—or: let's be frank about the furniture problems), *Népszabadság*, Vol. 27 (1969. 18th November) p. 5.

[41] — "Camera manufacturers", *The Oriental Economist*, Vol 36 (1968) No. 1 pp. 50–59.

[42] CAVES, R.: *American Industry: Structure, Conduct, Performance*, Englewood Cliffs: Prentice-Hall, 1964.

[43] CHAMBERLIN, E. H.: *The Theory of Monopolistic Competition*, Cambridge: Harvard University Press, 1956.

[44] COOK, G. J.: *Handbook of Textile Fibres*, Watford: Merrow, 1964.

[45] CYERT, R. M.—MARCH, J. G.: *A Behavioral Theory of the Firm*, Englewood Cliffs: Prentice-Hall, 1964.

[46] CSAPÓ, L.: "Gazdaságpolitika,mechanizmus, tervgazdálkodás—és kölcsönös összefüggéseik" (Economic policy, mechanism, planned economy—and their interralations), *Társadalmi Szemle* (Social Review), Vol. 21 (1966) No. 2, pp. 22–32.

[47] CSIKÓS-NAGY, B.: *Szocialista árelmélet és árpolitika* (Socialist price theory and price policy), Budapest, Kossuth, 1966.

[48] DANTZIG, G. B.–WOLFE, P.: "Decomposition principle for linear programs", *Operations Research*, Vol. 8. (1960) pp. 101–111.

[49] DANTZIG, G. B.–WOLFE, P.: "The decomposition algorithm for linear programs", *Econometrica*, Vol. 29. (1961) pp. 767–778.

[50] DEBREU, G.: *Theory of Value*, New York: Wiley, 1959.

[51] DENISON, E. F.: "Sources of postwar growth in nine Western countries," *American Economic Review, Papers and proceedings*, Vol. 57 (1967) pp. 326–336.

[52] DISCHKA, GY.–IZMAY, F.: "Szintetikus szálasanyagok általános minőség jellemzői" (General quality characteristics of synthetic fibres), *Magyar Textiltechnika* (Hungarian Textile Technics), Vol. 17 (1965) pp. 441–449.

[53] DORFMAN, R.: *Application of Linear Programming to the Theory of the Firm*, Berkeley and Los Angeles: University of California Press, 1951.

[54] DORFMAN, R.–SAMUELSON, P. A.–SOLOW, R. M.: *Linear Programming and Economic Analysis*, New York: McGraw-Hill, 1958.

[55] DUESENBERRY, J. S.: *Income, Saving and the Theory of Consumer Behavior*, New York: Oxford University Press, 1967.

[56] EINSTEIN, A.: *Ideas and Opinions*, New York: Crown, 1960.

[57] EISNER, R.: "Investment: Fact and fancy", *American Economic Review*, Vol. 53 (1963) pp. 237–246.

[57a] ERDEY-GRUZ, T. (ed): *Természettudományi Lexikon* (Lexicon of Natural Sciences), Budapest: Akadémiai Kiadó, 1964–1968.

[58] ERDŐS P.: *Adalékok a mai tőkés pénz, a konjunktúraingadozások és a gazdasági válságok elméletéhez* (Contribution to the theory of the present capitalist money, business-cycles and economic crises), Budapest: Közgazdasági és Jogi Könyvkiadó, 1966.

[59] EVAN, W. M.: "Indices of the hierarchial structure of industrial organizations", *Management Science*, Vol. 9 (1963) pp. 469–470.

[60] FEKETE, G: "Autók és szervizek" (Motor-cars and servicing), *Népszabadság*, Vol. 27. (1969, May 13) p. 3.

[61] FELLNER, W. et al.: *Ten Studies in the Tradition of Irving Fisher*, New York: Wiley, 1967.

[62] FORRESTER, J. W.: *Industrial Dynamics*, Cambridge: MIT Press, 1961.

[63] FRIEDMAN, M.: *Essays in Positive Economics*, Chicago: University of Chicago Press, 1953.

[64] FRIGYES, E.: *A munkások és alkalmazottak jövedelemeloszlásának elemzése és tervezési módszerei* (Analysis and planning methods of income distribution of workers and employees) (manuscript), Budapest: OT Tervgazdasági Intézet, 1966.

[65] FRISCH, R.: "Rational price fixing in a socialistic society", *Economics of Planning*, Vol. 6 (1966), pp. 97–125.

[66] FRISS, I.: (ed): *Magyar Tudományos Akadémia Közgazdaságtudományi Intézetének Évkönyve III. 1960–1961.* (Year-book of the Institute of Economics, Hungarian Academy of Sciences III), Budapest: Közgazdasági és Jogi Könyvkiadó, 1962.

[67] FRISS, I.: *Gazdasági törvények, tervezés, irányítás* (Economic laws, planning and control), Budapest: Kossuth, 1968.

[68] FRISS, I. (ed): *Reform of the Economic Mechanism in Hungary*, Budapest: Akadémiai Kiadó, 1969.

[68a] GALBRAITH, J. K.: "The disequilibrium system", *American Economic Review*, Vol. 32 (1947) pp. 287–302.

[69] GALBRAITH, J. K.: *American Capitalism*, Cambridge: Houghton-Mifflin 1959.

[70] GALBRAITH, J. K.: *The New Industrial State*, Boston: Houghton-Mifflin 1967.

[71] GLUSCHKOW, W. M.: *Theorie der abstrakten Automaten* (A theory of abstract automata), Berlin: Deutscher Verlag der Wissenschaften, 1963.

[72] GRAVES, R.–WOLFE, P. (ed): *Recent Advances in Mathematical Programing*, New York: McGraw-Hill, 1963.

[73] GRENIEWSKI, H.: "Kybernetik und Planung" (Cybernetics and plannings), *Wirtschaftswissenschaft* (1963), pp. 531–543.

[74] GRENIEWSKI, H.: "Kybernetisch-ökonomische Modelle" (Cybernetical economic models), [173], Vol. 2, pp. 327–377.

[75] GRILICHES, Z.: "Hybrid corn: An exploration in the economics of technological change", *Econometrica*, Vol. 25 (1957) pp. 501–525.

[76] GRILICHES, Z.: "Hedonic price indexes for automobiles: An econometric analysis of quality change", [28], pp. 103–130.

[77] GUETZKOW, H. (ed): *Simulation in Social Science*, Englewood Cliffs: Prentice-Hall, 1962.

[78] HAHN, F. H.–MATTHEWS, C. O.: "The theory of economic growth: A survey", [245], pp. 1–24.

[79] HALABUK, L.: "A magyar népgazdaság M–2/A statisztikai modelljéről" (On the M–2/A statistical model of the Hungarian economy), [182], pp. 282–294.

[80] HALEY, B. F. (ed): *A Survey of Contemporary Economics*, Homewood: Irwin, 1952.

[81] HALL, R. L.–HITCH, C. J.: "Price theory and business behavior", *Oxford Economic Papers*, (1939) No. 2. pp. 12–45.

[82] HANSEN, A. H.: *A Guide to Keynes*, New York: McGraw-Hill, 1953.

[83] HARSANYI, J. C.: "Measurement of social power, opportunity costs, and the theory of two-person bargaining games", *Behavioral Science* (1962) pp. 67–80.

[84] HAYEK, F. A. (ed): *Collectivist Economic Planning*, London: Routledge, 1935.

[85] HEADLEY, G.: *Nonlinear and Dynamic Programing*, Reading–Palo Alto–London: Addison Wesley, 1964.

[86] HEFLEBOWER, R. B.: "The theory and effects of nonprice competition", [14], pp. 177–201.

[87] HEGEDŰS, A.: "A tervgazdálkodás konkrét rendszeréről" (On the concrete

system of the planned economy), *Közgazdasági Szemle*, Vol. 7. (1960) pp. 1422–1438.

[88] HEISENBERG, W.: *Válogatott tanulmányok* (Selected studies), Budapest: Gondolat, 1967.

[88a] HILFERDING, R.: *Das Finanzkapital. Eine Studie über die jüngste Entwicklung des Kapitalismus*, Berlin: Dietz, 1947.

[89] HIRSCHMAN, A. O.: *The Strategy of Economic Development*, New Haven: Yale University Press, 1958.

[90] HOCH, R.: "Az indifferencia-felületek elméletének kritikai ismertetése" (A critical survey of the theory of indifference surfaces), *Közgazdasági Szemle*, Vol. 7 (1960) pp. 1305–1324.

[91] HOCH, R.: "Az indifferencia felületekről szóló tanítás elméleti alapjainak bírálata" (A critique of the theoretical foundations of the doctrine of indifference surfaces), [66], pp. 331–362.

[92] HÓDI, E.: *Matematikai érdekességek* (Mathematical curiosa), Budapest: Gondolat, 1969.

[93] HOGAN, T. M.: *A Preliminary Investigation of Four Planning Procedural Models* (mimeographed), Berkeley: University of California, 1968.

[94] HORVÁT, B.: "The optimum rate of investment", *Economic Journal*, Vol. 68 (1938) pp. 747–767.

[95] HORVÁTH, Á.: *A gépkocsi regénye* (The romance of the motor-car), Budapest: Zrinyi, 1968.

[96] HOUTHAKKER, H. S.: "The present state of consumption theory", *Econometrica*, Vol. 29 (1961) pp. 704–740.

[97] HURWICZ, L.: "Optimality and informational efficiency in resource allocation", [18], pp. 27–46.

[98] —: *International Statistical Year-book 1926*. Geneva: League of Nations, Economic and Financial Section, 1927.

[99] JÁNOSSY, F.: *A gazdasági fejlettség mérhetősége és új mérési módszere* (Measurability and a new measuring method of the economic development level), Budapest: Közgazdasági és Jogi Könyvkiadó, 1963.

[100] JÁNOSSY, F.: *A gazdasági fejlődés trendvonala és a helyreállítási periódusok* (The trend-line of economic development and the periods of reconstruction), Budapest: Közgazdasági és Jogi Könyvkiadó, 1966.

[101] JÁNOSSY, F.: "Gazdaságunk mai ellentmondásainak eredete és felszámolásuk útja" (The origin of present-day contradictions in our economy and the ways of their elimination), *Közgazdasági Szemle* (Review of Economics), Vol. 16. (1969) pp. 806–829.

[102] JEWKES, J.–SAWERS, D.–STILLERMAN, R.: *The Sources of Invention*, London: Macmillan, 1958.

[103] JOHANSEN, L.: "Labour theory of value and marginal utilities", *Economics of Planning*, Vol. 3 (1963) pp. 80–103.

[104] JORGENSON, D. W.–SIEBERT, C. D.: "Theories of corporate investment behavior", *American Economic Review*, Vol. 58 (1968) pp. 681–712.

[105] JORGENSON, D.W.–STEPHENSON, J. A.: "Investment behavior in U.S. Manufacturing, 1947–1960", *Econometrica*, Vol. 35 (1967) pp. 169–220.

[106] KATSENELINBOIGEN, A. I.–MOVSHOVICH, S. M.–OVSIENKO, Yu. V.: "Ob otnoseniakh i raspredelinia v sisteme optimalnogo funktsionirovanya sotsialisticheskosh ekonomiki" *Ekonomika i Matematicheskie Metody*, Vol. 4 (1968) pp. 551–556.

[107]   KATSENELINBOIGEN, A. I.–MOVSHOVICH, S. M.–OVSIENKO, J. V.: "Produktu
        dlitelnogo ispolzovanikha v sistema optimalnogo planirovanikha", *Ekonomika
        i matematicheskie metody*, Vol. 4 (1968) pp. 861–970.

[108]   KATSENELINBOIGEN, A. I.–MOVSHOVICH, S. M.–OVSIENKO, Yu. V.: "Kharak-
        teristiki rabotnikov i ikh deyatelnosti v modeli optimalnoi ekonomiki", *Eko-
        nomika i matematicheskie metody*, Vol. 4 (1969) pp. 183–194.

[109]   KADE, G.: *Die Grundannahmen der Preistheorie*, (The basic assumptions of
        price theory), Berlin und Frankfurt: Franz Vahlen, 1962.

[110]   KALMÁR, L.: *A matematika alapjai, I. kötet. 1. füzet* (The foundations of
        mathematics), Budapest: Felsőoktatási Jegyzetellátó Vállalat, 1956.

[111]   KANTOROVICH, L. V.: *Ekonomicheski raschet nailuchshego ispolzovaniya
        resursov* (Economic calculation of the optimal utilization of resources),
        Moskva: Izd. AN SSSR, 1959.

[112]   KARLIN, S.: *Mathematical Methods and Theory in Games, Programming and
        Economics*, Vol. I, Reading, Mass.: Addison-Wesley, 1959.

[113]   KATONA, G.: *The Powerful Consumer*, New York: McGraw–Hill, 1969.

[114]   KATONA, G.: *Psychological Analysis of Economic Behavior*, New York:
        McGraw–Hill, 1963.

[115]   KATONA, G.: *The Mass Consumption Theory*, New York: McGraw–Hill,
        1964.

[116]   KEMENY, J. G.–SNELL, J. L.–THOMPSON, G. L.: *Introduction to Finite
        Mathematics*, Englewood Cliffs: Debreu Prentice-Hall, 1957.

[117]   KENESSEY, Z. (ed): *Világgazdasági idősorok 1860–1960.* (World economic
        time series, 1860–1960), Budapest: Közgazdasági és Jogi Könyvkiadó,
        1965.

[118]   KEYNES, J. M.: *The General Theory of Employment, Interest and Money*,
        London: Macmillan, 1936.

[119]   KIRK, R. E.–OTHMER, D. F. (ed): *Encyclopedia of Chemical Technology*,
        New York: The Interscience Encyclopedia, 1953–1955.

[120]   KLEIN, L. R.: *An Econometric Model of the United States, 1929–1952*, Amster-
        dam: North–Holland, 1955.

[121]   KOCH, S. (ed) *Psychology: A Study of a Science*, Vol. 6., New York:
        McGraw–Hill, 1963.

[122]   KOO, A.Y.C.: „An empirical test of revealed preference theory", *Economet-
        rica*, Vol. 31 (1963) pp. 646–664.

[223]   KOOPMANS, T. C. „Intertemporal distribution and 'optimal' aggregate eco-
        nomic growth". [61], pp. 95–126.

[124]   KOOPMANS, T. C.: *Three Essays on the State of Economic Science*, New
        York: McGraw–Hill, 1958.

[125]   KOOPMANS, T. C.–MONTIAS, J. M.: "On the description and comparison of
        economic systems", in a volume edited by A. Eckstein, reporting on a *Confe-
        rence on Comparison of Economic Systems* held at the University of Michigan,
        Nov. 1968. (In press.)

[126]   KORNAI, J.: *Overcentralization in Economic Administration*, Oxford: Oxford
        University Press, 1959.

[127]   KORNAI, J.: "A Dantzig–Wolfe dekompozíciós eljárás közgazdasági értel-
        mezése és alkalmazásának problémái" (Economic interpretation and applica-
        tion problems of the Dantzig–Wolfe decomposition method), *Népgazdasági
        programozás (1966–70)* (National Economic Programming, 1966–1970), (1965)
        Bulletin No. 11.

[128] KORNAI, J.: *Mathematical Planning of Structural Decisions*, Amsterdam: North-Holland Publishing Company, 1967.

[129] KORNAI, J.: *A gazdaság működésének szimulációs modelljei* (Simulation models of the functioning of the economy) (mimeographed), Budapest: MTA Közgazdaságtudományi Intézet—KSH Információfeldolgozási laboratórium (Institute of Economics, Hungarian Academy of Sciences—Information Processing Laboratory, Central Statistical Office), 1966.

[130] KORNAI, J.: *Anti-equilibrium — Esszé a gazdasági mechanizmus elméleteiről és a kutatás feladatairól* (Anti-Equilibrium—An essay on the theories of the economic mechanism and on the tasks of research) (manuscript), Budapest: MTA Közgazdaságtudományi Intézet, 1967.

[131] KORNAI, J.: *Anti-Equilibrium — Esszé a gazdasági mechanizmus elméleteiről és a kutatás feladatairól* (Anti-Equilibrium—An essay on the theories of the economic mechanism and on the tasks of research) (mimeographed), Budapest: MTA Közgazdaságtudományi Intézet, 1967–1968.

[132] KORNAI, J.: "A többszintű népgazdasági programozás modellje" (The model of multi-level economy-wide programing), *Közgazdasági Szemle* (Review of Economics), Vol. 15 (1968) pp. 54–68.

[133] KORNAI, J.: "A többszintű népgazdasági programozás gyakorlati alkalmazásáról" (On the practical application of multi-level economy-wide programming), *Közgazdasági Szemle* (Review of Economics), Vol. 15 (1968), pp. 173–190.

[133a] KORNAI, J.: "Multi-level Programming—A First Report on the Model and on the Experimental Computation", *European Economics Review*, Vol. 1 (1969) No. 1. pp. 134–191.

[134] KORNAI, J.–DÖMÖLKI, B.: *A gazdasági mechanizmus szimulációja — Feljegyzés az 1. számú kísérletsorozat matematikai modelljéről* (The simulation of the economic mechanism—Memorandum on the mathematical model of the first experiment series) (manuscript), Budapest: MTA Közgazdaságtudományi Intézet, 1965.

[135] KORNAI, J.–LIPTÁK, T.: "A mathematical investigation of some economic effects of profit sharing in socialist firms", *Econometrica*, Vol. 30 (1962) pp. 140–161.

[136] KORNAI, J.–LIPTÁK, T.: "Two-level Planning", *Econometrica*, Vol. 33 (1965) pp. 141–169.

[137] KONDOR, GY.: *Az értékelés és a piac egyes kérdései nemlineáris modellekben* (Some problems of valuation and the market in non-linear models) (mimeographed), Budapest: MTA Közgazdaságtudományi Intézet, 1966.

[138] KOVÁCS, A.: "Nehéz emberek — A film hangszalagja" (The difficult men —Sound track of the film), *Valóság*, Vol. 8 (1965) No. 1. pp. 41–60.

[139] KÖZPONTI STATISZTIKAI HIVATAL FORGALOMSTATISZTIKAI FŐOSZTÁLYA (Department of Turnover Statistics, Central Statistical Office): *Az üzletek őszi ruházati áruellátása* (Autumn clothing supply in the shops). Budapest: Központi Statisztikai Hivatal, 1968.

[140] KÖZPONTI STATISZTIKAI HIVATAL FORGALOMSTATISZTIKAI FŐOSZTÁLYA (Department of Turnover Statistics, Central Statistical Office): *Az üzletek tavaszi ruházati áruellátása* (Spring clothing supply in the shops), Budapest: Központi Statisztikai Hivatal, 1969.

[141] KUENNE, R. E. (ed): *Monopolistic Competition Theory: Studies in Impact*, New York: Wiley, 1967.

26*

[142] KUENNE, R. E.: "Quality space, interproduct competition and general equilibrium" [141], pp. 219–250.

[143] KUH, E.: "Theory and institutions in the study of investment behavior", *American Economic Review*, Vol. 53 (1963) pp. 260–268.

[144] KURZ, M.: *The General Instability of a Class of Competitive Growth Processes* (mimeographed), Stanford: Stanford University, 1966.

[145] KÜNZI, H. P.–TAN, S. T.: *Lineare Optimierung grosser Systeme* (Linear optimization of large systems), Berlin–Heidelberg–New York: Springer, 1966.

[146] LANGE, O.: "On the economic theory of socialism", [153], pp. 57–142.

[147] LANGE, O.: *Politikai gazdaságtan* (Political economy), Budapest: Közgazdasági és Jogi Könyvkiadó, 1965, 1967.

[148] LANGE, O.: *Bevezetés a közgazdasági kibernetikába* (An introduction to economic cybernetics), Budapest: Közgazdasági és Jogi Könyvkiadó, 1967.

[149] LANZILOTTI, R. F.: "Pricing objectives in large companies", *American Economic Review*, Vol. 48 (1958) pp. 921–940.

[149a] LEDLEY, R. S.: *Programing and Utilizing Digital Computers*, New York: McGraw Hill, 1962.

[150] LENIN, V. I.: "Az imperializmus, mint a kapitalizmus legfelsőbb foka", *Lenin Válogatott Művei*, I. kötet, Budapest: Kossuth Kiadó, 1967.

[151] LEWIN, K.: *Principles of Topological Psychology*, New York: McGraw-Hill, 1936.

[152] LIGETI I.–SIVÁK, J.: *Nagyméretű lineáris programozási feladatok megoldásában felhasználható néhány dekompozíciós eljárás* (Some decomposition methods to be used in the solution of large-size linear programing problems) (mimeographed), Budapest: OT Tervgazdasági Intézet (Institute of Economic Planning, National Planning Office), 1969.

[153] LIPINCOTT, B. (ed): *On the Economic Theory of Socialism*, Minneapolis: University of Minnesota Press, 1938.

[154] LUCE, R. D.–RAIFFA, H.: *Games and Decisions*, New York: Wiley, 1958.

[155] MACHLUP, F.: "Theories of the firm: marginalist, behavioral, managerial", *American Economic Review*, Vol. 57 (1967) pp. 3–33.

[156] MAIMINAS, E. Z.: *Processy planirovanits v ekonomike informaionny aspekt*, Vilnus: Minthyis.

[157] MALINVAUD, E.–BACHARACH, M. O. L. (ed): *Activity Analysis in the Theory of Growth and Planning*, London–New York: Macmillan–St. Martin Press, 1967.

[158] MANSFIELD, E.: *Industrial Research and Technological Innovation*, New York: Norton, 1968.

[159] MARCH, J.–SIMON, H. A.: *Organizations*, New York: Wiley, 1958.

[160] MARSCHAK, J.: "Towards an economic theory of organization and information", [253], pp. 187–220.

[161] MARSCHAK, J.: "Theory of an efficient several-person firm" *American Economic Review*, Vol. 50 (1960) pp. 541–548.

[162] MARSCHAK, J.: "Economics of inquiring, communicating, deciding", *American Economic Review Papers and Proceedings,*, Vol. 58 (1968) pp. 1–18.

[163] MARSCHAK, J.–RADNER, R.: *Economic Theory of Teams* in print 1968.

[164] MARSCHAK, T.: "Centralization and decentralization in economic organizations", *Econometrica*, Vol. 27 (1959) pp. 399–430.

[165] MARSCHAK, T.: "Computation in organizations; Comparison of price mechanisms and other adjustment processes", [36].

[166] MARK, H. F.–GAYLORD, N. G.–BIKALES, N. M. (ed): *Encyclopedia of Polymer Science and Technology*, New York: Wiley, 1964–1968.

[166a] MARTOS, B.: *Nem-lineáris programozási módszerek hatóköre* (The range of effect of non-linear programing methods) (mimeographed), Budapest: MTA Közgazdaságtudományi Intézet, 1966.

[167] MARX, K.: *Das Kapital,*, Vol. 3, Berlin: Dietz, 1947.

[168] MARX, K.: *A Tőke*, Budapest: Szikra, 1955–1956–1961.

[169] MARX, K.: "Levél Annenkovhoz, 1846" (Letter to Annenkov, 1846), *Marx–Engels: Válogatott Művek, II. kötet*, Budapest: Kossuth Kiadó, 1963.

[170] MARX, K.: "Előszó a 'Politikai gazdaságtan bírálatához'" (Foreword to "The critique of political economy"), *Marx–Engels Művei, 13. kötet*, Budapest, Kossuth Kiadó, 1965.

[171] MARX, K–ENGELS, F.: *Selected Works*, Vol. II. Moscow: Foreign Languages Publishing House, 1949.

[172] MASSON, E. S.: "Monopolistic competition and the growth process in less developed countries", [141], 00. 77–104.

[173] —*Mathematik und Kybernetik in der Ökonomie, Internationale Tagung— Konferenzprotokoll* (Mathematics and cybernetics in economics—International session, protocole), Berlin: Akademie Verlag, 1965.

[174] McGUIRE, J. W.: *Theories of Business Behavior*, Englewood Cliffs: Prentice-Hall, 1964.

[175] McKENZIE, L. W.: "On the existence of general equilibrium for a competitive market", *Econometrica*, Vol. 27 (1959) pp. 54–71.

[176] MEGYERI, E.: "Nyereségérdekeltség, árrendszer és a termékek gazdaságossági rangsorolása az új mechanizmusban" (Profit sharing, price system and the rentability ranking of products in the new mechanism), *Közgazdasági Szemle* (Review of Economics), Vol. 14 (1967), pp. 1432–1443.

[177] MEGYERI, E.: "A vállalati beruházás gazdaságossági számítások néhány elvi-módszertani kérdése" (Some theoretical and methodological questions of the computation of investment efficiency in the firm), *Közgazdasági Szemle* (Review of Economics), Vol. 16 (1969) pp. 179–192.

[178] MESAROVIC, M. D. (ed.): *Views on General Systems Theory*, New York: Wiley, 1964.

[179] MIKLÓS, P.: "Bevezetés (a Helikon c. folyóiratnak a strukturalizmusról szóló különszámához)" (Introduction to the special issue of the periodical Helikon on structuralism), *Helikon*, Vol. 14. (1968) pp. 3–24.

[180] MILNOR, J.: "Games against Nature", [253] pp. 49–59.

[181] MILONOV, J. K. (ed): *A technika története* (A history of technics), Budapest: Kossuth, 1964.

[182] MÓD, A. (ed): *Korszerű statisztikai törekvések Magyarországon* (Modern statistical endeavours in Hungary), Budapest: Akadémiai Kiadó, 1968.

[183] MONCRIEFF, R. W.: *Man-made Fibres*, London: Heywood and Co., 1963.

[184] MORVA, T.: "A tervgazdaság Lange-féle piaci modellje" (Lange's market model of the planned economy), *Közgazdasági Szemle*, Vol. 13 (1966) pp. 156–170.

[185] NAGEL, E.: "Assumptions in economic theory", *American Economic Review*, Vol. 53 (1963) pp. 211–219.

[186] NAGY, T.: *A gazdasági mechanizmus reformja és a politikai gazdaságtan kategóriái* (The reform of the economic mechanism and the categories of political economy) (mimeographed), Budapest: Tudományos Ismeretterjesztő Társulat (Society for the Dissemination of Scientific Knowledge), 1966.

[187] NAYLOR, TH. H.–BALINTFY, J. L.–BURDICK, D. S.–KONG CHU: *Computer Simulation Techniques*, New York: Wiley, 1966.

[188] NÁDASY, L.: "A film tovább pereg" (The film goes on), *Új Írás* (New Writing), Vol. 7 (1967) No. 5 pp. 59–72, No. 6, pp. 65–72.

[189] NÁDASY, L.: "Érdemes-e folytatni"? (Is it worth-while to continue?) *Új Írás* (New Writing), Vol. 7 (1967) No. 12, pp. 84–102.

[190] NEUMANN, J.: "Az automaták általános és logikai elmélete" (General and logical Theory of automata), [247], pp. 55–114.

[191] NEUMANN, J.: *Válogatott előadások és tanulmányok* (Selected lectures and papers), Budapest: Közgazdasági és Jogi Könyvkiadó, 1965.

[192] NEUMANN, J.–MORGENSTERN, O.: Theory of Games and Economic Behavior, Princeton: Princeton University Press, 1953.

[193] NEYMANN, F. (ed): *Proceedings of the Second Berkeley Symposium on Mathematical Statistics and Probability*, Berkeley–Los Angeles: University of California Press, 1951.

[194] NYERS, R.: *25 kérdés és válasz gazdaságpolitikai kérdésekről* (Twenty-five questions and answers about problems of economic policy), Budapest: Kossuth, 1959.

[195] OKUN, A.: "Potential GNP: Its measurement and significance", *American Statistical Association Proceedings* (1962) pp. 98–104.

[196] OLIVER, J. W.: *History of American Technology*, New York: Ronald Press, 1956.

[197] ORCUTT, G. H.: *Research Strategy in Modelling Economic Systems* (mimeographed), Madison: University of Wisconsin, 1967.

[198] ORCUTT, G. H.–GREENBERGER, M.–KORBEL, J.–RIVLIN, A. M.: *Microanalysis of Socioeconomic Systems: A Simulation Study*, New York: Harper, 1961.

[199] ORCUTT, G. H–WATTS, S. H.–EDWARDS, J. B: "Data aggregation and information loss", *American Economic Review*, Vol. 58 (1968) pp. 773–787.

[200] PAPANDREOU, A. G.: "Theory construction and empirical meaning in Economics" *American Economic Review*, Vol. 53 (1963) pp. 205–210.

[201] PÉTER, GY.: *A gazdaságosság és jövedelmezőség jelentősége a tervgazdálkodásban* (The significance of rentability and profitability in the planned economy), Budapest: Közgazdasági és Jogi Könyvkiadó, 1956.

202] PICARD, F.: "Les voitures à moteur arrière. Point de vue européen" (Rear-engined cars — The European viewpoint), *Journal de la S.I.A.*, Vol. 30 (1957) pp. 265–280.

[203] PIGOU, A. C.: *Economics of Welfare*, London: Macmillan, 1920.

[204] POLTEROVICH, V. M.: "O nekotorykh absztrakhnykh modelakh funktsionirovanya yacheiek ierarkhii v upravlaemoi ekonomicheszkoi sistema", *Ekonomika i Matematicheskie Metodu*, Vol. 4 (1968) pp. 176–190.

[205] PORTES, R. D.: "The enterprise under central planning", *Review of Economic Etudies*, Vol. 36 (1969) pp. 197–202.

206] RADNER, R.: *The Evaluation of Information in Organizations* (mimeographed), Berkeley: Management Science Research Group, University of California, 1961.

[207] RADNER, R.: *Equilibrium of Spot and Future Markets under Uncertainty* (mimeographed), Berkeley: University of California, 1967.

[208] RADNER, R.: "Competitive equilibrium under uncertainty", *Econometrica*, Vol. 36 (1968) pp. 31–58.

[209] RÉNYI, A.: "A Barkochba játék és az információelmélet" (The game "Barkochba" and information theory), [92], pp. 269–286.

[210] ROBINSON, J.: *The Economics of Imperfect Competition*, London: Macmillan — St. Martin's Press, 1961.

[211] ROSEN, J. B.: "Convex partition programing", [72], pp. 159–176.

[212] RÖPKE, W.: *Die Lehre von der Wirtschaft* (The economic doctrine), Wien: Springer, 1937.

[213] SAMUELSON, P. A.: *Economics*, New York: McGraw-Hill, 1955.

[214] SAMUELSON, P. A.: *Foundations of Economic Analysis*, Cambridge: Harvard University Press, 1955.

[215] SAMUELSON, P. A.: "Problems of methodology—Discussion", *American Economic Review*, Vol. 53 (1963) pp. 231–236.

[216] SAMUELSON, P. A.: "The monopolistic competiton revolution", [141], pp. 139–146.

[217] SAUERMANN, H. (ed): *Beiträge zur experimentellen Wirtschaftsforschung* (Contributions to experimental economic research), Tübingen: Mohr, 1966.

[218] SAVAGE, L. J.: *The Foundations of Statistics*, New York: Wiley, 1954.

[219] SCARF, H.: "On the computation of equilibrium prices", [61] pp. 207–230.

[220] SCARF, H. "The core of an N person game", *Econometrica*, Vol. 35 (1967) pp. 50–69.

[221] SCHMIDBAUER, P. L.: *Information and Communication Requirements of the Wheat Market* (mimeographed), Berkeley: University of California, 1966.

[222] SCHNEIDER, E.: *Einführung in die Wirtschaftstheorie, II. Teil* (Introduction to economic theory), Tübingen: Mohr, 1956.

[223] SCHOELHAMMER, H.: "Corporate planning in France", *European Business*, (July 1969) pp. 22–31.

[224] SCHUMPETER, J. A.: *Kapitalismus, Sozialismus und Demokratie* (Capitalism, socialism and democracy), Bern: Francke, 1946.

[225] SCHUMPETER, J.A.: *Theorie der wirtschaftlichen Entwicklung* (Theory of economic development), Berlin: Duncker–Humblot, 1952.

[226] SHANNON, C. E.–MCCARTHY, J. (ed): *Automata Studies*, Princeton: Princeton University Press, 1956.

[227] SHAPLEY, L. S.–SHUBIK, M.: "Quasi-cores in a monetary economy with non-convex preferences", *Econometrica*, Vol. 34 (1966) pp. 805–827.

[228] SHUBIK, M.: "Simulation of the industry and the firm", *The American Economic Review*, Vol. 50 (1960) pp. 908–919.

[229] SHUBIK, M.: "Simulation of socio-economic systems", *General Systems*, Vol. 12 (1967) pp. 149–158.

[230] SIEGEL, S.: "Level of aspiration and decision-making", *Psychological Review*, Vol. 64 (1957) pp. 253–262.

[231] SIMAI, M.: "A fogyasztás fejlődésének fő irányai a világgazdaságban" (Main trends of consumption in the world economy), *Közgazdasági Szemle* (Review of Economics), Vol. 16 (1969) pp. 492–505.

[232] SIMON, GY.: *Népgazdasági programozás és árnyékárak* (National economic

388 REFERENCES

gazdaságtudományi Intézet (Institute of Economics, Hungarian Academy
of Sciences), 1968.
[233] SIMON, GY.: "Optimális tervezés reflektorprogramozással" (Optimal planning with reflector-programing) *Gazdasági fejlődés és tervezés*, (Economic
development and planning), Budapest: Közgazdasági és Jogi Könyvkiadó,
1969.
[234] SIMON, GY.–KONDOR, GY.: "A gazdasági optimumszámítások problémái
Kantorovics és Novozsilov műveiben" (The problems of economic optimum
computations in the works of Kantorovich and Novoshilov), *MTA Közgazdaságtudományi Intézetének Közleményei* (Publications of the Institute of
Economics, Hungarian Academy of Sciences), (1963) No. 2.
[235] SIMON, GY.–KONDOR, GY.: *Gazdasági hatékonyság, árnyékárak* (Economic
efficiency and shadow prices), Budapest: Közgazdasági és Jogi Könyvkiadó.
1965.
[236] SIMON, H. A.: "A behavioral model of rational choice", *The Quarterly Journal
of Economics*. Vol 69 (1955) pp. 99–118.
[237] SIMON, H. A.: "Theories of decision-making in economics and behavioral
science", *American Economic Review* Vol. 49 (1959) pp. 253–283.
[238] SIMON, H. A.: *Administrative Behavior*, New York: The Free Press, 1966.
[239] SIMON, H. A.—BONINI, C. P.: "The size distribution of business firms",
American Economic Review, Vol. 48 (1958) pp. 607–617.
[240] STARKE, P. H.: "Theorie stochastischer Automaten" (Theory of stochastic automata), *Elektronische Informationsverarbeitung und Kybernetik* (Electronic information processing and cybernetics), Vol. 1 (1965) pp. 5–32,
71–98.
[241] STARKE, P. H.: "Einige Bemerkungen über nicht-determinierten Automaten"
(Some remarks on non-determined automata), *Informationsverarbeitung und
Kybernetik* (Electronic information processing and cybernetics), Vol. 2
(1966) pp. 61–82.
[242] — *Statistical Year-book of the League of Nations 1931/32*, Geneva: League
of Nations, Economic Intelligence Service, 1932.
[243] — *Statistical Year-book of the League of Nations 1941/42*. Geneva: League
of Nations, Economic Intelligence Service, 1943.
[244] STREETEN, P.: "Unbalanced growth", *Oxford Economic Papers*, NS Vol. 11
(1959) pp. 170–181.
[245] — *Surveys of Economic Growth Vol. II.*, London–New York: Macmillan–St.
Martin's Press, 1967.
[246] SZABÓ, L.: Vállalati piackutatás (Market research in the firm), Budapest:
Közgazdasági és Jogi Könyvkiadó, 1969.
[247] SZALAI, S. (ed): *A kibernetika klasszikusai* (The classics of cybernetics),
Budapest: Gondolat, 1965.
[248] SZÉP, J.: *Analízis* (Calculus), Budapest: Közgazdasági és Jogi Könyvkiadó,
1965.
[249] STALIN, J. V.: A leninizmus kérdései (The problems of Leninism), Budapest: Szikra, 1953.
[250] TANKÓ, J.: *A gazdasági mechanizmus szimulációja. Az 1. számú kísérletsorozat modelljének és gépi programjának ismertetése* (Simulation of the
economic mechanism—Description of the model and computer program of
the first experiment series) (manuscript), Budapest: MTA Közgazdaság-

tudományi Intézet (Institute of Economics, Hungarian Academy of Sciences), 1966.

[251] TARJÁN, R.: *Gondolkodó gépek* (Thinking machines), Budapest: Bibliotheca, 1958.

[252] TARJÁN, R.: *Kibernetika* (Cybernetics), Budapest: Gondolat, 1964.

[253] THRALL, R. M.–COOMBS, C. H.–DAVIS, R. L. (ed): *Decision Processes*, New York: Wiley, 1954.

[254] THURSTONE, L. L.: "The indifference function", *Journal of Social Psychology*, Vol. 2. (1931) pp. 139–167.

[255] TIMÁR, M.: "Számadás év közben" (Rendering account in the course of the year), *Népszabadság*, Vol. 27 (September 14, 1969), p. 3.

[256] TINBERGEN, J.–BOS, H. C.: *Mathematical Models of Economic Growth*, New York: McGraw-Hill, 1962.

[257] TODA, M.–SLUFORD, E. H.: "Logic of systems. Introduction to a formal theory of structure", *General Systems*, Vol. 10 (1965) pp. 3–27.

[258] TŐKEI, F.: *Az "ázsiai termelési mód" kérdéséhez* (On the problem of the "Asian production method"), Budapest: Kossuth, 1965.

[259] TŐKEI, F.: *A társadalmi formák elméletéhez* (On the theory of social forms), Budapest: Kossuth, 1968.

[260] — *UN Statistical Yearbook 1956, 8th issue*, New York: Statistical Office of the UN Department of Economic and Social Affairs, 1957.

[261] — *UN Statistical Yearbook 1962, 14th issue*, New York: Statistical Office of the UN Department of Economic and Social Affairs, 1963.

[262] — *UN Statistical Yearbook 1968, 20th issue*, New York: Statistical Office of the UN Department of Economic and Social Affairs, 1969.

[263] UZAWA, H.: "Preference and rational choice in the theory of consumption", [18], pp. 129–148.

[264] VAN COURT HARE: *Systems Analysis: A Diagnostic Approach*, New York: Harcourt, Brace and World, 1967.

[265] VIETORISZ, T.: "Decentralization and project evaluation under economics of scale and indivisibilities", *Industrialization and Productivity*, Bulletin 12 (1968) pp. 25–58.

[266] VOZN'ESENSKI, N. A.: *A Szovjetunió hadigazdasága a Nagy Honvédő Háború éveiben* (The Soviet war economy in the Great Patriotic War), Budapest: Anonymus, 1948.

[267] WAELBROECK, J.: "La grande controverse sur la planification et la théorie économique mathématique contemporaine" (The great controversy on planification and present-day mathematical economic theory), *Cahiers de 'ISEA* (ISEA Brochures) (1964) No. 146, pp. 3–24.

[268] WALD, A.: "On some systems of equations of mathematical economics", *Econometrica*, Vol. 19 (1951) pp. 368–403.

[269] WALRAS, L.: *Elements of Pure Economics, or the Theory of Social Wealth*, London: George Allen, 1954.

[270] — "Wankel-motoros Mercedes" (Mercedes with Wankel engine), *Népszabadság*, Vol. 27 (September 9, 1969) p. 9.

[271] WARD, B. N.: *The Socialist Economy*, New York: The Random House, 1957.

[272] WEBER, M.: *Wirtschaft und Gesellschaft*, (Economy and society), Köln–Berlin: Kiepenhauer und Witsch, 1964.

[273] WEITZMAN, M.: *Iterative Multi-Level Planning with Production Targets* (mimeographed), New Haven: Cowles Foundation, 1967.

[274] WILLE, H. H.: *A kocsitól a gépkocsiig* (From coach to motor-car), Budapest: Táncsics, 1967.

[275] WILLIAMSON, O. E.: "Managerial discretion and business behavior", *American Economic Review*, Vol. 53 (1963) pp. 1032–57.

[276] WINTER, S. G.: "Economic natural selection and the theory of the firm", *Yale Economic Essays*, (1964) pp. 225–272.

[277] WINTGEN, G.: "*A kibernetikai rendszer fogalma és alkalmazása a közgazdaságtanban*" (The concept of the cybernetic system and its application in economic theory), *Szigma*, Vol. 2 (1969) pp. 89–99.

[278] — *Workshop on Analytic Techniques for the Comparison of Economic Systems* (mimeographed), Berkeley: University of California, 1967.

[279] ZAUBERMAN, A.: *Reforms Formalized* (mimeographed), London: London School of Economics, 1969.

[280] ZAUBERMAN, A.: "The rapprochement between East and West in mathematical-economic thought", *The Manchester School*.

[281] ZELLNER, A. (ed): *Economic Statistics and Econometrics*, Boston: Little, Brown and Co., 1968.

AUTHOR INDEX

Abadie, J. M. 86, 377
Adelman, I. 215, 377
Albach, H. 370, 377
Alchian, A. A. 188, 377
Amos, A. J. 377
Andorka, R. 285, 377
Antal, I. 278
Aoki, M. 365, 377
Arrow, K. J. XVI, 21, 25, 130, 131, 135, 179, 299, 300, 347, 348, 355, 364, 366, 369, 377, 378
Ashby, W. R. 51
Augustinovics, M. 71, 378

Bacharach, M. O. L. 384
Bain, S. 200, 372, 378
Balderston, F. E. 223, 228, 370, 378
Balintfy, J. L. 386
Baran, P. A. 357, 378
Barone, E. 350, 358, 378
Bauer, T. XVI
Baumol, W. J. 19, 131, 365, 378
Beer, S. 51, 378
Bellman, R. 188, 378
Benedek, P. XVI
Benson, P. H. 135, 378
Bentham, J. 13
Berei, A. 378
Berge, C. 38, 378
Bikales, N. M. 385
Billington, A. E. 377
Blau, P. M. 84, 378
Bonini, C. 197, 370, 378, 388
Borch, C. 378
Bos, H. C. 106, 389
Boulding, K. E. 351, 378
Bródy, A. XVI. 359, 378
Brus, W. 54, 378
Burdick, D. S. 386
Buzási, J. 290, 378

Caves, R. 378
Chamberlin, E. H. 368, 379
Cook, G. J. 379
Coombs, C. H. 389
Cournot, F. 348, 360
Cyert, R. M. 89, 92, 93, 379
Czétényi, P. 241

Csapó, L. 53, 379
Csikós-Nagy, B. 379

Dantzig, G. B. 86, 87, 379, 382
Dányi, D. 285, 377
Davis, R. L. 389
Deák, A. XVI, 241
Debreu, G. 7, 8, 18, 19, 23, 25, 38, 46, 76, 122, 131, 230, 347, 348, 354, 355, 366, 369, 377, 379
Denison, E. F. 215, 379
Dischka, G. 379
Dorfman, R. 349, 379
Dömölki, B. XVI, 52, 383
Duesenberry, J. S. 141, 379

Edwards, J. B. 386
Einstein, A. 9–11, 379
Eisner, R. 379
Endrei, W. 268
Engels, F. 385
Enthoven, A. C. 21, 377
Erdey-Grúz, T. 379
Erdős, P. XVI, 59, 379
Evan, W. M. 84, 379

Fekete, G. 240, 379
Fellner, W. XVI, 379
Forrester, J. W. 370, 379
Frey, T. XVI
Friedman, M. 7, 379
Frigyes, E. 370, 379

SUBJECT INDEX

In the compilation of the index we adopted – in addition to the usual stand-points – a few special principles.

1. Several of the entries in the index are concepts, which are explained in the book under a Definition of a certain serial number. The page number indicating the place of the Definition in the book is set in italics, to make it different from the other page numbers. E.g., the page numbers following the entry "Disequilibrium" begin like this: 25, *253*, 256 ... etc., i.e. the definition of disequilibrium is to be found on p. 253.

2. There are certain *main concepts* among the entries, which occur repeatedly throughout the whole book, e.g. organization, economic system, etc. It would not be a help to the Reader if all occurrences were shown in the Index. Instead, we did the following.

The most frequently used main concepts are set in italics in the Index. Only the place of their definition is shown, if a definition is given at all.

Decision algorithm 113–21, *116*, 129, 161, 180, 182, 196, 208, 209, 225, 234, 327, 374

Decision alternatives XVIII, 12, 95, 102–10, 114, 118, 120–4, 127, 133, 139, 142, 145, 149, 151, → Acceptable ∼, Eligible ∼, Implementable ∼, Possible ∼

Decision distribution XIX, 109, 110, 112, 144, 163, 164

Decision-maker
∼ collecting information 66, 114, 143, 145, 149, 339
learning of ∼ 143, 145, 149, 152, 161, 297, 336
consistency of ∼, → consistent ∼, restrictedly consistent ∼, steadily inconsistent ∼, steadily consistent ∼

Decision problem XIX, 14, 66, *101*, 102, 106, 114, 116, 138, 154, 171, 172, 233, 234, 339, 371

Decision process 36, 97, 98, 100–76, 203, 222, 232–4, 243
complex ∼ *101*, 102, 171, 226
elementary ∼ 100–13, *101*, 114, 115, 145, 154, 156, 161, 169, 171, 222
extensive indices of ∼ 159–65, *164*, 170, 374
fundamental ∼ 117–21, 129, 149, 152, 164, 182, 201, 339–41
standard ∼ 117–21, 129, 148, 164, 182, 184, 201, 339, 340

Decision theory 12–4, 121, 343

Demand 20, 24, 27, 56, 170, 193, 221, 225, 229–39, 257–60, 282, 292, 299, 300, 308, 309, 315, 329–32, 342, 348, 356, 361, 366

Deminishing returns 29, 349

Desideratum, ∼ of an economic system *210–7*

Directive *77–9*, 105, 106

Directive ordering *79*

Direct reflection *60*, 63, 72, 105, 183, 184, 336

Disequilibrium 25, *253*, 256, 260, 285, 287, 290, 292, 295, 297, 299, 300, 301, 303, 309, 311, 313, 323, 327, 332, 333, 342

Dualistic description *39*, 42, 56, 75, 108, 230, 352–5

Duality 352–5, 365

Econometrics 15, 16, 206, 214, 257, 258, 284, 371, 373

Economic role of the state, government intervention, central control 6, 28, 54, 84, 105, 107, 110, 111, 132, 135, 151–3, 158, 177, 180, 186, 187, 189, 190, 298, 300–2, 308, 313, 316–9, 322, 323, 328, 329, 331, 334–43, 350–6, 362, 365, 372, 375

Economic system 50–52
characteristics of ∼ *51*, 53, 217
performance of ∼ 213–7, *214*, 286, 301, 303, 308, 327, 337, → Adaptive property, Autonomous function, Desideratum, Higher function, Multilevel structure
stochastic-causal description of ∼ 57, 147
structure of ∼ 36, 51

Economic systems theory 3, 5, 7, 16, 17, 30, 31, 35, 36, *51–3*, 71, 75, 88, 188, 197, 198, 202, 217, 230, 313, 332, 347, 351, 362, 367, 370–6

Elementary buying intention *234*
degree of maturity of ∼ 234, 237

Elementary decision process 100–13, *101*, 114, 115, 145, 154, 156, 161, 169, 171, 222

Elementary investment decision *317*
validity period of ∼ *317*

Elementary selling intention XIX, 233–7, *234*, 238, 239,
degree of maturity of ∼ 234–8

Eligible decision alternatives XIX, *108*, 109, 111, 112, 118, 120, 162, 163, 235, 238, 239

Empirical observations, investigation 5, 8–17, 36, 58, 71, 72, 75, 85, 97, 125, 129, 130, 135, 136, 147, 151, 158, 159, 161, 172, 197, 228, 252, 257–9, 282, 307, 332, 357, 361, 374

Equilibrium XV, 24–7, 29, 31, 73, 186, 216, 229, 239, *253*, 260, 293–6, 299, 300, 303, 307, 309–15, 323, 330, 333, 349, 350, 352, 356, 359, 365 ∼ theory → GE-theory

Evaluable series of decision *128*, 134–6, 138, 151

Explored decision alternatives XVIII, *105–14*, 118, 120, 121, 145, 149, 162, 235, 238, 239

402   SUBJECT INDEX

Static character 19, 30, 124, 125, 133–5, 185, 366
Stationary character 19, 20, 22, 23, 30, 124, 125, 185, 199, 366
Steadily consistent decision-maker *135*, 136, 146, 151
Steadily inconsistent decision-maker *135*, 136
Stochastic response function 47, 50, 57, 147, 207
Stock XIX, 21, 39, *43*, 47, 48, 51, 64, 92, 118, 178–81, 183, 185, 217, 225, 228, 229, 232, 233, 237, 251, 259, 261, 287, 296, 299, 304, 307, 310, 314, 315, 322, 323, 326, 366
Sub-system 41, 51, 73, 77, 80, 89, 98, 99, 132, 161, 173, 189, → Control sub-system
Subordinated organization *80*
Suction 240–9, 252–62, *253*, 286–330, 332, 335, 338, 341, 362, 371, 375
Superordinate organization *80*–2, 106
Supply 20, 24, 27, 56, 221, 225, 229–39, 258–61, 299, 300, 309, 329–32, 342, 348, 356, 361, 366
Survival 94, 95, 142, 305, 308
System → Economic system
Systems theory → Economic ∼, Mathematical ∼

Technical development, progress 22, 28, 64, 92, 137–9, 146, 151, 173, 185, 187, 188, 197, 200, 210, 216, 259, 262–85, 287–93, 310, 311, 323, 325, 327, 328, → Q-activities, Quality, Product development
Tension
    degree of ∼ → Aspiration ∼, Decision ∼

∼ of aspiration XIX, *160*, 161, 164, 175, 197, 244–9, 252–4, 297, 303, 304, 307, 308
∼ of decision XIX, *160*, 164, 197
Theoretical structure 15–7
Threshold of sensation 191, 192, 194, 196, 198, 201, 211
Time of decision preparation *101*, 106, 110, 114, 155
Time horizon → Anterior time horizon, Posterior time horizon
Transmitted reflection → Indirect reflection

Uncertainty 23, 29, 66, 74, 88, 129, 133, 134, 137, 142–6, 296, 302, 322, 334, 339, 359, 366, 367
Unfilfelment ratio 247, 315
Unit 38, 50
Upper level 77, 81, 84
Utility, 24, 122, 123, → Utility function
Utility function XIX, 22, 107, 112, 121–53, *123*, 173, 258, 348, → Utility

V-activities *283*, 284, 286, 287, 298
Vertical chain *81*, 82
Vertical flow of information *83*, 85, 87, 106, 115, 184, 336
Vertical ordering 80
Volume XIX, 42, 118, 170, 206, 223, 262–71, 283–7, 300, 302, 311, 316, 321, 325, 326, 338, 350
Volume indices, V-indices 269, 283–5

Wage 72, 93, 170, 313, 315, 316, 319, 321, 322, 339
Welfare function 351, 352
World standard, qualitative 281, 290, 291